IP Routing Protocols

IP Routing Protocols

Fundamentals and Distance-Vector Routing Protocols

James Aweya

CRC Press
Taylor & Francis Group
Boca Raton London New York

CRC Press is an imprint of the
Taylor & Francis Group, an **informa** business

First edition published 2021
by CRC Press
6000 Broken Sound Parkway NW, Suite 300, Boca Raton, FL 33487-2742

and by CRC Press
2 Park Square, Milton Park, Abingdon, Oxon, OX14 4RN

CRC Press is an imprint of Taylor & Francis Group, LLC

© 2021 Taylor & Francis Group, LLC

The right of James Aweya to be identified as author of this work has been asserted by him in accordance with sections 77 and 78 of the Copyright, Designs and Patents Act 1988.

Reasonable efforts have been made to publish reliable data and information, but the author and publisher cannot assume responsibility for the validity of all materials or the consequences of their use. The authors and publishers have attempted to trace the copyright holders of all material reproduced in this publication and apologize to copyright holders if permission to publish in this form has not been obtained. If any copyright material has not been acknowledged please write and let us know so we may rectify in any future reprint.

Except as permitted under U.S. Copyright Law, no part of this book may be reprinted, reproduced, transmitted, or utilized in any form by any electronic, mechanical, or other means, now known or hereafter invented, including photocopying, microfilming, and recording, or in any information storage or retrieval system, without written permission from the publishers.

For permission to photocopy or use material electronically from this work, access www.copyright.com or contact the Copyright Clearance Center, Inc. (CCC), 222 Rosewood Drive, Danvers, MA 01923, 978-750-8400. For works that are not available on CCC please contact mpkbookspermissions@tandf.co.uk

Trademark notice: Product or corporate names may be trademarks or registered trademarks and are used only for identification and explanation without intent to infringe.

Library of Congress Cataloging-in-Publication Data
Names: Aweya, James, author.
Title: IP routing protocols / James Aweya.
Description: First edition. | Boca Raton : CRC Press, 2021. | Includes bibliographical references and index. | Contents: v. 1. Fundamentals -- v. 2. Link-state and patth-vector routing protocols. | Summary: "This two-volume book describes the most common IP routing protocols used today (RIPv2, EIGRP, OSPFv2, IS-IS, and BGPv4), explaining the underlying concepts of each protocol and how the protocol components and processes fit within the typical router"-- Provided by publisher.
Identifiers: LCCN 2020052474 (print) | LCCN 2020052475 (ebook) | ISBN 9780367709624 (v. 1 ; paperback) | ISBN 9780367710415 (v. 1 ; hardback) | ISBN 9780367709631 (v. 2 ; paperback) | ISBN 9780367710361 (v. 2 ; hardback) | ISBN 9781003149040 (v. 1 ; ebook) | ISBN 9781003149019 (v. 2 ; ebook)
Subjects: LCSH: TCP/IP (Computer network protocol)
Classification: LCC TK5105.585 .A94 2021 (print) | LCC TK5105.585 (ebook) | DDC 004.6/65--dc23
LC record available at https://lccn.loc.gov/2020052474
LC ebook record available at https://lccn.loc.gov/2020052475

ISBN: 978-0-367-71041-5 (hbk)
ISBN: 978-0-367-70962-4 (pbk)
ISBN: 978-1-003-14904-0 (ebk)

Typeset in Times
by SPi Global, India

Contents

Preface xv
Author xxi

Chapter 1   Introduction to IP Routing Protocols 1

    1.1 Why We Need Routing Protocols 1
    1.2 Routing Methods 1
        1.2.1 Directly Connected Interface 2
        1.2.2 Static Routing 3
        1.2.3 Default Routing 4
            1.2.3.1 Default Route for an IP Host 4
            1.2.3.2 Default Route in a Router 4
        1.2.4 Dynamic Routing Protocols 5
            1.2.4.1 Routing Updates 8
            1.2.4.2 Periodic versus Triggered Routing Updates 8
            1.2.4.3 Routing Information Authentication 9
            1.2.4.4 Routing Information and Network Convergence 10
    1.3 Autonomous System 11
        1.3.1 What Is a Network Prefix and a Route? 11
        1.3.2 Autonomous System Numbers (ASNs) 12
        1.3.3 Multiple Routing Domains in an Autonomous System 13
        1.3.4 Types of Autonomous Systems 14
    1.4 Routing Metrics and Costs 15
        1.4.1 Hop Count 15
            1.4.1.1 Network Diameter: Maximum Hop Count 16
            1.4.1.2 "Infinity" Hop Count as a Signaling Mechanism for Network Failures 17
            1.4.1.3 Limitations of the Hop Count as a Routing Metric 17
        1.4.2 Bandwidth 17
        1.4.3 Delay 18
        1.4.4 Traffic Load 18
        1.4.5 Reliability 18

		1.4.6	Cost	19
			1.4.6.1 Example: OSPF Cost	19
			1.4.6.2 Cost Based on Interface Bandwidth	19
	1.5	Classification of Routing Protocols		20
		1.5.1	Interior versus Exterior Routing Protocols	21
	1.6	Least-Cost Routing		22
	Review Questions			24
	References			24
Chapter 2	Types of Dynamic Routing Protocols			25
	2.1	Introduction		25
	2.2	Distance-Vector Routing Protocols		25
		2.2.1	Basic Characteristics of Distance-Vector Routing Protocols	25
		2.2.2	Distance-Vector Routing Protocol Operations	30
		2.2.3	What Is a Rooting Loop?	31
		2.2.4	Routing Loops and Workarounds – Enhancing the Performance of Distance-Vector Routing Protocols	32
			2.2.4.1 Initial Full Routing Table Update and Periodic Updates	33
			2.2.4.2 Route Maintenance and Invalidation Timers	34
			2.2.4.3 Holddown Timers	36
			2.2.4.4 Triggered Updates	38
			2.2.4.5 Count-to-Infinity (Maximum Hop Count)	39
			2.2.4.6 Poison Reverse	41
			2.2.4.7 Split Horizon	42
	2.3	Link-State Routing Protocols		44
		2.3.1	OSPF versus IS-IS Metrics	45
		2.3.2	Basic Characteristics of Link-State Routing Protocols	46
		2.3.3	Link-State Routing Protocol Operations	47
			2.3.3.1 Neighbor Discovery	48
			2.3.3.2 Link-State Flooding	50
			2.3.3.3 Link-State Database	51
			2.3.3.4 Link-State Routing Process	52
			2.3.3.5 Calculating the Shortest Paths and Constructing the Routing Table	53
			2.3.3.6 Areas	56
	2.4	Path-Vector Routing Protocols		57

		2.4.1	Why an IGP Is Not Recommended for Routing between Routing Domains or Autonomous Systems 58
		2.4.2	Using an EGP between Routing Domains: Path-Vector Routing Protocol 60
		2.4.3	BGP: A Path-Vector Routing Protocol 62
			2.4.3.1 Internal and External BGP Peering............. 63
			2.4.3.2 Basic Characteristics of BGP Routes 63
			2.4.3.3 BGP Autonomous System Path Advertisement 65
			2.4.3.4 Loop-Free Paths within an Autonomous System 65
			2.4.3.5 Manually Configured BGP Connections over TCP 65
		2.4.4	BGP and Path Attributes 66
	2.5	The IP Routing Table and Selection of Best Paths 68
		2.5.1	Path Metrics and Routing Protocols 68
			2.5.1.1 Equal-Cost Multipath (ECMP) Routing 69
		2.5.2	Administrative Distance and Route Selection 69
			2.5.2.1 Administrative Distance Use Case Example 70
		2.5.3	Prefix Length and Longest Prefix Matching Lookup 71
	Review Questions 73
	References 74

Chapter 3	Routing and Forwarding Tables in Routing Devices 77
	3.1 Introduction 77
	3.2 Functional Components of an IP Router 77
	3.2.1 IP Control Engine (or Route Processor) 77
	3.2.2 IP Forwarding Engine 79
	3.3 High-Level Router Architectures 80
	3.3.1 Router Architectures with Centralized Forwarding Engine 80
	3.3.2 Router Architecture with Multiple Centralized Forwarding Engines 81
	3.3.3 Router Architecture with Distributed Forwarding Engines 82
	3.3.4 Control Plane Redundancy 84
	3.4 IP Routing and Forwarding Tables 85
	3.4.1 Routing Table 85

		3.4.1.1	Routing Table Entries 85
		3.4.1.2	Routing Tables in a Router with Multiple Protocols 88
		3.4.1.3	Types of Unicast Routing Tables 90
		3.4.1.4	Aggregate or Summary Routes in the Routing Table 91
	3.4.2	Forwarding Table 91	
3.5	A Note on Layer 2 Adjacency Table 92		
3.6	IP Forwarding Operations 92		
	3.6.1	Handling Special Addresses During Packet Forwarding 94	
3.7	Redistributing Routing Information and Routing Metric Translation 95		
	3.7.1	The Need for Route Redistribution 95	
	3.7.2	Filtering Inbound and Outbound Routing Information 95	
	3.7.3	The Need for Routing Metric Translation 96	
Review Questions 97			
References 98			

Chapter 4 Static Routes in the Routing Table 99

4.1	Introduction 99
4.2	Benefits of Dynamic Routing Protocols 99
4.3	Benefits of Static Routing 100
4.4	Configuring Dynamic Routing versus Static Routing 103
4.5	Standard Static Route 104
	4.5.1 Concept of Qualified Next Hop 105
4.6	Default Static Route 106
4.7	Summary Static Route 108
4.8	Floating Static Route 111
Review Questions 113	
References 113	

Chapter 5 Routing Information Protocol (RIP) 115

5.1	Introduction 115
5.2	Routing Protocols and Their Databases 115
5.3	RIP Overview 117
5.4	RIPv2 Message Format and Other Characteristics 117
	5.4.1 RIPv2 Message Format 118
	5.4.2 Interpreting the Address Family Identifier (AFI) Field in RIPv2 119
	5.4.3 RIPv2 Routing Table 120

Contents ix

		5.4.4	RIPv2 Timers	120
		5.4.5	RIPv2 Request Message	120
		5.4.6	RIPv2 Response Messages	121
		5.4.7	Sending and Receiving RIPv2 Request and Response Messages	122
	5.5	RIPv2 Authentication		122
		5.5.1	Plaintext Authentication	123
		5.5.2	Cryptographic Authentication	124
			5.5.2.1 RIPv2 Authentication Message Generation	126
			5.5.2.2 RIPv2 Authentication Message Reception	127
	5.6	High-Level RIP Router Architecture, Processes, and Databases		128
		5.6.1	The RIP Process	129
		5.6.2	The Management Process	129
			5.6.2.1 The Route Store Process	131
			5.6.2.2 The Interface Manager	132
			5.6.2.3 The Sockets Manager	134
			5.6.2.4 The Redistribution Manager	137
		5.6.3	The Routing Table Manager Process	137
	5.7	Filtering Routing Updates in RIP		140
		5.7.1	Configuration of Passive Interface to Prevent or Restrict Routing Updates	140
		5.7.2	Filtering Routes in Incoming and Outgoing Routing Updates	141
			5.7.2.1 Distribute-list In	141
			5.7.2.2 Distribute-list Out	142
	5.8	Summary of RIPv2 Features		142
	Review Questions			144
	References			144
Chapter 6	Enhanced Interior Gateway Routing Protocol (EIGRP)			147
	6.1	Introduction		147
	6.2	EIGRP Overview		147
	6.3	EIGRP Concepts		149
		6.3.1	Reliable Transport Protocol	149
		6.3.2	Main EIGRP Databases	151
			6.3.2.1 Neighbor Table	151
			6.3.2.2 Topology Table	152
			6.3.2.3 Routing Table	153
			6.3.2.4 Other EIGRP Concepts	154
		6.3.3	Neighbor Formation	154

6.4	EIGRP Message Types	155
	6.4.1 HELLO Packets	157
	6.4.2 UPDATE Packets	159
	6.4.3 QUERY Packets	159
	6.4.4 REPLY Packets	160
	6.4.5 REQUEST Packets	161
	6.4.6 EIGRP TLVs	161
	6.4.7 EIGRP Flags Field	163
6.5	EIGRP Metrics	165
6.6	Feasible and Reported Distance	167
	6.6.1 Feasible Distance	167
	6.6.2 Reported (or Advertised) Distance	167
6.7	Successor and Feasible Successor	168
	6.7.1 Successor	168
	6.7.2 Feasible Successor	170
6.8	Route States: ACTIVE and PASSIVE States	172
	6.8.1 PASSIVE State	172
	6.8.2 ACTIVE State	173
	6.8.3 Comments on Feasible Successors when a Route Is in the PASSIVE or ACTIVE State	174
6.9	Feasibility Condition	175
6.10	EIGRP Diffusing Update Algorithm (DUAL)	176
	6.10.1 High-Level Description of DUAL	176
	6.10.2 Message Types Used by DUAL	177
	6.10.3 Some Behaviors of DUAL	178
	6.10.4 Stuck-In-Active (SIA) and the Use of SIA-QUERY and SIA-REPLY Messages	179
	6.10.4.1 Stuck-In-Active (SIA)	179
	6.10.4.2 SIA-QUERY	180
	6.10.4.3 SIA-REPLY	181
6.11	EIGRP Neighbor Discovery and Maintenance	181
	6.11.1 Neighbor Hold Time	182
	6.11.2 Use of HELLO Packets	182
	6.11.3 Use of UPDATE Packets	183
	6.11.4 Use of QUERY Packets	183
6.12	Building the EIGRP Topology Table	184
	6.12.1 Route Management	184
	6.12.1.1 Internal Routes	184
	6.12.1.2 External Routes	185
	6.12.2 Use of Split Horizon and Poison Reverse	186
6.13	Initial Neighbor and Route Discovery	187
6.14	EIGRP Load Balancing	189
6.15	EIGRP Route Redistribution	191
6.16	EIGRP Route Summarization	193

Contents   xi

        6.16.1   Auto-summarization 193
        6.16.2   Manual Summarization 194
  6.17   EIGRP Authentication 194
        6.17.1   Simple Password Authentication 195
        6.17.2   MD5 Authentication 195
  6.18   High-Level EIGRP Router Architecture, Processes, and
        Databases 197
        6.18.1   Neighbor Table 198
        6.18.2   Topology Table 198
        6.18.3   Routing Table 199
        6.18.4   Determining Successors and Feasible
                Successors 200
        6.18.5   Populating and Maintaining the Neighbor Table 201
                6.18.5.1   Understanding the SRTT, RTO,
                            and Q Cnt Parameters 202
        6.18.6   Populating and Maintaining the Routing Table 203
        6.18.7   Handling the Loss of a Route to a Network
                Destination 204
        6.18.8   Handling Queries for a Route to a Network
                Destination 204
        6.18.9   Note on Route States, Successors, and
                Feasible Successors 205
  6.19   Summary of EIGRP Features 206
  Review Questions 207
  References 207

Chapter 7 Network Path Control and Factors That Affect Routing
Table Properties 209

  7.1   Introduction 209
  7.2   Running Multiple Routing Protocols 209
        7.2.1   Running Multiple Overlapping Routing
               Protocols 210
        7.2.2   Running Multiple Non-Overlapping Routing
               Protocols 210
  7.3   The Need for Network Path Control Tools 212
        7.3.1   What Is a Routing Policy? 212
        7.3.2   Implementing Routing Policies 215
               7.3.2.1   Routing Policy Control Points 217
               7.3.2.2   Packet Filter Policy Control Points 218
        7.3.3   Routing Policies and BGP Attribute
               Manipulation 219
  7.4   What Is Policy-Based Routing (PBR)? 219
  7.5   Route Summarization 221

		7.5.1	Using of VLSM and CIDR 221
		7.5.2	RIP Route Summarization 222
		7.5.3	EIGRP Route Summarization 223
		7.5.4	OSPF Route Summarization 224
		7.5.5	IS-IS Route Summarization 224
		7.5.6	Static Route Summarization 225
	7.6	Route Redistribution 226
		7.6.1	One-Point Route Redistribution 229
		7.6.2	Multipoint Route Redistribution 231
	7.7	Path Control Tools 234
		7.7.1	The Need for Path Control Tools 235
		7.7.2	Route Redistribution Configuration Tools 236
			7.7.2.1 Redistributing Routes into RIP 236
			7.7.2.2 Redistributing Routes into OSPF 237
			7.7.2.3 Default Metric for RIP, OSPF, or BGP 238
			7.7.2.4 Redistributing Routes into EIGRP 238
			7.7.2.5 Default Metric for EIGRP 239
			7.7.2.6 Redistributing Routes into BGP 239
			7.7.2.7 Redistributing Directly Connected Networks and Static Routes into a Routing Protocol 240
		7.7.3	Route Metric: Route Redistribution and the Seed Metric 240
			7.7.3.1 Configuring Seed Metrics 240
		7.7.4	Administrative Distance and Path Control 243
			7.7.4.1 Administrative Distance as a Path Control Tool 243
		7.7.5	Route Tagging 246
		7.7.6	Passive Interfaces 248
		7.7.7	Static Routes 249
		7.7.8	Default Routes 252
			7.7.8.1 Setting Default Routes Dynamically 252
			7.7.8.2 Setting Default Routes Statically 252
			7.7.8.3 Configuring Default Routes 253
			7.7.8.4 Generating Default Routes in RIP 254
			7.7.8.5 Generating Default Routes in EIGRP 254
			7.7.8.6 Generating Default Routes in OSPF 254
			7.7.8.7 Generating Default Routes in IS-IS 255
			7.7.8.8 Generating Default Routes in BGP 256
		7.7.9	Route Maps 256
			7.7.9.1 Route Map Applications 256
			7.7.9.2 Defining a Route Map 258
		7.7.10	Distribute Lists 261
			7.7.10.1 Filtering Incoming Routing Updates 261

		7.7.10.2	Filtering Outgoing Routing Updates 262

- 7.7.11 Prefix Lists 263
- 7.7.12 Using Policy Based Routing (PBR) for Path Control 265
- 7.7.13 Offset Lists 267
- 7.7.14 IP Service Level Agreement (SLA) Probes 268
  - 7.7.14.1 When to Use Cisco IOS IP SLA Probes ... 269
  - 7.7.14.2 Workings of Cisco IOS IP SLA Probes 270
- 7.8 Special Focus: Path Control Tools in BGP 274
  - 7.8.1 BGP Route Filtering and Path Attribute Manipulation 274
    - 7.8.1.1 Identifying BGP Routes 275
    - 7.8.1.2 Accepting/Rejecting BGP Routes 275
    - 7.8.1.3 BGP Path Attribute Manipulation 275
  - 7.8.2 BGP Route Filtering 276
    - 7.8.2.1 BGP AS-Path Filter Lists 276
    - 7.8.2.2 BGP Prefix Lists 277
    - 7.8.2.3 BGP Distribute Lists 278
    - 7.8.2.4 BGP Route Maps 278
  - 7.8.3 Injecting Routing Information into BGP 278
    - 7.8.3.1 Injecting Routes Statically into BGP 279
    - 7.8.3.2 Injecting Routes Dynamically into BGP 280
    - 7.8.3.3 Route Injection and the BGP ORIGIN Attribute 281
  - 7.8.4 BGP AS-Path Prepending: AS-Path Attribute Manipulation Using Dummy Entries 281
  - 7.8.5 Route Aggregation in BGP 283
    - 7.8.5.1 Performing BGP Route Aggregation 284
    - 7.8.5.2 Route Aggregation without AS_SET 286
    - 7.8.5.3 Route Aggregation with AS_SET 286
    - 7.8.5.4 Changing the BGP Attributes of an Aggregate Route 287
    - 7.8.5.5 Advertising the Aggregate Route Only, while Suppressing the More-Specific Routes 288
    - 7.8.5.6 Advertising the Aggregate Route Plus All of the More-Specific Routes 288
    - 7.8.5.7 Advertising the Aggregate Route Plus a Subset of the More-Specific Routes 288
  - 7.8.6 Default Routes in BGP 289
  - 7.8.7 IGP Routes versus BGP Routes: Looking at Backdoors Routes 290
- 7.9 Unnumbered Interfaces 290

	7.9.1	Conserving IP Addresses with Unnumbered Interfaces 290
	7.9.2	Configuring IP Unnumbered Interfaces 292
	7.9.3	Limitations of IP Unnumbered Interfaces 293
	7.9.4	Receiving Routing Updates over IP Unnumbered Interfaces 293
	7.9.5	Forwarding IP Packets over IP Unnumbered Interfaces 294
7.10	Routing Protocol Timers 294
Review Questions 296
References 296

Index 299

Preface

IP routing is a general term used to represent the collection of methods and protocols that determine the paths across multiple internetworks that a piece of user data (called a packet) takes in order to travel from its source to its intended destination. Each packet is routed hop-by-hop through a series of routers, and across multiple networks from its source to the destination. Each hop in this case represents a routing device (or router). Routers use IP routing protocols (set of procedures and rules) to communicate with each other and to exchange routing information about network reachability.

Each router uses this reachability information to build a network topology map and a local Routing Table which contains the network destination addresses, and their associated next hop addresses plus outbound interfaces. A next hop IP address is an entry in the router's Routing Table, which specifies the next best or closest router that lies on the most optimal path to the packet's destination. The most common unicast routing protocols used in today's internetworks include RIP (Routing Information Protocol), EIGRP (Enhanced Interior Gateway Routing Protocol), OSPF (Open Shortest Path First), IS-IS (Intermediate System – Intermediate System), and BGP (Border Gateway Protocol).

Typically, a router uses an IP Forwarding Table for IP packet forwarding, which is a reduced/optimized table containing only the most relevant information distilled from the Routing Table. The contents of the IP Forwarding Table are directly relevant to packet forwarding and mirror the main packet routing information in the Routing Table such as the destination addresses, outgoing interfaces, and next hop IP addresses.

When a router receives an IP packet to be forwarded, it parses the destination IP address in the IP packet header, and consults its Forwarding Table to determine the next hop and outgoing interface for the packet's destination, and forwards the packet appropriately (after performing some required IP packet header field updates). The next router receives the packet and repeats this process using its own IP Forwarding Table, until the packet reaches its final destination. At each hop (router), the IP destination address in the packet header provides the basic routing instructions used to determine the next hop and outgoing interface.

For the purpose of routing and management, internetworks such as the Internet, are divided into smaller routing and administrative zones called Autonomous Systems. An Autonomous System can be loosely defined as a group of routers that are under the control and management of a single administrative authority, and exchange routing information using a common clearly defined routing policy. For example, a corporate network or an ISP network is usually structured as an individual Autonomous System. The global public Internet can be viewed as a collection of different Autonomous Systems.

Typically, an Interior Gateway Protocol (IGP) determines best/optimal routes within a single Autonomous System. RIP, EIGRP, IS-IS and OSPF are well-known IGP examples. The IGP enables routers on different networks within the single Autonomous System to discover and to send data to one another. Also, when an

Autonomous System provides transit services, an IGP can be used to allow data to be forwarded across the transit Autonomous System from ingress to egress. Routes are distributed between Autonomous Systems by an Exterior Gateway Protocol (EGP). BGP is currently the only EGP used in today's internetworks. The EGP enables routers within an Autonomous System to choose the best egress point on an Autonomous System when they have data for external destinations.

The IGPs and EGP running within each Autonomous System cooperate to route data across internetworks. The EGP determines the Autonomous Systems in the internetwork that data must travel over, in order to get from its source to the destination, while the IGP determines the best/optimal path within each Autonomous System that data must travel over to get from the ingress point (i.e., the data source) to the egress point (i.e., the final destination).

As discussed above, routing protocols run in routers and provide the intelligence that guide how routers communicate with each other to determine the best/optimal paths for routing user data. Routing protocols are a key component of a router. Many books discuss routing protocols without sufficiently linking them to the other key components of a router, and how all these components interact to allow a router to perform its key functions. As will be discussed in greater detail in this two-volume book, the key functions of a router are:

- communicate with other routers to determine optimal paths to destinations in a network,
- react to network topology and state changes, and recompute optimal paths,
- receive data packets and perform packet verifications,
- perform IP Forwarding Table lookups to determine the next hop and outgoing interface for packets,
- perform a number of IP packet header updates,
- and finally, forward the packet out the outgoing interface to the next hop.

Volumes 1 and 2 of this two-part book, describe the most common IP routing protocols in used today (RIPv2, EIGRP, OSPFv2, IS-IS, and BGPv4) by explaining the underlying concepts of each protocol, and describing how the protocol components, and processes fit within the typical router. Each routing protocol uses a number of databases to perform its functions. This two-volume book also describes the types of databases each routing protocol uses, how the databases are constructed and managed, and how the various protocol components and processes relate and interact with the databases. The description of the routing protocols is from a systems perspective, recognizing the most important routing and packet forwarding components and functions of a router.

A majority of existing books tend to be vendor-focused, dealing instead with how a particular vendor's router and the routing protocols running on it are set up and configured, and other matters related to troubleshooting. This two-volume book appeals more to readers who want a detailed discussion of routing protocols, yet do not want to be tied down or distracted by the often, lengthy discussions on configuration and troubleshooting instructions and routines, found in a majority of IP routing protocols books. The presentation is in a style that makes it appealing to undergraduate and graduate level students, research and practicing engineers and scientists, IT personnel, and network engineers.

Volume 1 focuses on the fundamental concepts of IP routing and distance-vector routing protocols (RIPv2 and EIGRP), while Volume 2 focuses solely on link-state routing protocols (OSPF and IS-IS), and the only path-vector routing protocol in use today (BGP). BGP is the only path-vector EGP in use today. Traditional distance-vector routing protocols such as RIPv1 and RIPv2 determine the best routes to a network destination based on a distance metric such as the number of routers (i.e., the hop count) each candidate route has to that destination.

Routers running link-state routing protocols advertise routing information about the network topology (which includes information about their directly connected links and the state of those links) to all link-state routers in the routing domain until all the routers have identical information about the network. The link-state routers exchange routing information using multicast addresses and triggered routing updates. The routers calculate the best path to each network destination based on constraints such as minimum available path bandwidth, maximum delay, and other path related parameters.

For each destination network, a router in a (traditional) distance-vector routing protocol network constructs a distance-vector, which is a one-dimensional array or vector that presents or indicates the destination network on the router's least-cost tree of network prefixes. The routing information is structured in the form of vectors or arrays with elements (distance, direction) where "distance" in a vector is a route metric (or cost), such as the hop count to reach the destination network, and "direction" is the next-hop IP router to be used to reach that destination. To prevent routing loops, speed up network convergence, and improve network stability, the routers use various protocol timers and loop prevention mechanisms such count to infinity, split horizon, and poison reverse.

Newer distance-vector routing protocols such as EIGRP use a composite routing metric that takes into account path hop count, delay, bandwidth, path Maximum Transmission Unit (MTU), plus other factors such as traffic load, and path reliability. EIGRP's composite routing metric is designed to be flexible in such a way that these route attributes (some of which are factors that reflect traffic on a given route) can be selected as needed for a given EIGRP routing domain. Whereas RIPv1/2 uses the Bellman–Ford algorithm to find the least cost path to network destinations, EIGRP uses the Diffusing Update Algorithm (DUAL). Both RIPv1/2 and EIGRP support routing loop prevention mechanisms such as count-to-infinity, split horizon, and poison reverse. It may be argued if EIGRP is truly a distance-vector routing protocol, but this book stays away from that debate.

Chapter 1 describes the various methods used by routers to learn routing information. In this chapter, we introduce the main concepts of static routing and dynamic routing including their benefits and limitations. We discuss the different sources of routing information, the classification of the different dynamic routing protocols, and the routing metrics or costs the routing protocols use to determine the best paths to network destinations.

Chapter 2 describes the different categories of dynamic routing protocols in addition to their main distinguishing features (distance-vector routing protocol, link-state routing protocols, and path-vector routing protocols). RIPv2 and EIGRP are classified as distance-vector routing protocols, while OSPFv2 and IS-IS fall under the category of link-state routing protocols. BGPv4 is so far the only path-vector routing protocol in use today. The discussion includes the characteristics of the different dynamic routing protocols, and how they differ in design and operation. Understanding the different routing methods available (whether static or dynamic), is important and key to making informed decisions about which method to use for routing in a network. This allows a network engineer to determine which routing method is most suitable for a particular network environment.

A routing device, in general, maintains two key databases, the IP Routing Table and IP Forwarding Table, which hold the routing information required for forwarding packets in a network. Chapter 3 discusses these two key databases, their distinguishing features, and how they are used for packet forwarding. The chapter describes the difference between the routing and forwarding planes in an IP router. The chapter also explains the processing steps involved in forwarding IP packets through an IP router to their destinations.

Chapter 4 contrasts static routing and the widely implemented dynamic routing methods. The discussion covers, in addition, the different methods used for configuring static routes in Routing Tables. This chapter includes a description of different types of static routes such as default static routes, summary static routes, and floating (or backup) static routes. Many of today's networks of all sizes, use a combination of static and dynamic routing. Static routing is very appealing and widely used mainly because it does not require the same amount of processing, memory and routing information messaging overhead as in dynamic routing protocols. The advantages and disadvantages of static routing are also discussed.

Chapters 5 and 6 provide detailed descriptions of the most common distance-vector routing protocols RIPv2 and EIGRP, respectively. The discussion covers their most identifying characteristics, operations, and the databases they maintain. Each routing protocol maintains a number of databases which hold information about the local router's neighbor routers, and the routing information it has learned from other routers in the network. Each database type is used for specific operations as defined by the particular routing protocol. For each routing protocol, we discuss the main components, data structures, routing protocol messages, and best path computation algorithm. Each chapter also covers a high-level router architecture, processes, and databases for the particular routing protocol being discussed.

Chapter 5 begins with a review of distance-vector routing protocols and a discussion on the main features of RIPv2. The chapter then describes the packet formats used by RIPv2, RIPv2 authentication mechanisms, and RIPv2 router high-level architecture, processes, and databases. The discussion includes a high-level view of the RIPv2 router and components, plus and overview of its inbound and outbound message processing.

Chapter 6 starts by providing an overview of the main features of EIGRP. The chapter then provides a detail description of the different databases used by EIGRP. The chapter includes a detail discussion on EIGRP packets, protocol message processing, EIGRP routing information generation and maintenance, and EIGRP authentication. The discussion includes a high-level architecture of the EIGRP router and components, plus the inbound and outbound message processing.

Traffic within networks and between networks flow according to the routes discovered by routing protocols. So, altering the behaviors of the routing protocols, and path parameters and attributes, can translate into changes in routes and traffic flow to network destinations. Chapter 7 discusses the various mechanisms IP routers use for controlling routing in networks. This chapter discusses the different factors that affect Routing Table properties such as route summarization, route redistribution, and path control tools such static routes, Administrative Distance (or Route Preference), and route filtering (using methods such as IP prefix lists, route maps, distribute lists, and policy-based routing [PBR]). This chapter covers various methods for processing policies for route import/export between the routing protocols running in a router.

This two-volume book presents an illustrated discussion of IP routing protocols using real-world network examples. The entire discussion is presented from a practicing engineer's perspective, linking theory and fundamental concepts to common practices and real-world examples. It is hoped that the approach taken would contribute to a better understanding of IP routing protocols.

Author

James Aweya, PhD, is a chief research scientist at the Etisalat British Telecom Innovation Center (EBTIC), Khalifa University, Abu Dhabi, UAE. He was a technical lead and senior systems architect with Nortel, Ottawa, Canada, from 1996 to 2009. He was awarded the 2007 Nortel Technology Award of Excellence (TAE) for his pioneering and innovative research on Timing and Synchronization across Packet and TDM Networks. He has been granted 68 US patents and has published over 54 journal papers, 40 conference papers, and 43 technical reports. He has authored four books including this one and is a senior member of the Institute of Electrical and Electronics Engineers (IEEE).

1 Introduction to IP Routing Protocols

1.1 WHY WE NEED ROUTING PROTOCOLS

Each router in a network needs to be provisioned with the relevant routing information to enable it correctly forward packets from their sources to their respective network destinations. The information used in packet forwarding is maintained in a Routing Table and can be configured manually or automatically (and dynamically) through various means. To enable Layer 3 (or IP) forwarding of packets across internetworks, routers share the routing information they have learned with other routers usually via dynamic routing protocols. A routing protocol is a set of procedures that govern how a router communicates with other routers in order to share routing information about the state and reachability of network destinations.

Dynamic routing protocols not only allow routers to determine single or multiple paths to network destinations, but also the best paths to those destinations which in turn can be installed in the Routing Tables. The routing protocols also support update functions that allow the router to determine the next-best path if an existing best path to a destination fails or becomes unavailable. Dynamic routing protocols support various mechanism that allow them to determine network topology changes, and to compensate for such changes with the goal of keeping the network operational and network destinations reachable, as much as possible. This capability in particular makes the use of dynamic routing protocols more important and advantageous over static routing.

This chapter describes the various methods (manual and dynamic) used by routers to discover routing information. We introduce the main concepts of dynamic routing protocols and their benefits. We discuss the classification of the different dynamic routing protocols in use today, and the routing metrics (or costs) each protocol uses to determine the best paths to network destinations.

1.2 ROUTING METHODS

The primary responsibility of the routing function (in a switch/router or router) is to forward Layer 3 packets (IP packets) from one network, subnetwork, or Virtual Local Area Network (VLAN), to another network, subnetwork, or VLAN. Routers must therefore be configured with routing information to enable communication between different networks. This section discusses the mechanisms used to configure routing information in routers.

All hosts within a network that want to communicate with other hosts in other networks (subnetwork or VLAN) must have access to a router. Furthermore, wherever a router has an interface that connects to a network, that interface must have an IP address assigned to it. This assigned interface IP address serves as the address that indicates to other routers (via routing protocols), reachability information about the networks the router is connected to.

Also, when a router receives routing information from another router, the source IP address allows the receiving router to know the router (interface) that sent the routing update. The router interface must support the relevant Physical and Data Link Layer protocols that allows it to connect to the network. These Physical and Data Link Layer protocols also allow the router to be aware of the "up" or "down" state of the attached link and network.

A number of mechanisms exist for a router to construct and maintain its Routing Table. These routing information discovery mechanisms include the use of directly connected interfaces, static routing, default routing, and dynamic routing. All of these serve as information sources that provide a router with the network information necessary to build and maintain its Route Table. The different routing methods and mechanisms are described in this section.

1.2.1 Directly Connected Interface

A directly connected interface (also referred to as directly connected route or network) is a local interface on a router that is directly attached to a network, subnet, or host (Figure 1.1). A router always adds networks that are directly connected to its interfaces to the Routing Table. These routes are appropriately identified in the Routing Table as directly connected, and no routing is needed from the local router to reach such destinations. The directly connected interface has to be configured with an IP address, and the interface must be in the "up" state for the router to be able to forward all packets destined to all hosts in the attached network, directly.

When an IP address and subnet mask (or simply, network mask or netmask) are configured on a router interface, that interface becomes an IP host on the network associated with that network prefix (i.e., the prefix to which the interface is attached). Once the interfaces to the directly connected networks are appropriately configured, the router immediately learns and adds them in the Routing Table. The router installs the network address prefix (resulting from the IP address and network mask) of the

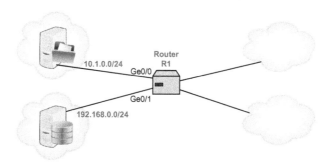

- Networks that are directly connected to router R1's interfaces are added to its Routing Table.
- The two active interfaces, Ge0/0 and Ge0/1 have been configured with IP addresses and are currently in the up-up state, so router R1 adds these networks to its Routing Table.
- Router Ri will be able to route all packets destined to all hosts in the networks that are directly connected to its active interfaces.

FIGURE 1.1 Directly Connected Routes

Introduction to IP Routing Protocols

interface, in addition to the interface type and number, into the Routing Table as a directly connected network.

When a router forwards a packet, for example, to a video, email, or web server that is directly connected to one of its local interfaces, that host is considered to be on the same network as the router's connecting interface. A directly connected network is inherently identified through the router's configured local interface to which the network is attached. Such networks are immediately recognizable by the router, and packets are forwarded direct to these networks without requiring any assistance from dynamic routing protocols.

Directly connected interfaces or networks have an Administrative Distance value of 0 (the lowest value), and always take precedence over static routes or routes discovered through dynamic routing protocols (see discussion in Chapter 2). When several routes to the same destination (discovered by different routing methods) exist, the Administrative Distance (also called the Route Preference) is the criterion that a router uses to select which route among those routes to install in its Routing Table. Since directly connected networks have the lowest possible (default) Administrative Distance value of 0, they are always installed in the Routing Table.

A directly connected route is always the best paths to a destination because the router will always clearly recognize when a packet is destined to such a destination. The router does not have to rely on some other routing methods such as via static configuration or dynamic routing protocols, to learn such a route. However, when packets are destined to other (remote) networks not directly attached to the router's interfaces, then other routing methods are needed.

1.2.2 STATIC ROUTING

The mechanism used for configuring routing information in a router can be either static or dynamic. A static mechanism requires manual configuration by the user and the routing information does not change until the user changes it. A dynamic mechanism on the other hand, involves using routing protocols that facilitate dynamic exchange of routing information between routers. This allows all routers in a network to learn and adapt to changes in the network topology and state. Network changes may occur for a number of reasons, including failure of a router, introduction of a new link, failure of a link, and change of link parameters. The dynamic routing protocols allow information in the Routing Tables of all routers to be updated when there is a network topology change.

Static routes are paths to network destinations that have to be manually entered into the router's Routing Table by a network administrator (see Figure 1.2). The information associated with a static route is the IP address of the next-hop router, and the local outgoing interface on the router. The local router uses this information when forwarding packets to a particular destination that is reachable over the static route. Because a static route is a fixed entry in the Routing Table to a destination, it does not change or adapt to dynamically changing conditions in the network. If the router itself or the interface associated with the static route becomes unavailable or fails, the static route defined to that destination also fails. The old static route entry has to be manually reconfigured if a new static route is required.

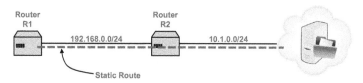

- We assume router R1 is directly connected to router R2, which in turn is is directly connected to the network 10.1.0.0/24.
- Since network 10.1.0.0/24 is not directly connected to router R1, we can manually configure a static route on router R1 to route packets through R2 to network 10.1.0.0/24.

FIGURE 1.2 Static Routes

1.2.3 Default Routing

Default routes can be configured to provide an IP host a means to communicate out of its local network to hosts on other networks. A default route can also provide a router with a route of last resort to forward a packet if no other route specifically matching its destination exists in the router's Forwarding Table.

1.2.3.1 Default Route for an IP Host

IP hosts do not usually maintain local Routing Tables that hold routing information to other networks, although have operating systems that support the capability. Instead, IP hosts on a network rely on local routers (also called the default gateways or routers) to forward packets to hosts on other (remote) networks. Thus, for an IP host in one network to be able to communicate with other IP hosts on different networks, it must be configured with the IP address of at least one local router. The host may be statically or dynamically configured (via the Dynamic Host Configuration Protocol [DHCP]) to discover its default gateway's IP address to enable it communicate with others outside the local network.

When a host wants to communicate with another host outside its local network, it sends its packets to the default gateway for forwarding outside the network. The default gateway is the point of exit and/or entry connecting local hosts to other hosts outside the local network (see Figure 1.3). Without the default gateway (local router), hosts are limited to communicating with hosts connected to the same (local) network.

1.2.3.2 Default Route in a Router

When a router receives a packet, the destination IP address is parsed and then compared to each entry in the Forwarding Table until a best route match is found. If a best match is found in the Forwarding Table, the router updates the packet's Time-To-Live (TTL) and IP checksum, and readdresses the packet's MAC addresses using the outgoing interface's MAC address as the source MAC address, and the next-hop router's receiving interface's MAC address as the destination MAC address.

After updating the Ethernet frame checksum, the router transmits the packet out the outgoing interface to the next-hop router. When the router does not find any matching entry in its Forwarding Table, the packet may be sent to a default route (if one is configured), or the packet may be discarded and an ICMP error is sent to the packet's originator.

Introduction to IP Routing Protocols

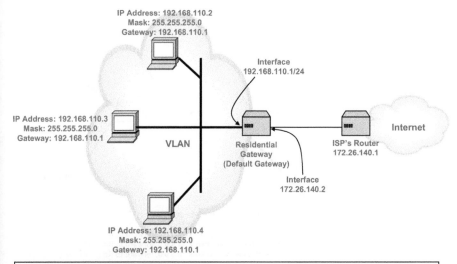

- The default gateway provides a default route or path packets will use when there is no explicit route from a host in the VLAN to external destinations.
- Without the default gateway, each host in the VLAN would have to use a Routing Table containing explicit routes to destinations on the Internet, a solution that is not workable.

FIGURE 1.3 Default Route for an IP Host

A router may use a default route as a last resort when all other routes (directly connected, static, or dynamic) in the Forwarding Table have not produced a match for a destination address (see Figure 1.4). If a directly connected, static, or dynamic route exists for the packet's destination, the router forwards the packet out the appropriate outgoing interfaces to the next-hop router.

If the packet's destination (i.e., next-hop router and outbound interface) is unknown, that is, no routing method has produced a learned route, the router has no other choice but to use a default route if one is configured. Typically, a network administrator would implement default routes on point-to-point link (i.e., a link interconnecting two routers) for example, between a company's network to the outside world.

Generally, configuring a default route on a router is not necessary because the router should already have the capability to learn routes and forward packets by consulting its Forwarding Table for known network destinations. However, if the router has no learned route to a destination, it may use the default route (the route of last resort).

1.2.4 Dynamic Routing Protocols

A routing function (in a router or switch/router) learns about (or discovers) other networks through the use of dynamic routing protocols or via static configuration by a network manager. A routing protocol is the special set of rules a router uses when it communicates with other routers in order to share information about the status and reachability of networks. Routers use routing protocols in addition to other

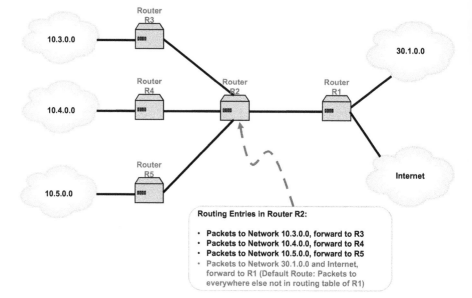

FIGURE 1.4 Illustrating How Default Routing Works

mechanisms to build and maintain routing information, and to communicate this to other routers.

Dynamic routing protocols operate by distributing network topology information and routing updates between the various routing functions without manual intervention anytime network topology changes occur. After discovering or mapping out the network topology, each routing function tries to select the best loop-free path in a network, on which each arriving packet should be forwarded to get to its destination.

A network or subnet that is not directly attached to any one of a router's local interfaces is considered a remote network. A router can only reach a remote network by sending packets to another router, that is the next-hop IP router. Remote networks are discovered and added to the Routing Table by static route configuration or via dynamic routing protocols. As discussed above, static routes are those configured manually to other networks by a network administrator. A router also learns dynamic routes to remote networks automatically using dynamic routing protocols.

Even though the primary responsibility of a dynamic routing protocol is to automatically learn best paths to network destinations, and detect and adapt to changes in network topology and state, this is done at the price of increased router memory and CPU processing time, in addition to higher network bandwidth. In practice, a dynamic routing protocol becomes a good choice only after the user determines that static routing is not a practical solution for the network under consideration. Obviously, the network manager will not incur so much administrative overhead to implement static routing in a smaller network or one that has a hub-and-spoke topology, but this is far from what real-world networks look like.

Introduction to IP Routing Protocols 7

The various dynamic routing protocols are discussed in more detail in Chapter 2. To summarize, the various routing protocols discussed are responsible for the following:

- **Sending and Receiving Routing Information from Other Routers**: Each router supports procedures for advertising and receiving reachability information about network destinations.
- **Computing the Best Paths to Network Destinations and Installing These Routes in the Routing Table**: Each router supports procedures for computing optimal or best routes to network destinations based on the reachability information it has received, and for installing these routes in its Routing Table.
- **Detecting and Reacting to Changes in the Network State and Topology, and Communicating These to Other Routers**: Each router supports procedures that defines how the router (and consequently, the network) reacts to, compensates for, and advertises network topology and state changes.

Each routing protocol supports, at a minimum, built-in mechanisms for best path determination, assigning costs or metrics to network routes, routing loop prevention, facilitating faster network convergence, and load balancing across multiple paths. Each protocol supports its own mechanisms for ensuring that other routers will receive routing update information while at the same time preventing such routing updates from looping, that is, circulating endlessly in the network.

As networks evolved to handle more network nodes and users, and consequently became larger and more complex, newer routing protocols have been developed over time. RIPv2 was developed as an improvement to RIPv1, and includes improved addressing and security features that allow it to work better in today's network environments. However, RIPv2 still does not scale well for implementation in larger networks. So, advanced routing protocols such as EIGRP (designed by Cisco), OSPF, and IS-IS were developed with the goal of addressing the needs of larger and more complex networks.

Furthermore, because of the need to interconnect and provide routing between different large and complex internetworks, BGPv4 (Border Gateway Protocol) was developed. With the advent of IPv6, newer versions of the routing protocols were also developed for IPv6, such as IPv6 RIPng, EIGRP for IPv6, OSPFv3, IS-IS for IPv6, and Multiprotocol BGP (MBGP).

Newer routing protocols such RIPv2, EIGRP, OSPFv2, and IS-IS are all classless routing protocols and include the network mask in their routing updates. Classless routing or Classless Inter-Domain Routing (CIDR) **[RFC1517] [RFC1518] [RFC1519] [RFC4632]** allows the use of Variable Length Subnet Masks (VLSM) **[RFC1878]** and provides better and efficient route summarization.

In spite of the availability of different routing protocols with different capabilities, selecting the right protocol for a network is still a complex task. Determining which routing protocol (whether IGP or EGP) is suitable for a particular network is often influenced by a number of factors. At least, the following factors have to be considered when selecting a routing protocol for a network: The size and complexity of the network; support for VLSMs; the expected traffic levels in the network; the expected maximum latency characteristics of the network; routing protocol security needs; reliability expectations of the network; desired organizational routing policies for the network. Many of these factors also influence the Routing Table properties and, the type of path control tools that can be used with each routing protocol.

1.2.4.1 Routing Updates

Routers (via their routing protocols) exchange routing updates that allow them to learn about networks and routes to network destinations. Routing updates carry network state and reachability information, and routers rely on these to construct and maintain their Routing Tables. A router running routing protocols such as RIPv1 transmit routing updates in the form of broadcast messages to all other routers on the same network, while those running protocols such as RIPv2 **[RFC2453]** and OSPF **[RFC2328]** transmit updates as multicast messages addressed to routers belonging to a multicast group on the network.

- RIPv1 broadcast routing updates addressed to the all-broadcast address 255.255.255.255. These updates are received and processed by all routers (including hosts) on the local network. This can create unnecessary traffic processing overhead for IP hosts that do not function as routers.
- RIPv2 transmits routing updates to all other RIP2-aware routers on a network using the IPv4 multicast address 224.0.0.9.
- EIGRP **[RFC7868]** transmits routing updates to all other EIGRP-aware routers on a network using the IPv4 multicast address 224.0.0.10.
- OSPF transmits Hello packets to all OSPF routers on a network using the All OSPF Routers IPv4 address 224.0.0.5.
- OSPF transmits OSPF routing information to designated routers on a network using the All Designated Routers (DR) IPv4 address 224.0.0.6.

Sending routing updates as multicast messages reduces the network traffic, and also helps in reducing the processing overhead in routers which are not target of a routing update message type (an OSPF router that is not a Designated Router for a broadcast network segment will not have to worry about listening to the IPv4 address 224.0.0.6). Only the routers that are a target of a routing update type would join the multicast group to which the routing update is addressed.

1.2.4.2 Periodic Versus Triggered Routing Updates

Some routing protocols (e.g., RIPv1 and RIPv2) use period timers to determine the instance of transmission of routing updates. These timers control how often routers send routing updates to other routers. Upon startup, a router, irrespective of the routing protocol it is running, would transmit its entire Routing Table to its neighbor routers. Routers are considered neighbors if they can communicate (reachable) over a common data link (Layer 2) protocol and are running the same routing protocol. A routing protocol transmits its routing updates to neighbors, and in turn relies on those routers to propagate this routing information on to their neighbors.

At startup, each router only knows about its directly connected networks or routes. Then through the routing protocols, the routers would then start exchanging

Introduction to IP Routing Protocols

information about their directly connected routes. Thereafter, the router would send routing updates (depending on the routing protocol type it is running), either driven by periodic timers (i.e., periodic updates), or when a recognizable network event takes place (i.e., triggered updates).

When a router receives periodic routing updates from a neighbor, it compares the routes received to the routes it has already installed in its Routing Table. If the Routing Table already contains a better route (e.g., one with a lower Administrative Distance), the router does not enter the new route into the Routing Table. New and better routes learned from neighbors are always installed in the Routing Table. When the router sends its next routing update, it will advertise the better routes in its Routing Table plus the new routes it has just learned from its neighbors.

In the case where updates are event driven (triggered updates), the router does not send periodic routing updates, but instead would send an update, for example, only when a change in the network occurs – this event causes or triggers an update to be sent. EIGRP does not send periodic updates like RIPv1 and RIPv2, but does so only when there is a change in the network topology. RIPv2 was designed to support triggered updates unlike RIPv1 which supports only periodic updates. Routing protocols that transmit periodic routing updates to their neighbors (solely driven by periodic timers), do so even when no change in network topology or state has occurs, creating network bandwidth wastage.

1.2.4.3 Routing Information Authentication

Routing updates when not properly secured, can easily be intercepted by unauthorized entities in a network when exchanged between routers. Securing a network includes securing not only the user data that flow through the network, but also flowing between the various routing and control functions in the network. The exchange of routing updates must be secured to ensure that the routing information the routers enter into the Routing Tables is valid, and does not come from a malicious source intending on disrupting the operations of the network.

A malicious source may try to introduce invalid routing updates during the routing exchange process with the intent of seriously degrading the performance of the network, or to fool a router into forwarding data to a wrong network destination. To prevent this, and to allow for secure exchange of routing updates between routers, modern routing protocols support various forms of authentication mechanisms.

Older routing protocols like RIPv1 do not support mechanisms for authenticating a neighbor router before routing information is exchanged. Some routing protocols support only plain text authentication using simple clear-text passwords, which offers some degree of routing security but can be breached by hackers scanning the network with just simple protocol analysis tool. Other routing protocols RIPv2, EIGRP, OSPFv2, IS-IS, and BGPv4, support more sophisticated security mechanism such as the use of cryptographic authentication.

Two different types of authentication are supported in RIPv2, which are plain text authentication (using clear-text passwords) and Message Digest 5 (MD5) authentication **[RFC2453] [RFC4822]**. RIPv1 does not support any form of authentication. OSPF supports three different types of authentication, namely, null authentication (where the OSPF packet header carries no authentication information), plain text authentication (which uses simple clear-text passwords), and MD5 authentication (which uses MD5 cryptographic passwords) **[RFC2328]**.

1.2.4.4 Routing Information and Network Convergence

Convergence is the process describing how long it takes for all routers (running a specific routing protocol within a routing domain) to agree on reachability information to network destinations. Convergence is achieved when all the routers have agreed on the optimal routes to network destination and thereby have completed the update of their Routing Tables. Routing convergence takes place as a result of changes in the network state or topology, for example, when a network link fails or becomes available.

When a link goes down, fails or becomes available in a network, routers send routing updates to other routers in the network describing the changes that have occurred. Each router then runs a routing algorithm (specific to the routing protocol it is running) to recompute the best routes based on the routing information received, and installs these new routes in the local Routing Table. When all routers in the network have received and processed routing updates, and all have a consistent view of the network topology and best routes, then they (and the network) can be deemed to have converged. The process of adapting to the network changes and arriving at a consistent result is referred to as convergence.

The router can still forward packets before convergence is completely achieved. Each router is not even aware of this process, and so, will not stop forwarding packets it receives. However, forwarding packets when convergence has not been achieved may lead to routers using less than optimum routes, which may also result in routing loops being created in the network, or packets being dropped because destinations are deemed unreachable.

Distance-vector routing protocols such as RIP have slow convergence and therefore routing domains running such protocols are highly susceptible to routing loops. Routing protocols like EIGRP, OSPF, and IS-IS have faster convergence making the occurrence of routing loops less likely.

Specifically, convergence time is the time it takes for all routers in the routing domain to receive routing updates and arrive at a consistent view of the network and best routes. It is a measure of how fast a group of routers process routing updates, modify their Routing Tables and agree on a consistent reachability information to network destinations.

Convergence time is a very important performance measure for routing protocols, and has become one of the key design goals for modern routing protocols. The faster the routers in the network are able to share routing information and reach a state where they all have a consistent knowledge of the network state or topology, the more preferable the protocol. As discussed above, routing loops can occur when the routers have inconsistent knowledge of the network and the Routing Tables are not correctly updated due to slow convergence when changes in the network occur.

Newer routing protocols implement several mechanisms that are designed to enable all routers in a network running that protocol to converge quickly and reliably. The size of the network (i.e., the routing domain) also definitely plays an important role in the convergence time. For the same routing protocol, a smaller network will converge faster than a larger one.

Introduction to IP Routing Protocols

Generally, the speed of convergence is directly proportional to the size of the network. The speed of convergence is influenced by how fast the routers propagate information about a change in the network state or topology to their neighbors, and how fast the routers calculate best routes using the new routing information they have received. A network is generally not completely stable and correctly operational, until convergence is achieved, making routing protocols with shorter convergence times preferable.

RIPv2 has very slow convergence even in a network with a small number of routers; the network can take a couple of minutes to converge. Thus, in the event a new route is being advertised, routers can use triggered updates to speed up the convergence of RIPv2. However, if a router wants to flush a route that previously existed in its Routing Table, this process can take a long time due to the time delay imposed by the Holddown timers used by RIPv2 (see discussion in Chapter 2). EIGRP and OSPF are faster-converging routing protocols, and a network consisting of a small of OSPF routers can converge in the order of just seconds.

1.3 AUTONOMOUS SYSTEM

Within the Internet, an Autonomous System (AS) is a group of interconnected routers and network address prefixes owned or under the control of one or more network operators (e.g., organizations) but are managed by a single administrative entity (e.g., Internet service provider [ISP]). Furthermore, this interconnection of routers and network prefixes share a common routing policy or plan, and are presented by the administrative entity to the Internet under this common, clearly defined routing policy. The entire Autonomous System is viewed by the outside world as a single entity.

A routing policy here refers to how routing decisions are made within the administrative entity/domain. Using its routing policy, the Autonomous System presents a consistent and coherent view of the network destinations that can be reached through it to other Autonomous Systems. A routing or network prefix represents, here, a group of IP addresses that can be reached through the network of the administrative entity, for example, the ISP's network.

1.3.1 What Is a Network Prefix and a Route?

A network address prefix is a contiguous set of the most significant bits in an IP address and represents, collectively, a set of systems within a network. In CIDR, the network prefix represents the network portion of an IP address. In IPv4 networks, the network portion of an IP4 address is referred to as the "network part", while the remaining bits of the address make up the "host part" of the address. For a given IPv4 address and network part, the host addresses are selected from the host part. In IPv6 networks, the network portion of an IPv6 address is referred to as the "subnet prefix", while the remaining bits represent the "interface identifier". The network prefix length is the number of contiguous bits that make up the network prefix portion of an IP address. For example, the network part of the IPv4 address 10.10.126.0/17 is 17 bits wide, that is, the network prefix length is 17.

A route is defined with respect to a specific next hop to which packets can be sent on their way toward their destinations (as defined by a destination network address prefix). A route is the basic unit of information about a specific network destination discovered by the routing protocols, and is a candidate for the Routing Table of a router. In general, a route is expressed as the n-tuple <IP address prefix, Next Hop [...other routing or non-routing protocol attributes...]> **[RFC4098]**.

1.3.2 Autonomous System Numbers (ASNs)

The public Internet is made of tens of thousands of Autonomous Systems that span the whole world (Figure 1.5). If an Autonomous System connects to other systems or the public Internet using a routing protocol such as BGP, then it must be assigned a unique Autonomous System Number (ASN). To be connected to the Internet, each Autonomous System is assigned an ASN which uniquely identifies it globally. An ISP might support multiple organizations (each with its own network prefixes) but the ISP will use its own unique routing policy to connect these network prefixes to the Internet.

Each organization might have multiple private ASNs and then use BGP to connect these ASNs to an ISP, which in turn will connect all the organizations it supports to the Internet. The routing policy seen by the Internet is that of the ISP and not that of the individual organizations or Autonomous Systems behind it even though the ISP

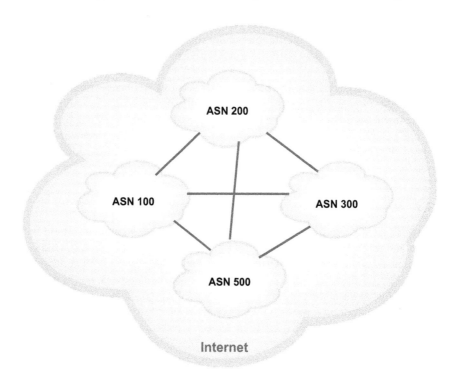

FIGURE 1.5 Internet as a Group of Autonomous Systems

may be supporting multiple small Autonomous Systems. An ISP is officially assigned and identified by a registered ASN. Each Autonomous System that communicates with the public Internet has a unique ASN which it uses in BGP routing.

There are two types of ASNs:

- **Public ASNs**: This is used when an Autonomous System is required to exchange routing information with other Autonomous Systems on the public Internet. The routes exchanged will be visible on the Internet along with their public ASNs.
- **Private ASNs**: This is used if an Autonomous System only needs to communicate with a single provider or peer system via BGP without making this visible on the public Internet. The routes exchanged along with their private ASNs will not be visible on the Internet.

An Autonomous System requires an officially registered ASN to be visible to other Autonomous Systems on the Internet and when it needs to exchange routing information with multiple Autonomous Systems on the public Internet. However, if an Autonomous System only needs to communicate privately with a peer or an ISP via BGP, then a private ASN can be used which makes the Autonomous System not visible on the public Internet. The Internet Assigned Numbers Authority (IANA) allocates ASNs in blocks to Regional Internet registries (RIRs) and each RIR in turn assigns the block assigned by the IANA to entities (e.g., ISPs) within its designated area.

Each ISP uses its ASN and BGP to exchange network prefix information with other ISPs in the (public) Internet. As will be seen in Chapter 2, BGP is the standard protocol for exchanging routing information among Autonomous Systems in the Internet. Routers use the ASNs and network prefixes to identify the originating or destination Autonomous System of Internet routing information. These identify where a particular Internet routing information comes from, and where in the Internet, a piece of routing information (originating from inside an ISP's network) should be sent to. The ASNs are used to control routing within an ISP network and to exchange routing information among ISPs.

1.3.3 Multiple Routing Domains in an Autonomous System

An Autonomous System could run a single Interior Gateway Protocol (IGP) (see more about IGP below), or could be a collection of routing domains each running a different IGP and working together to provide interior routing for the Autonomous System (see Figure 1.6). A routing domain represents a collection of routers running a common routing protocol and under a common administrative control. The role of the IGP in the routing domain is to provide routing connectivity between the routers within the routing domain. The routers running the common IGP also run a common best path selection algorithm to determine the best paths to each network destination.

In addition to advertising network reachability information within a given routing domain, an IGP can be used to advertise routing information from its routing domain to another routing domain running a different IGP – a process known as route

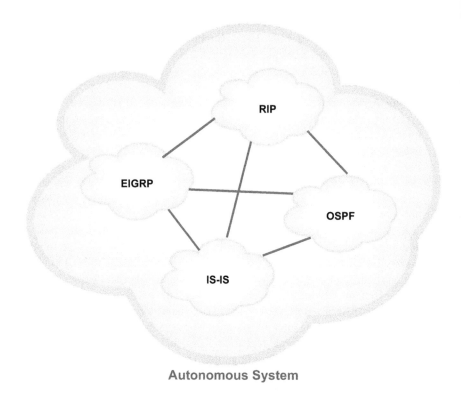

FIGURE 1.6 Different Routing Domains in an Autonomous System

redistribution. Route redistribution is the process by which distinct IGPs running in different routing domains, exchange routing information between themselves in order to tie the multiple routing domains within the Autonomous System together when intra-Autonomous System connectivity is required. Route redistribution allows the different routing domains and IGPs to learn and advertise routing information, and to build Routing Tables that contain entries for all routes (network prefixes) for the entire Autonomous System.

Routers outside the Autonomous System see it as a single entity although it might be a collection of smaller routing domains each running a different IGP. The concept of Autonomous Systems allows each one Autonomous System to run its own set of IGPs independent of the IGPs used in other Autonomous Systems. Each Autonomous System could implement its own set of routing protocols and policies that uniquely distinguishes its networks and services from those of other Autonomous Systems.

1.3.4 Types of Autonomous Systems

There are three types of Autonomous Systems:

- **Stub Autonomous System**: This type of Autonomous System has a single connection to one other Autonomous System. Communication to or from a

Introduction to IP Routing Protocols

destination outside the Autonomous System is done over this single connection. A stub may have peering or private connections to other Autonomous Systems that are not visible on the public Internet but the stub will appear to have only a single connection to the public Internet.
- **Transit Autonomous System**: This type of Autonomous System connects one Autonomous System to another and allows communication between them to pass through it. An ISP is an example of a transit Autonomous Systems if it offers other Autonomous Systems access to other Autonomous Systems in the Internet.
- **Multihomed Autonomous System**: This type of Autonomous System maintains connections to two or more Autonomous Systems, but it does not allow traffic from one Autonomous System to pass through on its way to another Autonomous System. This allows the (multihomed) Autonomous System to maintain connectivity to the Internet even if one connection fails. Traffic received over one of these connections will not be forwarded out of the Autonomous System to another. This type of Autonomous System does not provide a transit service to other Autonomous Systems. A multihomed Autonomous System is similar to a stub Autonomous System, except that the ingress and egress paths for traffic traveling to or from the Autonomous System can be selected from one of these (multiple) connections, depending on which connection offers the best route to the ultimate destination. Many large corporate networks are normally designed as multihomed Autonomous Systems.

1.4 ROUTING METRICS AND COSTS

When multiple paths exist to a network destination, a router must have a way of determining the best or optimum path to that destination. Routing metrics are numeric values that routing protocols (RIP, EIGRP, OSPF, etc.) use to determine the best path to a network destination. Each routing protocol defines its own set of metrics which it uses to decide which path among multiple paths is the preferred or best path to a particular destination (Figure 1.7).

Each route to a destination is assigned a metric to provide the routing protocol a means of ranking the multiple routes to that destination, from best (or most preferred) to worst (or least preferred). Note that a routing metric is different from the Administrative Distance or Route Preference (discussed in Chapter 2). The most common routing metrics are hop count, bandwidth, delay, traffic load, reliability, and cost. Some routing protocol use a single metric only while others may use a combination of metric, or multiple metrics, as desired by the user. These routing metrics are described below.

1.4.1 HOP COUNT

A hop count is a metric used to measure the distance from a particular router to a network destination and is based on the number of (intermediate) routers a packet would traverse to get to that destination. Each router a packet crosses counts as a single hop. A routing protocol that uses hop count as its primary metric views the

16 IP Routing Protocols

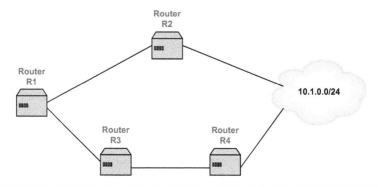

- When a router discovers two different routes to the same network using the same routing protocol, it has to decide the best route among the two to be installed in its routing table. RIP, for example, uses hop counts as its metric.
- In this figure, router R1 has two paths to reach the network 10.1.0.0/24. The path through router R2, has the lowest hop count and will be the one installed in the routing table of router R1.

FIGURE 1.7 Routing Metric

preferred or best path to a network destination (among multiple paths) to be the path with the smallest hop count. The underlying assumption when a single hop count metric is used, is that, the path with the least number of hops to a network destination is the best path. A network that is directly connected to a router interface is assigned a hop count of zero.

RIP uses a routing metric of hop count to determine the best route to a network destination when multiple routes are available. Each router in a RIP network uses the total number of hops on each route between itself and a given network destination as the routing metric when determining the best route to install in its IP Routing Table. The route having the fewest number of hops to the destination is considered the best route: the lowest-cost route.

1.4.1.1 Network Diameter: Maximum Hop Count

RIPv1 and RIPv2 are limited in scope, that is, they have a maximum network diameter which limits the travel distance of routing messages (measured from transmitter to receiver). The maximum network diameter or scope (also referred to as the maximum hop count in RIP) specifies the maximum distance a routing message can travel before the routing protocol considers the network destination to be unreachable. When the maximum distance is exceeded, the routing message will be discarded. The main reason for specifying a maximum hop count is avoid routing loops; this prevent packets from looping around unceasingly.

Both RIPv1 and RIPv2 support just a maximum hop count value or metric of 15. This means, any RIP-aware router that is more than 15 hops away is considered as unreachable. As routing messages traverse the network, RIP-aware routers receive them and increment the hop count by one before forwarding them to the next router. When a routing message reaches the 16th RIP-aware router, that router would discard the message because a hop count vale of 16 or greater is considered out of scope by RIP; a network with a hop count of 16 is considered unreachable. The router that discards the routing message would generate an ICMP message to be sent back to the source router indicating that the network destination is unreachable.

1.4.1.2 "Infinity" Hop Count as a Signaling Mechanism for Network Failures

A router may also use the maximum hop count value carried in a routing message as a mechanism for signaling failures in a network. When a network link on a router fails, the affected router can transmit information about this failure in its next routing update to other routers. The router communicates this failure by writing a hop count value that is one higher than the maximum value (i.e., 16 for RIPv1 and RIPv2), which indicates to the receiving routers that the affected network is unreachable. Routers receiving this routing update will be appropriately informed and will remove this (failed) route from their Routing Tables.

1.4.1.3 Limitations of the Hop Count as a Routing Metric

The routing metric of hop count ignores differences in link speed, link utilization, and all other factors, many of which can be important in selecting the best path to a destination. Routing protocols that use only a hop count as their metric do not necessary provide the best path to a network destination. The path with the smallest number of hops to a destination does not necessarily make it the best path (if multiple paths exist). The path with the least hop count may contain the slowest links (10 Mb/s) while the path with the most hops may contain the fastest links (1,000 Mb/s). However, routing protocols such as RIP use only hop count and were not designed to consider other metrics in their routing decisions. The small hop count range of 0 to 15 also makes RIP unsuitable for large networks.

1.4.2 Bandwidth

A routing protocol can use the bandwidth or capacity of an interface or link (measured in bits per second) as a routing metric. In this case, links that support higher bandwidth (Gigabit Ethernet) are preferred over lower bandwidth links (100 Mb/s Ethernet). To determine the best paths, the routing protocols considers the bandwidth of each link along the path up to the network destination. The best path is then taken as the path with the overall higher bandwidth to that destination.

Unlike using the hop count metric, a path with a higher number of hops may have a higher overall bandwidth while a path with fewer hops may have lower overall capacity. In this case, the path hop count is irrelevant in the best path decision-making process. A routing protocol using solely bandwidth as a metric would choose a higher-bandwidth path over a lower-bandwidth path regardless of the path state (e.g., delay, traffic load, reliability, etc.).

Bandwidth by itself may not be a suitable routing metric because it does not incorporate other useful information about the path, such as when it is heavily loaded with traffic and when it is lightly loaded. This is because a higher-bandwidth path when chosen as the best path, could be heavily loaded and have a higher end-to-end delay.

1.4.3 Delay

The delay over a path is a measure of the time (in microseconds) it takes a unit of data (a packet) to traverse that path. A routing protocol that uses path delay as a routing metric would choose the path with the smallest end-to-end delay (among multiple paths to the same network destination) as the best path. Path delay may include the node processing delay, transmission delay (i.e., the time it takes to place data on the transmission medium), queuing delay, and propagation delay (which is the travel time over the transmission medium). Most often, the delay along the path is dominated by factors such as router processing latency and queuing delay.

A router may implement mechanisms for delay measurements, or the path delay may be not measured at all, but instead it may be static quantity defined for the path. The delay may be an estimate based on the type of links that make up the end-to-end path starting from the originating interface on the router. The performance of the best path selection process of the routing protocol depends very much on how accurate the path delay quantities are.

1.4.4 Traffic Load

This routing metric is a measure of the amount of traffic utilizing the links that make up a path to a network destination. In this case, the routing protocol that uses this metric would choose the path with the lowest load (among multiple paths to the same network destination) as the best path. The load metric can be a measure of the amount of traffic occupying the slowest link on the path over a measurement time period, and expressed as a percentage of the link's total bandwidth.

Unlike the hop count and bandwidth metrics, the traffic load on a path is dynamic and changes from time to time. Therefore, the traffic load metric also changes, a factor the routing protocol must take into account when determining best paths. This means the routing protocol must handle frequent metric changes careful to avoid route flapping. Route flapping occurs when a router alternates the advertisement of a destination network from one route to another and then back to the first route in quick (alternating) sequences.

The load metric used by a routing protocol may be manually configured as a static value by a network administrator for a path, or it may be dynamically measured, allowing the routing protocol to adapt to traffic changes in the network. Additionally, a routing protocol may measure the traffic load to recognize when a path to a destination is becoming heavily loaded or congested, and use an alternate path to that destination, if available.

1.4.5 Reliability

The reliability metric reflects the degree to which a particular path to a network destination can be dependent on to be operational and useable. The reliability metric assigned to a path can be either fixed or variable and changes depending on other factor network factors. A fixed reliability routing metric is generally based on a defined value (as determined by the network administrator) assigned to a path that

Introduction to IP Routing Protocols

reflects the quality of the links that make up that path. The routing protocol will choose the path (among multiple paths to the same network destination) with highest reliability as the best path.

A variable reliability routing metric for a path can be based on the number of times a link (on the path) has failed, or the number of transmission and data errors it has experienced within a specified time period. Routers running a routing protocol that uses a reliability metric may observe its attached interfaces and links to record relevant error statistics and problems, such as interface errors, lost packets, link failures, etc.

The router would then consider links experiencing more problems to be less reliable than those experiencing less – the higher the reliability of the constituent path links, the better the path for routing. The routing protocol may rank the paths that contain links with more problems as less desirable paths. If the reliability metric is a measured quantity, then given that network conditions are continuously changing, the path reliability metric will change accordingly.

1.4.6 Cost

Cost is a generic term that holds the same meaning as metric. It is almost pointless to debate the differences since both terms are measures that a routing protocol uses to decide which path (among multiple paths to a given destination) is the best one to forward packets on. In general, cost represents the overhead required for a router to forward packets across a certain interface/link or path.

1.4.6.1 Example: OSPF Cost

OSPF assigns a cost to each interface (or equivalently, link) on a router that can be used to reach a network destination. OSPF routing metric based on cost and it is the routing metric that OSPF uses in its link-state calculations.

Routes (to a particular network destination) with lower total path costs are preferred (or considered to be best paths) compared to those with higher path costs. To understand how OSPF interface costs are used, we describe how the Shortest Path First (SPF) algorithm works. The SPF algorithm is used to construct and calculate the shortest path to all known network destinations and is based on the Dijkstra algorithm.

Upon OSPF initialization on a router, or when there is the need to communicate any change in routing information to other routers, an OSPF router generates a Link-State Advertisement (LSA) to be sent to those routers (see discussion on link-state routing protocol and OSPF in Chapter 2). The LSA contains all link-states on the originating router and is propagated to other OSPF routers in the network. Each OSPF router that receives this LSA would write a copy of the contents in its local link-state database (LSDB) and then propagate this LSA to other OSPF routers in the network.

1.4.6.2 Cost Based on Interface Bandwidth

After OSPF has updated the LSDB, it runs the SPF algorithm to select the best route to each destination from all available routes learned. The SPF algorithm uses a cost to compute and select the best route for installation in the Routing Table. Given that a packet will take less time in crossing a lower bandwidth 10 Mb/s link than crossing a higher bandwidth 100 Mb/s link, OSPF uses this understanding to calculate the cost for each path. The cost is taken as inversely proportional to the link bandwidth, meaning a lower bandwidth interface/link has a higher cost while higher bandwidth one has a lower cost. OSPF defines the following expression for calculating the cost:

Cost = Reference bandwidth / Interface bandwidth in bps.

Reference bandwidth in this expression is defined as an arbitrarily chosen value for OSPFv2 **[RFC2328]**. Different OSPF implementations can use their own reference bandwidth values, but Cisco implementations use reference bandwidth value of 100 Mb/s (10^8). Using this reference bandwidth, the cost equation becomes:

Cost = 10^8 / interface bandwidth in bps

For example, a 10 Mb/s link has a cost of 10 while a 100 Mb/s link has a cost of 1. Some key points to note about the cost computation are that the cost is a positive integer value meaning any resulting decimal value computed is rounded to the nearest positive integer. Also, any value that is below 1 is always rounded up or considered as 1. The reference bandwidth can be changed to handle higher speed links.

1.5 CLASSIFICATION OF ROUTING PROTOCOLS

As shown in Figure 1.8, routing protocols can be categorized into different groups according to where they operate (within an Autonomous System or between Autonomous Systems), or their characteristics from a design and operational point of

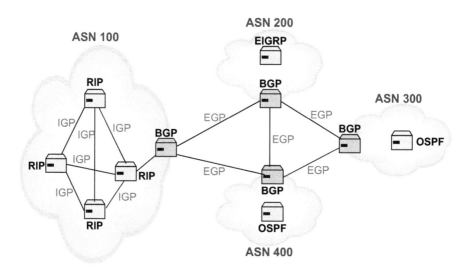

FIGURE 1.8   Illustrating IGP Versus EGP Routing

Introduction to IP Routing Protocols

view (distance-vector, link-state, or path-vector routing). These classifications are discussed in this section using the most common IP routing protocols.

1.5.1 INTERIOR VERSUS EXTERIOR ROUTING PROTOCOLS

The standard-based dynamic routing protocols commonly used in today's networks are RIPv2, OSPF, IS-IS, and BGP. The various classes and types of routing protocols are further illustrated in Figure 1.9. The Enhanced Interior Gateway Routing Protocol (EIGRP) is proprietary routing protocol developed by Cisco and mainly on Cisco routers. The various IGPs and EGPs shown in Figure 1.9 use different algorithms for routing computation and best path selection.

The routing protocols RIP, EIGRP, OSPF, and IS-IS are known as Interior (or Internal) Gateway Protocols (IGPs), and are designed and optimized for exchanging routing information within an Autonomous System. The IGPs allow a router to learn about routes to networks that are within or internal to an Autonomous System. The IGP is responsible for constructing, maintaining, and distributing routing information within a single Autonomous System. An Autonomous System may be running one or more IGPs to map out the routes to networks or subnets within that Autonomous System.

When routers in an Autonomous System run the same IGP (e.g., OSPF), they only share routing information with other routers in the system running that IGP. Each IGP (RIP, EIGRP, OSPF, or IS-IS) running in a network within an Autonomous System represents a separate IGP routing domain. Thus, a router running multiple

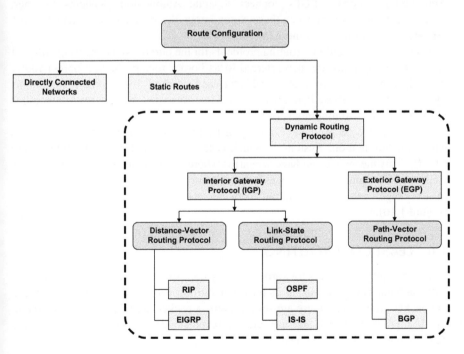

FIGURE 1.9 Types of Unicast Dynamic Routing Protocols

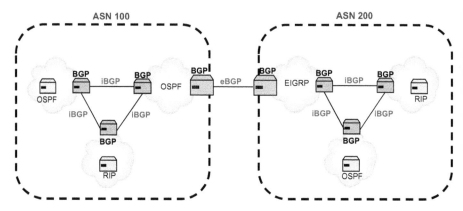

FIGURE 1.10 Illustrating iBGP Versus eBGP Routing

IGPs, for example, OSPF and RIP, is considered a member of two separate IGP routing domains in the Autonomous System. Such multirouting protocol routers are called border routers because they are located on the border between the two separate IGP routing domains.

BGP is an Exterior (or External) Gateway Protocols (EGP) which is used for exchanging routing information between Autonomous Systems (Figure 1.10). In this case, BGP is referred to as an external BGP (eBGP). EGPs, such as BGP, are typically deployed to serve as entry/exit points for communication between different Autonomous Systems. EGPs connect separate Autonomous Systems together, providing transit paths between these Autonomous Systems, thereby facilitating the forwarding of traffic across these internetworks.

Routers that run multiple routing protocols for the purpose of connecting different Autonomous Systems are also referred to as border routers. Such a border router typically runs an IGP over the local interfaces attached to its home Autonomous Systems, and an EGP (BGP) over the external interfaces that connect this Autonomous System to remote Autonomous Systems.

BGP can also be used for exchanging routing information between two peers within an Autonomous System, in which case, it is referred to as an internal BGP (iBGP). This means we can have scenarios where we have one or multiple IGPs, as well as, iBGP running within a single Autonomous System, and eBGP sessions used to exchange routing information between the various Autonomous Systems (Figure 1.10).

1.6 LEAST-COST ROUTING

During routing, a packet is forwarded, hop by hop, from its source to its destination, with each hop (or router) using its local Forwarding Table to decide how to forward the packet. Most often, there are several paths a packet can take from the source to the destination, and each hop or router must independently determine which is the best route the packet should take.

Introduction to IP Routing Protocols

A network can be modeled as a weighted graph, in which each edge is assigned a cost (or metric). A router selects the best route to a destination to be the one has the least total cost among all the routes that exist to that destination. A least-cost tree can be described as a tree with the source router placed at the root, with the tree visiting all other nodes (routers and destinations), plus the requirements that the path between the root and any other node be the shortest. In such a network, there is only one shortest-path tree for each node.

In distance-vector routing (see Section 2.2), each router in the network (or routing domain) first constructs its own least-cost tree with the rudimentary or minimum routing information it has received from its neighbor routers. Each router then exchanges its incomplete tree or Routing Table with its neighbors (via routing protocol updates) to allow all routers in the network to eventually construct more complete trees that represent the whole network.

Each distance-vector router uses the Bellman–Ford algorithm to find the least cost path between itself and a destination network through a number of intermediary nodes. For each destination network, a router constructs a distance-vector which is a one-dimensional array or vector that represent the destination network on the router's least-cost tree. Upon startup, each router creates a very rudimentary tree with the information describing its directly attached networks.

After a distance-vector router has created its Routing Table, it will transmit a copy of this (via routing updates) to all its neighbor routers. When a router receives a Routing Table from a neighbor, it updates its own Routing Table by running the Bellman–Ford algorithm. Also, after a router updates its Routing Table, it will immediately transmit its updated Routing Table to all its neighbors. To prevent routing loops, speed up network convergence, and improve network stability, the routers use mechanisms such count-to-infinity, split horizon and poison reverse (see Chapter 2).

In link-state routing, the term link-state is used to define the characteristic of an edge, or equivalently, a link in a network. The cost associated with a link defines the state of the link in the network. In order to construct a least-cost tree based on link-state routing, each link-state router requires a complete topological map of the network. This means each router needs to know the state of each link in the network. Each router learns and collects the states of all links in the network and maintains this in a link-state database (LSDB), as discussed in Chapter 2.

Each link-state router sends out network reachability information via LSA messages to all of its immediate neighbor routers with the goal of gathering information from each neighbor. The information includes the identity of each neighbor router and the cost of their directly connected link. Each router sends this information in link-state messages out of each of its active interface to all neighbors allowing all to have a common view of the network topology and identical LSDBs. Each link-state router then runs the well-known Dijkstra algorithm to create a least-cost tree for itself using the information in the common LSDB. The router selects itself as the root of the tree, creating a least-cost tree originating from the root, and sets the total cost of each router in the network based on the information in the LSDB.

REVIEW QUESTIONS

1. What is the difference between static routing and dynamic routing?
2. Explain briefly the difference between a default route in an IP host and a default route in an IP Router.
3. What are the main functions of a dynamic routing protocol?
4. What is a routing update?
5. What is the difference between periodic updates and triggered updates?
6. Why do dynamic routing protocols use authentication mechanisms?
7. What is the meaning of convergence in the operation of a routing protocol?
8. What is an Autonomous System in IP routing?
9. What is an Autonomous System Number?
10. What is the difference between Public Autonomous System Numbers (ASNs) and Private ASNs?
11. What routing metric does RIP use?
12. Why does RIP define a maximum network diameter or hop count?
13. What are the limitations of using hop count as a routing metric?
14. What does sending a hop count of 16 signify in RIP?
15. What routing metric does OSPF use?
16. What is the difference between an Interior Gateway Protocol (IGP) and an Exterior Gateway Protocol (EGP)?

REFERENCES

[RFC1517]. R. Hinden, Ed., "Applicability Statement for the Implementation of Classless Inter-Domain Routing (CIDR)", IETF RFC 1517, September 1993.
[RFC1518]. Y. Rekhter and T. Li, "An Architecture for IP Address Allocation with CIDR", IETF RFC 1518, September 1993.
[RFC1519]. V. Fuller, T. Li, J. Yu, and K. Varadhan, "Classless Inter-Domain Routing (CIDR): An Address Assignment and Aggregation Strategy", IETF RFC 1519, September 1993.
[RFC1878]. T. Pummill and B. Manning, "Variable Length Subnet Table For IPv4", IETF RFC 1878, December 1995.
[RFC2328]. J. Moy, "OSPF Version 2", IETF RFC 2328, April 1998.
[RFC2453]. G. Malkin, "RIP Version 2", IETF RFC 2453, November 1998
[RFC4098]. H. Berkowitz, E. Davies, Ed., S. Hares, P. Krishnaswamy, and M. Lepp, "Terminology for Benchmarking BGP Device Convergence in the Control Plane", IETF RFC 4098, June 2005.
[RFC4632]. V. Fuller, T. Li, "Classless Inter-Domain Routing (CIDR): The Internet Address Assignment and Aggregation Plan", IETF RFC 4632, August 2006.
[RFC4822]. R. Atkinson and M. Fanto, "RIPv2 Cryptographic Authentication", IETF RFC 4822, February 2007.
[RFC7868]. D. Savage, J. Ng, S. Moore, D. Slice, P. Paluch, and R. White, "Cisco's Enhanced Interior Gateway Routing Protocol (EIGRP)", IETF RFC 7868, May 2016.

2 Types of Dynamic Routing Protocols

2.1 INTRODUCTION

Routers in a network need routing protocols to maintain the Routing Tables they use to route traffic to network destinations. The Routing Tables must be maintained such that they always contain the most current network reachability information. How this is done, depends on the type of routing protocol used (distance-vector, link-state, or path-vector), and the specific mechanisms supported within the particular routing protocol (RIP, EIGRP, OSPF, BGP).

In this chapter, we describe the different categories of dynamic routing protocols in use today, and their main distinguishing features. The discussion includes the distinguishing characteristics of the different dynamic routing protocols, and how they differ in design and operation. Understanding the different routing methods available (whether static or dynamic), is important and key to making informed decisions about which routing method to use in a particular network. This allows a network engineer to determine which routing method is most appropriate for a particular network environment.

2.2 DISTANCE-VECTOR ROUTING PROTOCOLS

Distance-vector routing protocols determine the best path to a network destination (when multiple paths exist) based on routing metrics that indicate how far the destination is from the router making the decision. The routing metric used to reflect "distance" can be a simple hop count (as in RIP), or a combination of several variables that represent a distance value (as in EIGRP). EIGRP typically uses a combined routing metric that is based on bandwidth and delay **[RFC7868]**. While EIGRP can use a combined metric based on bandwidth, delay, traffic load, and reliability, the extra metric are generally not used.

2.2.1 BASIC CHARACTERISTICS OF DISTANCE-VECTOR ROUTING PROTOCOLS

A router running a distance-vector routing protocol advertises routing information to its neighbors with the information structured in the form of vectors or arrays with elements (distance, direction). The element "distance" in a vector, is a routing metric (or cost) such as the hop count to reach a destination network, and "direction" is the next-hop IP router to be used to reach that destination. Note that the next-hop router from the advertising router is associated with an outbound local interface in the direction leading to the destination network plus the IP address of the receiving interface of the next-hop router.

Basically, a router running a distance-vector routing protocol does not know the entire path to a network destination, but instead knows only the local interface (i.e., direction) on which packets to a destination should be forwarded, and the distance (i.e., how far it is) to that destination. A distance-vector router is only aware of the IP addresses assigned to its local interfaces and the addresses of the remote networks it can reach through its neighbor routers. The router does not possess any broader knowledge of the entire topology of the network it is operating in. Essentially, the routers running the distance-vector routing protocol are not aware of the entire network topology.

The routing information that a router receives from its neighbor are stored ("as is") in a local routing database (i.e., a "route store") that the distance-vector routing protocol maintains. The distance-vector routing protocol then uses a distance-vector algorithm (Bellman–Ford algorithm) to calculate the best (and possibly) loop-free paths/routes to each destination, if multiple paths exist (Figure 2.1). The best paths are then installed in the IP Routing Table, and are also advertised as routing information to each neighbor router. Thus, once a router determines the best paths to all known destinations, it advertises its entire IP Routing Table containing these best paths to each directly connected adjacent router.

The above process shows that each router running a distance-vector routing protocol learns routes from its neighbors' perspectives, and then calculates and

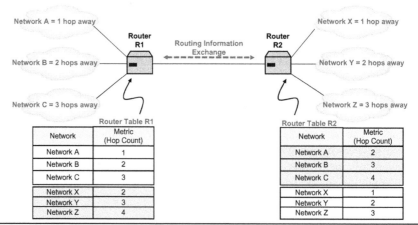

FIGURE 2.1 How Distance-Vector Routing Protocols Work

Types of Dynamic Routing Protocols

advertises the best routes from its own view of how to get to those destinations (the local router's own perspective). For example, a router A may send a routing update to a neighbor stating that "It takes 4 hops to reach Destination Y, in the direction of next-hop router E". The distance of 4 hops away is just the distance, router A has derived from other routing updates received from its neighbors, but routers A has no way of knowing the complete path to the destination or the individual routers that are at each hop.

Each router simply advertises routing information to its neighbors consisting only of the distance information it has gleaned from routing updates, but not from a complete knowledge of the network topology map identifying potential routes to network destinations. The router just creates a local perspective of getting to the destination from the routing information that it has received (Figure 2.2).

This behavior prevents a distance-vector routing protocol from having a complete map of the whole network up to any given destination network. Instead, the Routing Table maintained by the protocol reflects simply how best a neighbor router knows how to reach a particular destination network based on how far that neighbor thinks it is from that destination network. The local router does not know how many hops (or how many other routers) are on the best path leading to any of those destination networks. This behavior has prompted distance-vector routing to be sometimes referred to as "routing by rumor". Each router learns from its neighbors how best to reach a particular destination, routing information which the neighbors in turn may have inferred from their neighbors, and so on.

Distance-vector routing protocols generally have slow convergence and poor scalability, and are generally suitable for only small networks. Protocols such as RIPv1 and RIPv2 also have limitations because neighbors exchange routing updates as if they only have unidirectional connectivity as illustrated in Figure 2.3. The advantage of distance-vector routing protocols, however, is that they are less CPU/computationally intensive, require less memory (which can run easily run out in low-end routing platforms), and have simpler implementation and maintenance.

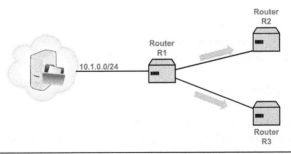

- We assume Router R1 is directly connected to Network 10.1.0.0/24. Router R1 sends routing updates to Routers R2 and R3.
- The routing updates contain the network IP address, network mask and metric for this route.
- Router R2 and R3 receive this routing update and add the route to their respective routing tables. Both routers list the metric of 1 because Network 10.1.0.0/24 is only one hop away.
  Note that the maximum hop count for a RIP route is 15 and any route with a higher hop count is considered to be unreachable.

FIGURE 2.2 Sending RIP Routing Updates

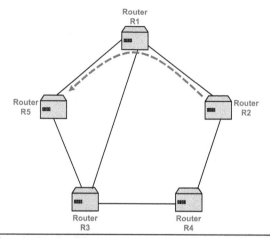

- Distance vector routing protocols such as RIP networks are limited by their unidirectional connectivity. RIP cannot ensure bidirectional connectivity because it processes routing information based solely on the receipt of routing table updates. On the other hand, link state routing protocols such as OSPF establish bidirectional connectivity with a three - way handshake.
- Let us assume Routers R1 and R4 flood their routing table information to Router R2 describing routes to Router R5. Let us assume also that Router R2 has data destined to a network connected to Router R5.
- Because the path from Router R2 to Router R5 through Router R1 has the smallest number of hops, Router R2 installs that route in its Forwarding Table.
- Now, let us assume Router R1 is able to transmit traffic to but not receive traffic from Router R2 because of an unavailable linkor invalid routing policy.
- Given that Router R2's forwarding table has listed the only route to Router R5 to be the one through Router R1, any traffic destined for Router R1 will be lost, because bidirectional connectivity was never established.

FIGURE 2.3 Illustrating the Limitations of Unidirectional Connectivity in Distance-Vector Routing Protocols

However, EIGRP which is sometimes regarded as a hybrid protocol between distance-vector and link-state routing, has fast convergence, medium resource usage, and higher scalability (suitable for large networks). EIGRP uses a number of features typically found in link-state routing protocols. The main features of RIP **[RFC2453]** and EIGRP **[RFC7868]** are summarized as follows (see also Table 2.1):

RIP:

- Uses hop count as routing metric.
- If the hop count is greater than 15, RIP considers the network to be unreachable.
- Routing updates are broadcast (RIPv1) or multicast (RIPv2) every 30 seconds (default setting).
- RIPv2 sends triggered updates as soon as a change in the reachability of a network is detected, instead of waiting for the Update timer interval to expire. RIPv2 immediately sends an update about this change to neighbor routers.
- RIPv2 supports CIDR **[RFC1517] [RFC1518] [RFC1519] [RFC4632]** and includes the network mask (VLSM) **[RFC1878]** in routing updates. RIPv2 supports VLSM while RIPv1 does not.
- RIPv2 supports route summarization which reduces the size of Routing Tables and allows the routers in a network to handle more routes. This improves efficiency and scalability in large networks.

Types of Dynamic Routing Protocols

TABLE 2.1
Characteristics of Distance-Vector Routing Protocols

Characteristic		Routing Protocol			
		RIPv1	RIPv2	IGRP	EIGRP
Route Updates	Broadcasts	√		√	
	Multicasts		√		√
	Update sends entire Routing Table	√	√	√	
	Triggered updates		√		√
	Incremental updates				√
	Periodic timer	30 seconds	30 seconds	90 seconds	5 seconds (See Note 1)
Metrics	Hop count	√	√		
	Composite metric (bandwidth, delay, etc.)			√	√
VLSM			√		√
Type of Service (ToS) capable routing (See Note 2)				√	√
Load Balancing	Equal Cost	√	√	√	√
	Unequal Cost			√	√
Maximum network diameter (hops)		15 hops	15 hops	255 hops (See Note 3)	See Note 3
Authentication			√		√

Note 1: This interval relates to the rate at which a router running EIGRP transmits Hello messages to neighbors. The default Hello interval setting is 60 seconds for low-speed, point-to-point links (T1/E1) and non-broadcast multiple access network (NBMA) networks, and 5 seconds for other types of networks. Note that in an NBMA network or point-to-point link, messages are transmitted and travel only directly from one node to another over the communication medium (i.e., a link, virtual circuit/connection or across a network/medium that emulates point-to-point communication).

Note 2: This refers to routing protocols that are capable of learning and advertising routing information while accounting for the ToS ("differentiated service [DS]") of traffic carried over the learned routes. Such routing protocols are capable of making routing decisions based on specially defined ToS related bits contained within the IP header of packets. These well-defined bits patterns are typically set by network access devices and by trusted end systems as a way of requesting a specific level of service for packet forwarding.

Note 3: Although these protocols generally observe the notion of a maximum network diameter, this is not used as a metric of distance value in best path computation and selection. EIGRP has no specific limit on maximum network diameter which directly relates to how many neighbors the particular implementation/instantiation of the protocol (Internet Operating System [IOS]) can support. Generally, the true maximum network diameter is these cases are determined by a number of factors such as the particular platform running the protocol processor, memory devoted to the protocol, number of routers neighbors that are stub, number of routes are advertised to neighbors and received from these neighbors. EIGRP being a significant enhancement over IGRP, can work with a large number of neighbors. In Cisco IOS, the **metric maximum-hops** command, with command syntax, **metric maximum-hops** *hops-number*, where *hops-number* specifies a maximum hop count in decimal, can be used to configure EIGRP to advertise a route as unreachable when the hop count is greater than the value assigned to the *hops-number* argument. The maximum number of hops (*hops-number*) that can be specified is 255 and the default value is 100. Setting a maximum hop count provides a safety mechanism in EIGRP that can be used to prevent any potential count-to-infinity problems in a network (similar to the maximum hop count used in RIPv1/v2.

- RIPv1 and RIPv2 supports load balancing over up to 6 equal cost paths (default is 4 equal cost paths).
- RIPv2 supports authentication and key management mechanisms to secure routing updates **[RFC4822]**.

EIGRP:

- Although EIGRP propagates route updates in a distance-vector manner, it uses a composite metric developed from bandwidth, delay, load and reliability, but in practice, only bandwidth and delay are used.
- Uses the Diffusing Update Algorithm (DUAL) for shortest path calculations which allows fast convergence.
- Does not send periodic routing updates as in RIP but instead, updates are sent only when there is a network topology change (triggered updates). IGRP (which is now obsolete), broadcasts routing updates every 90 seconds (default setting).
- EIGRP sends triggered and bounded updates which are updates sent only to those routers which are affected by a network topology change (route changes, route addition/removal, or changes in the metric of a route). These are non-periodic incremental (or partial) routing updates instead of sending the entire Routing Table only to the routers affected by the change. Using bounded updates means the partial or incremental updates are propagated in the network such that only those routers that need the new routing information are updated (i.e., such updates are automatically limited or bounded to only those routers who need them).
- Routing updates are sent as multicast messages instead of as broadcasts messages.
- EIGRP maintains the routes learned from neighbor routers in an EIGRP Topology Table (and these include routes that not the best routes). Information about EIGRP neighbor routers are stored in an EIGRP Neighbor Table while the best routes are stored in an IP Routing Table. EIGRP establishes adjacencies with neighbor routers using the EIGRP Hello protocol.
- DUAL allows an EIGRP router to install backup routes in its Topology Table to be used when the primary route fails. This is a local procedure (carried out in the router) which allows switchover to the backup route to be immediate and not involve action from any other routers.
- When multiple paths exist to a network destination, EIGRP can perform equal or unequal cost load balancing across these paths.
- EIGRP supports CIDR and VLSM while IGRP does not.
- EIGRP supports VLSM and route summarization and allows the creation of hierarchically structured large networks.

2.2.2 Distance-Vector Routing Protocol Operations

A router running a distance-vector routing protocol (RIPv1 and RIPv2) is required to inform its neighbor routers, periodically, and in RIPv2, also when changes in the

Types of Dynamic Routing Protocols

network topology. Each router sends out routing updates periodically where all of the router's Routing Table is transmitted to all neighbor routers. Each router broadcasts (in RIPv1) or multicasts (in RIPv2) its entire Routing Table to every neighbor at every routing update period. Even if the network topology has not changed, a RIP router still continues to send periodic updates to all neighbors.

RIP routers receiving these periodic routing updates, must process the entire update to discover any relevant routing information and then discard the rest. The information transmitted with the complete Routing Table, includes the IP addresses of the destination networks, the hop count for each destination network (i.e., the distance to all known networks), and possibly other information related to routing update authentication (in RIPv2).

Once a router receives its routing information updates from neighbors, it modifies its own Routing Table to reflect the changes transmitted to it, and then advertises these changes to its own neighbor routers. Each router updates its Routing Table, and sends this to its neighbors causing distance information (e.g., hop counts) to propagate across the routing domain. With this process, eventually each router obtains distance information about all reachable network destinations in the routing domain.

Each router relies solely on the routing information provided to it by other routers and does not assess or map out the network topology by itself. As described above, this process is referred to as "routing by rumor" because each router relies on the routing information it receives from other routers but the receiver cannot reliably determine if the routing information is actually valid and true.

2.2.3 What Is a Rooting Loop?

A routing loop is a network condition in which a packet is continuously circulated (endlessly) amongst a group of routers in a network without having the chance of reaching its intended destination network. A routing loop can occur when more than two routers (or group of routers) in a network have routing information that incorrectly point to only other routers in that group as having a valid path to a network destination. The packet ends up being caught up in that group and is circulated endlessly among them without getting to its destination. A routing loop can significantly degrade the performance of a network or even cause network downtime.

Generally, a routing loop may be created as a result of the following:

- When static routes are incorrectly configured in the Routing Table.
- When Routing Tables that are inconsistent in the network, are not being updated with correct routing information due to slow network convergence. A routing loop can be created by periodic updates sent by routers during the period a network is converging.
- When Routing Tables are incorrectly configured with static or dynamic routes that have been discarded/removed by other Routing Tables.
- When route redistribution between different routing protocols are incorrectly configured. Route redistribution is the process of transferring routing information from one routing protocol to another. Route redistribution allows routes that are learned by other routing methods (e.g., by another routing protocol,

directly connected, or routes static routes) to be passed onto another routing protocol.

The consequences of a routing loop in a network can be following conditions:

- The CPU in a router's control plane could be stressed due to looping of routing updates in the network.
- A router could be preoccupied with forwarding routing updates that could further negatively impact the convergence of the network.
- The router's data plane itself could be stressed with forwarding useless packets that will never get to their intended destinations.
- Precious network bandwidth could be used for forwarding traffic that end up looping endlessly between routers in the network.
- Routing updates may eventually be discarded (because of lifetime limits), get lost, or not reach their intended targets to be processed in a timely manner. Such conditions could end up introducing additional routing loops, thereby, creating more problems in the network, even additional routing loops.
- Routing updates and end user data may end up lost in "black holes" (see discussion below).

2.2.4 Routing Loops and Workarounds – Enhancing the Performance of Distance-Vector Routing Protocols

Networks that use distance-vector routing protocols are relatively more prone to routing loops and have slow convergence. A router running RIP broadcasts or multicasts routing updates every 30 seconds (which is the default setting of the periodic timer). This periodic update interval (an elapse time during which a lot can happen) may be longer than required for a network, thereby causing slow convergence. During these intervals (or periods between routing updates), the routers in the routing domain may not learn about network topology changes in a timely manner. The routers in the routing domain may be using routing information that is incorrect or outdated.

Slow convergence can lead to routing loops being created in the network, causing packets to circulate endlessly amongst a series of routers if not detected early – a situation that causes the routers to start a "count-to-infinity" (as discussed below). As already discussed above, routing loops can be created in the network when outdated or bad routing information exists in the Routing Table.

Thus, the problem of slow convergence and routing loops stems mainly from the reliance of routers in the routing domain on periodic routing updates (over longer timer periods) to learn and propagate routing information. Recognizing this problem, a number of mechanisms have been proposed that distance-vector routing protocols can take advantage of to avoid or minimize the impact of routing loops in the network. These mechanisms have been developed to enhance the stability and accuracy of distance-vector routing protocols.

Types of Dynamic Routing Protocols

Typically, RIP and EIGRP use the following techniques (or some combinations) to minimize the communications of incorrect routing information between routers, and to avoid routing loops; count-to-infinity, split horizon, poison reverse, and the use of Holddown timers. Split horizon and poison reverse also help reduce the amount of routing protocol traffic sent by distance-vector routing protocols, as well as allow more efficient transmission of routing information in a network. The actual implementations and settings of these routing enhancement mechanisms are very much vendor-dependent.

2.2.4.1 Initial Full Routing Table Update and Periodic Updates

Periodic updates are driven by a period timer and are sent by a router at the end of a define time period. RIP sends periodic updates every 30 seconds while IGRP does so every 90 seconds (default settings). The setting of periodic updates in routing protocols has to be in such a way that updates are not sent too frequently to cause/aggravate network congestion, and not too infrequently to cause slow network convergence. A compromise has to be struck between these two extremes, so most routing protocols allow the periodic update timers be configurable by specifying an update time range.

When a router first starts up on a network, it knows nothing about the network topology and so, has to discover other active routers, and also announce its presence. The starting router discovers its directly attached hosts and networks, and sends updates about these known networks out all of its interfaces. A RIP router directly attached to the network 10.1.1.0/24, for example, will send an update about this network out all other interfaces with a metric of 1. When a RIP router adds directly connected networks and static routes into its Routing Table, they each take a default seed metric value (hop count) of 1 (see Chapter 7). A neighbor router that receives an update about this network will add it to its Routing Table but will increment its hop count to 2 (as described in Figure 2.1). The overall network is assumed to have converged when all routers know about all the hosts and networks that are directed attached to all their neighbors.

RIPv1 routers send their updates to the IPv4 broadcast address 255.255.255.255 while RIPv2 send to the IPv4 multicast address 224.0.0.9. Other neighboring routers running the same routing protocol will receive these broadcasts or multicasts messages and then process them accordingly. Routing protocols such as RIP transmit their entire Routing Table, and neighbors that receive these updates (containing the complete table) will simply extract the information they need and discard the rest.

2.2.4.1.1 Sending Asynchronous Routing Updates

Through the use of asynchronous (periodic) routing updates, the likelihood of routers in a network transmitting their scheduled periodic updates at the same time is minimized. The periodic update timers are adjusted by small random timing jitters. This avoids routers in a network receiving and processing their updates, and then passing on these updates at the same time, thereby leading to synchronized update timers and operations **[CCIEDOYDEH]**. The delays related to generating, transmitting, and processing of routing updates in the routers in the network creates a situation where the different update timers become synchronized. Routers may maintain asynchronous updates using one of the following two methods [**CCIEDOYDEH**]:

- Each time the router sets a (scheduled) update period, it adds a small random time, or timing jitter, to that period as an offset.
- Each router implements and runs its update timer independently of the routing process it is running and, therefore, ensuring that it is not driven and affected by the routing information processing loads it is carrying.

For example, if the periodic route update interval is set to 30 seconds, a router can add a small random timing jitter to the update timer value each time the timer is reset. When all routers in the network implement their update timers with this added random time, network congestion that can occur if all routers send routing updates to their neighbors simultaneously, can be prevented.

The Cisco IOS uses a specific variable called *RIP_JITTER* (in its implementation of RIP) to prevent the synchronization of scheduled (periodic) updates between routers in a network running RIP. A router subtracts a variable amount of time from the specified periodic routing update interval, and the resulting interval is used to schedule the next routing update. This random length of time or jitter ranges from 0% to 15% of the specified/configured periodic update interval. Using this jitter range and a default 30-second periodic update interval, the actual update interval takes random values in the range from 25 to 30 seconds.

2.2.4.2 Route Maintenance and Invalidation Timers

When a network or a link attached to a router goes down, the router in its next routing update, can simply flag this network or link as unreachable, and pass on this information to other routers in the network. However, when the router itself fails, other routers will still have entries in their Routing Tables about the networks that are reachable via the failed router, even though this information is no longer valid. This is because there is no way the failed router can inform the other routers about its condition and the unreachable networks.

The other functioning routers will unknowingly continue to forward packets to the unreachable networks via the failed router. Other routers are still unaware that these networks are unreachable because they did not receive any routing update about the routes to these networks (via the failed router). The failed router in this case then becomes a "black hole" in the network because the neighbor routers still see these networks as still reachable.

The failed router creates a black hole in the internetwork which significantly degrades the performance of the network. This problem is addressed by allowing a router to set a route invalidation timer (generally referred to as *route invalid timer*) for each entry in its Routing Table. When a router first learns about a particular network and enters the information into its Routing Table, it also sets a timer for that route.

Whenever the router receives regular scheduled updates from other routers about this network, it discards this already-known information about this network. However,

Types of Dynamic Routing Protocols

whenever the router is refreshed with information about this network, it also resets the route invalid timer for that network or route. With this, if any router leading to this particular network goes down, the router will no longer receive updates about the network, and the timer will expire. The router will then flag this network/route (via the failed router) as unreachable and will pass this information to other routers in its next routing update.

Typical settings for the route invalid timer range from three to six times the periodic update interval, also called the route update period (i.e., 90–180 seconds), but the default is usually 180 seconds. A setting longer than the route update period is based on the reasoning that a router should not just invalidate a route when one or two updates have been missed, because this could have been the result of some lost or corrupted routing update messages, or due to a long network delay. At the same time, setting the route invalid timer value too large can cause network reconvergence to be excessively long and slow.

Other than the Holddown timer (discussed below) and several timers specific to EIGRP (which are different), distance-vector routing protocols used the following basic timers (Figure 2.4).

2.2.4.2.1 Route Update Timer

As discussed above, this drives the periodic update interval and refers to how often a router has to send routing updates to its neighbors – the amount of time the router waits since the last update before transmitting again its entire Routing Table (default is 30 seconds for RIP).

2.2.4.2.2 Route Invalid Timer

A router refreshes the age of routing information in its Routing Table each time a routing update is received. If a router does not receive a routing update for a particular route within the route invalid time period (i.e., does not receive a refresh update), then it will declare the route as invalid by setting its hop count metric to 16. The

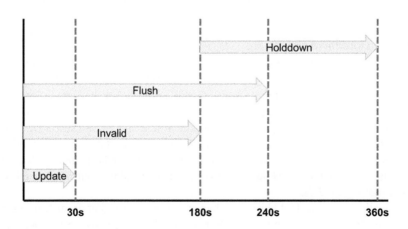

FIGURE 2.4 RIP Timers

default setting of the route invalid timer is 180 seconds for RIP. The setting of the route invalid timer should be at least three times the route update timer value.

When the timer expires for a route, the router places the route in a Holddown state (see discussion below), where it is tagged as inaccessible, and also advertised to neighbor routers as unreachable. The router will distribute the reachability information for this route in routing updates to neighbors (with metric of 16 or greater). The router will still retain the unreachable route in its Routing Table until the route flush timer expires (see below).

As discussed below, a router places a route in the Holddown state when it receives an update that indicates that the route is down or unreachable. The route is simply marked as inaccessible and advertised as unreachable, however, the router may still use this route to forward packets. However, when the Holddown timer expires (see below), routing updates about this route will be accepted, and the route will be declared as no longer inaccessible.

2.2.4.2.3 Route Flush Timer

The setting of the route flush timer specifies the amount of time a router must wait before removing a route from its Routing Table. When a router waits this amount of time without getting any routing update for a particular route, that route is simply removed from the Routing Table. The setting of the flush time interval should be greater than the sum of the invalid time interval plus the Holddown time. The default value is 240 seconds for RIP. If a routing update is not received to refresh a route (after it has been installed in the Routing Table) by the specified invalid time interval, the route is marked as invalid. The router then removes this route from its Routing Table after the Holddown period expires.

If the flush time interval is set to be less than this total value (i.e., invalid time interval plus the Holddown time), then the actual (desired) Holddown time interval cannot elapse, which can cause the router to accept a new route before the Holddown time interval expires. The router notifies its neighbor routers that the flushed route will be removed from the Routing Table before its actual removal. This means also the flush time interval has to be large enough to allow this communication to happen.

2.2.4.3 Holddown Timers

Holddown is a timer-based method used by distance-vector routing protocols to prevent routing loops from occurring in a network. When a router receives a regular (periodic) routing update about a route that is unreachable, it starts a Holddown timer. The router ignores all other routing updates from other routers for that route until the Holddown timer expires (RIP has a default setting of 180 seconds). During that period, the router would accept only updates sent from the router that originally advertised the unreachable route. If the originating router sends an update advertising that route as reachable, the Holddown timer is stopped and the routing update is processed (see Figure 2.5).

Basically, the route Holddown mechanism prevents a router from learning and using information from other routers about a route it already knows is unreachable or has failed. The Holddown timers allows a router to learned that a route is

Types of Dynamic Routing Protocols

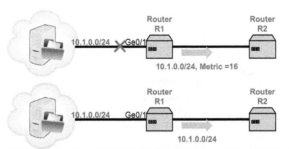

- Router R1 has advertised that it is directly connected to Network 10.1.0.0/24.
- After some period of time, interface Ge0/1 on Router R1 fails and R1 sends a routing update with metric equal to 16 to Router R2 (which is the infinite metric indicating that 10.1.0.0/24 is unreachable).
- Router R2 receives the routing update, marks the route as unreachable and starts the holddown timer.
  - During the holddown period (180 seconds by default in RIP), Router R2 will discard routing updates from other routers that indicate the route is reachable. Router R2 ignores all updates from other routers about that route to prevent routing loops from forming.
  - Only routing updates about the unreachable route will be processed by Router R2, the same router that originally advertised the route.
- If Router R1 detects that interface Ge0/1 is back up, it again advertises the route (but this time with a metric indicating reachability).
- Router R2 receives and process that update even if the holddown timer is still running and has not expired, because the routing update is sent by the same router (Router R1) that originally advertised the route.

FIGURE 2.5   Holddown Timer

unreachable, and at the same time, ensure that this route will not be inadvertently reinstalled in its routing table by a routing update received from another router that has not yet discovered that this route is unavailable.

The benefit of using a Holddown timer is that it helps to stabilize the propagation of routing information in the network, and prevents the formation of routing loops during periods when the network is converging on new routing information. Once a router marks a route as unreachable, that route is held in Holddown state long enough for all other routers in the network learn about the unreachable route.

Typically, a router will set the Holddown timer value to be greater than the typical total convergence time of the network, thereby allowing all routers sufficient time to learn, consolidate, and propagate accurate routing information through the network. During the Holddown timer period, the router ignores any new routing updates advertising routes it already knows to be unreachable or invalid that have similar or less favorable routing metric value than what it has learned. This prevents the router from inadvertently reinstalling an unreachable route, but would allow it instead to accept and reinstall previously declared invalid routes only if it receives a new routing update with a better metric than what it has learned, or when the Holddown timer has expired.

Distance-vector routing protocols implement Holddown timers to allow unreachable routes to recover, or the routers to switch to the next best routes available to the same destination. This mechanism is typically useful in the scenario where routes go down and come back up rapidly (a process called route flapping). A router entering one of such (flapping) routes in and out of the Routing Table repeatedly, can cause routing loops in the network, preventing it from converging. Holddown timers also prevent a router from immediately making changes to a route entered in its Routing Table that was recently declared failed or unreachable.

The use of Holddown timers involves a number of trade-offs. It reduces the likelihood of entering wrong routing information into the Routing Table, but this can also lengthen network reconvergence time. If the Holddown time for a route is set too short, it may not be effective as intended, and if it is set too long, it could prevent genuine and useful routing information from being received and entered into the Routing Table. This means Holddown timers must be set carefully to avoid creating undesirable conditions in the network and eventually defeating the purpose of Holddown timers.

2.2.4.3.1 How Holddown Timers are Used

The following summarizes how Holddown timers are used:

1. A router receives a scheduled (periodic) routing update from a neighbor router indicating that a route that was previously available is now no longer available.
2. The router then marks this route in its Routing Table as possibly unavailable, and starts its Holddown timer. For the route identified as possibly unavailable, the router ignores for a specified amount of time (i.e., the Holddown period), any other routing information received indicating the same status for that route, or worse.
3. If the router receives a routing update with a better metric for that route from any neighbor router during the Holddown period, the router reinstates that route and the Holddown timer is stopped.
4. If the router receives a routing update from any other neighbor router during the Holddown period with the same or worse metric for that route, the router ignores that update. The goal during this process is to allow more time for the routing information about the network change to be propagated throughout the routing domain. This means that routers will leave a route marked as unavailable in that state for a period of time that is long enough for routing updates sent to propagate and reach all routers in the routing domain. This allows all routers to be informed about the most current information so that they can appropriately update their Routing Tables.
5. A router will still forward packets on a route that is placed in Holddown state (i.e., marked as possibly down). This prevents the router from reinstalling routes in the Routing Table that have problems with intermittent connectivity (route flapping). If a route truly is down and yet packets are forwarded, it is possible that a black hole routing situation can be created in the network that can last until the Holddown timer expires.

2.2.4.4 Triggered Updates

A router will immediately send out a triggered update (or flash update) as soon as it is aware of a network topology change without waiting for its (periodic) route update timer to expire. This allows network reconvergence to occur faster than it would if all routers in the network had to wait to send periodic routing updates. This also greatly reduces complete reliance on mechanisms such as count-to-infinity (see below) and so on.

Types of Dynamic Routing Protocols

A router may still schedule regular (periodic) updates along with triggered updates since the latter are only event driven. Sending only triggered updates by themselves cannot guarantee that such updates would reach every other router in the network to effect appropriate Routing Table updates immediately. Using only Triggered updates (without additional routing mechanisms) is not often sufficient enough to effectively handle problems such as the following:

- Routing update messages may be corrupted or dropped in the network, resulting in the correct and most recent routing information not getting to other routers in the network.
- Triggered updates sent by a router may not reach other routers instantaneously, meaning it is possible that other routers that have not yet received or received at all the triggered update, will still be using incorrect routing information. This could cause unreachable routes to be still retained in the Routing Tables of neighbor routers that have not been informed through the recently sent triggered update.

Essentially, the use of triggered updates (on top of the other routing mechanisms) adds responsiveness to a network that is in the process of reconverging. A router that detects a network topology change will immediately transmit a triggered update to neighbor routers. The receiving routers, in turn, will send out triggered updates to notify their neighbors of the network change. Generally, a router sends triggered updates when one of the following events occur:

- A link changes state from up to down or vice versa
- A route changes state from unreachable to reachable or vice versa
- A new route is installed in a router's Routing Table

By providing routers immediately with correct routing information, triggered updates help to prevent the situation where a router might receive incorrect information about a route from another router (via regular updates) that has not yet reconverged. This helps to prevent or minimize routing errors that may occur while a network is in the process of reconverging.

To further improve the effectiveness of triggered updates, an update can include only the networks that actually caused the triggered update to be sent, rather than the router's entire Routing Table. The reduced routing update size helps to reduce the processing time of routing updates at the receiving routers, as well as, the amount routing traffic on the network.

2.2.4.5 Count-to-Infinity (Maximum Hop Count)

Distance-vector routing protocols place a maximum limit on the number of routers that can increment the distance metric value (hop count) in routing updates before they are considered invalid and have to be discarded. This essentially limits the distance (or hops) a routing update may traverse before it is invalid. This allows routers to automatically discard a routing update when the distance metric exceeds the maximum, for example, if a routing loop exists within the network topology.

The fact that distance-vector routing protocols learn routes to network destinations "by rumor", and have slow convergence, can lead to routing loops being created in the network. RIP has a maximum hop count limit of 15, and Interior Gateway Routing Protocol (IGRP), which is now obsolete but precedes EIGRP, has a maximum of 255.

Count-to-infinity happens when routers in a routing domain increment the metric of a route to "infinity" based on inaccurate routing updates (e.g., due to a routing loop). Any route with infinity metric is considered to be no longer reachable. Protocols such as RIP define a maximum metric value ("infinity") to prevent routers from endless incrementing the routing metric of a route, as it is propagated and circulates around in a network (see Figure 2.6). For example, RIP defines infinity as a hop count that is greater than 15 (i.e., 16 hops and more), that is, a route that is unreachable. Once a router "counts" the hop count of a route to "infinity", it marks that route as unreachable.

- Count-to-infinity (sometimes called counting-to-infinity) is phenomenon, but is sometimes used to refer to a mechanism for preventing routing updates from traveling in loops. This mechanism sets a maximum hop count value, that, when exceeded, corresponds to an unreachable destination (or regarded as a distance of infinity hops). Any value above the maximum hop count for RIP of 15 (and IGRP of 255) is considered infinity and serves as a way of signaling "destination unreachable" to other routers.

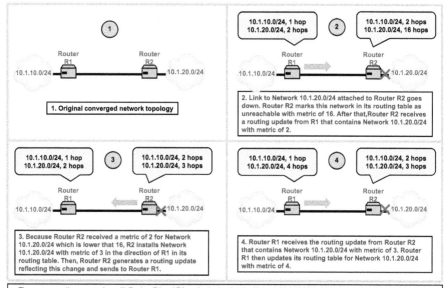

- The process continues on and on with Routers R1 and R2 continuing to advertise Network 10.1.20.0/24 with higher and higher hop counts until 16, the infinite metric, is reached. When this happens, the network is considered unreachable and the route will be eventually timed out of the routing tables. This process is referred to as the count-to-infinity problem which takes some time to manifest.
- Note that during the count-to-infinity process, the route from Router R1 to Network 10.1.20.0/24 points to Router R2. Router R2 thinks the opposite (thinking the route points to R1) which means there is a **routing loop** between R1 and R2 for the duration of the count-to-infinity problem.
- Speeding up the network convergence time is one of the reasons the maximum hop count of RIP is set to 15 (and 16 for unreachable).

FIGURE 2.6 Illustrating Count-to-Infinity

Types of Dynamic Routing Protocols

- The distance-vector routing protocols use the maximum limit as a mechanism for stopping the incessant increment of the hop count of a route (count-to-infinity) as it is propagated through a network. Recall that each router increments the distance routing metric of a route when it is being propagated. If a routing loop exists in a network, this process could go on endlessly if there is no mechanism for stopping it. Distance-vector routing protocols specify a maximum value that signifies an infinity value that, when reached, causes an endlessly circulating routing update to be discarded.
- Without this mechanism in place, excessively or unrealistically long routes, and eventually incorrect routing information, can spread throughout a network. Without a maximum hop count in place, incrementing the hop count (during routing loops) will cause some routes to have very long distances, and look like they are unreachable. But when in place, such routes will be removed from the Routing Table, thereby causing the routing loop to be resolved.

2.2.4.6 Poison Reverse

Another method used by distance-vector routing protocols for preventing routing loops is poison reverse, also referred to as route poisoning. With poison reverse or route poisoning, a router uses any value greater than the maximum hop count to signal to other routers to stop using a route in order to prevent a routing loop from occurring in the network. When a router loses a route or finds a route to be unavailable, it can advertise that route by sending a routing update with a hop count value greater than the maximum hop count (see Figure 2.7).

The router receiving that routing update would learn that destination network to be unreachable, and would in turn advertise this route to others. The receiving router will also send the routing update back toward the source router to ensure that the failed route is now "poisoned" throughout the entire network. The process of also updating the source router about the failed route is called poison reverse.

The benefits of poison reverse are as follows:

- **Allow Routers to Be Immediately Informed of Routes That Are Unavailable**: For example, when a router detects that one of its directly connected routes has failed, it can send a routing update for that route with the hop count metric value set to "infinity" (colloquially, "poisoning the route"). Any router that receives this routing update will interpret this advertisement as reporting that route as failed, and would remove the route from its Routing Table. Poison reverse prevents routers from propagating routing updates with inconsistencies throughout a network. Routers that have learned a route with a better hop count metric to the network destination advertised will ignore that routing update indicating destination unreachable.
- **Prevent the Propagation of Routing Updates with Inconsistencies**: Poison reverse allows a router to send routing updates on the same interface on which that routing information was learned (thereby violating the split horizon rule as discussed below), but the router "poisons" these learned route by advertising them with a routing metric value greater than the maximum hop count. The router receiving this particular routing update/advertisement will be informed that the "poisoned" routes are routes it had sent itself.

IP Routing Protocols

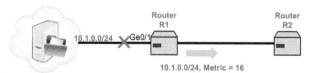

- Router R1 sends a routing update to Router R2 advertising its directly connected network 10.1.0.0/24.
- When interface Ge0/1 on Router R1 fails, Router R1 sends a routing update to Router R2 indicating that Network 10.1.0.0/24 is unreachable.
- The routing update has a metric of 16 (infinite metric) for the failed route, which is more than the RIP's maximum hop count of 15. This indicates to Router R2 that the route to Network 10.1.0.0/24 is definitely unreachable.
- Router R1 sends a routing update with an infinite metric 16, "poisoning the route". R2 will receive this routing update and will consider the route as unreachable and remove it from its routing table.
- The poison reverse technique helps to optimize the transmission of routing information and improve the time to reach network convergence.

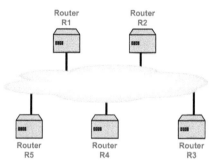

- Router R1 learns through one of its interfaces that routes to Routers R3, R4, and R5 are unreachable.
- Router R1 readvertises those routes out the same interface as unreachable.
- This advertisement informs Router R2 that Routers R3, R4, and R5 are definitely not reachable through Router R1.
  o Thus, when any router detects that one of its directly connected routes has failed, it will advertise a failed route with an infinite metric 16 ("poisoning the route"). Routers that receive this routing update will consider the route as unavailable and will remove it from their routing tables.
  o If Router R1 learns about unreachable routes through one of its interfaces, it advertises those routes as unreachable (hop count of 16) out the same interface.

FIGURE 2.7 Explaining Route Poisoning

- **Prevent the Propagation of Further Information Regarding Unreachable Routes**: Poison reverse also helps to optimize the communication of distance-vector routing protocol information in a network, and improve its convergence time. If a router receives a routing update from one of its interfaces and learns about a route that it already knows to have failed, or is unreachable, it will send a routing update (with hop count value of 16) advertising that route as unreachable out the same interface, thereby preventing the propagation of further information regarding that route.

2.2.4.7 Split Horizon

As discussed above, networks running distance-vector routing protocols generally have slow convergence and are susceptible to routing loops. Split horizon is one of the mechanisms that distance-vector routing protocols use to prevents routing loops from occurring. With split horizon, when a router learns routing information from a particular interface, it cannot advertise that information back on that interface (Figure 2.8).

The router should never advertise a route back onto the interface on which routing information originated since doing so could create a routing loop. The route that leads back to the router from which the routing information originates, is referred to as a reverse route. Split horizon is therefore a method that prevents the creation of reverse routes between two routers.

Types of Dynamic Routing Protocols

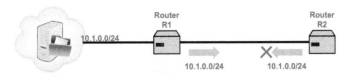

- Router R1 sends a routing update to Router R2 advertising a route to Network 10.1.0.0/24.
- Router R2 receives the routing update and enters the route in its routing table. Router R2 knows that the routing update for that route has come from Router R1, so it sees that it is unnecessary to readvertise that route back to Router R1. This technique is referred to as split horizon which helps to reduce extra traffic plus prevent routing loops by eliminating this type of route advertisements.
    - This is because, if the Network 10.1.0.0/24 goes down, Router R1 could receive a routing update from Router R2 advertising a route to Network 10.1.0.0/24. This will cause Router R1 to think that Router R2 has a route to that network and R1 would send packets destined to Network 10.1.0.0/24 to Router R2.
    - Router R2 would receive the packets from Router R1 and in turn send them back to Router R1, because Router R2 thinks that Router R1 has a route to Network 10.1.0.0/24, thereby creating a routing loop.

FIGURE 2.8 Explaining Split Horizon

When a change in the network occurs such as a link failure, the router noticing that change would send routing updates advertising that change to neighbor routers. The routers that receive these routing updates would in turn only advertise that change in one direction, meaning routing updates should be sent out on all other interfaces except the one from which the change was learned. Doing so prevents routing information from being transmitted back in the direction from which that information was received. With split horizon, the receiving interface on each router for an update becomes a starting point and the routing information is only propagated on the other interfaces.

Split horizon also helps to limit the amount routing protocol traffic sent by distance-vector routing protocols by allowing routers to eliminate routing information that the router and other neighbor routers on the receiving interface have already learned. If a router receives a set of routing updates on a particular interface, the router knows that those updates do not need to be advertised back on the same interface thereby reducing traffic on that interface.

There are two ways of implementing split horizon, which are, simple split horizon, and split horizon with poison reverse:

- **Simple Split Horizon**: The router should not advertise routes back to the routers from whom the routes were learned. In this method, when a router sends routing updates on any one of its interfaces, it must not include routes that were learned from updates received on that interface.
- **Split Horizon with Poison Reverse**: In this method, when a router sends routing updates on any one of its interfaces, the router can also designate some routes that were learned from routing updates received on that interface as unreachable by the router performing route poisoning (i.e., by sending a routing update with a hop count value greater than the maximum hop count). The router advertises reverse routes but with routing update carrying an unreachable or infinity hop count value.

Unlike simple split horizon, split horizon with poison reverse is a modification that provides more helpful routing information that is more constructive for routing loop prevention and faster network convergence. Split horizon with

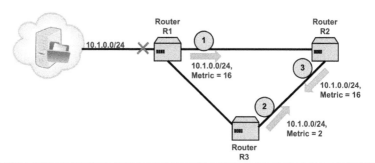

FIGURE 2.9 Split Horizon with Poison Reverse

poison reverse allows a router to explicitly inform other routers to ignore a route which is better than not informing them at all about the route in the first place. For these reasons, most modern implementations of distance-vector routing protocols use split horizon with poison reverse.

Essentially, the poison reverse feature allows routers to break the split horizon rule by allowing routers to send updates advertising (failed/unavailable and already known) routing information learned from an interface out of the same interface. However, this can only be done if the router performing poison reverse sends updates advertising these failed/unavailable routes learned from the receiving interface back on that interface with a value greater than the maximum hop count, indicating a destination unreachable, or equivalently, "poisoning" the failed/unavailable route.

When both poison reverse and split horizon are implemented in a distance-vector routing protocol, poison reverse takes precedence. When a router loses a route, or finds a route to be unavailable, it can override split horizon by sending a routing update that advertises the failed route with a destination unreachable (or equivalently an infinity hop count) distance metric value on all the local interfaces including the one on which the failed route was reported.

2.3 LINK-STATE ROUTING PROTOCOLS

Unlike distance-vector routing protocols, link-state routing protocols determine the best path to a network destination using relatively more complex methods and routing metrics that can take into account link variables, such as bandwidth, delay, traffic load, and reliability. Both OSPF **[RFC2328]** and IS-IS **[ISO10589:2002] [RFC1195]** routers assign a cost to each of their interfaces, which is a routing metric that the router uses in its link-state calculations. Routes with lower total path costs to a network destination are preferred over those with higher path costs.

2.3.1 OSPF versus IS-IS Metrics

OSPF is designed to use a cost metric that can be based on several network parameters but, in typical implementations, the metric is based on bandwidth only. The original IS-IS standard **[ISO10589:2002]** defined several different 6-bit metrics referred to as old-fashion or narrow metrics: default metric, expense metric (reflecting the monetary cost of sending data over a link), delay metric (reflecting transit delay), and error metric (accounting for the residual error probability associate with a link). In practice, many IS-IS implementations use only the 6-bit default metric for IP routing with the other metrics generally not implemented. A router uses these metrics to calculate a path's cost when running the shortest path first (SPF) algorithm.

Unlike OSPF, a router running IS-IS does not take into account link bandwidth when it sets the link's metrics. IS-IS routers do not have to calculate the routing metric based on bandwidth link or delay as done in other routing protocols like EIGRP **[RFC7868]** or OSPF **[RFC2328]**. IS-IS routers performing IP routing, typically, support only the default metric where the default metric of a link or interface is always set to 10, no matter the speed of the interface. A 10 Gb/s interface is assigned the same default metric as a slower 100 Mb/s link. This means, in practice, IS-IS will behave similar to RIP, and the best path to be used is the one with the least number of hops.

Using the default metric approach, a router (via the network administrator) will configure all its interfaces with a metric of 10 (by default) **[ISO10589:2002]**. This means in a network with links of varying speeds, this simplistic approach of assigning routing metric can result in suboptimal routing. Typically, using what is generally referred to as a narrow metric (which applies to the IS-IS default, delay, expense, and error metrics), the IS-IS metrics that a router assigns to an interface can take a value from 1 to 63 ($=2^6 - 1$). The router then calculates the total cost to a given network destination to be the sum of the metrics on all outgoing interfaces along a particular path from the source router to the destination, and the paths with the least costs are preferred.

The total path metric an IS-IS router can evaluate and set on an interface using the narrow metric is limited to a default value (*MaxLinkMetric*) of 1023 ($=2^{10} - 1$). This maximum metric value (*MaxPathMetric*) is considered in many cases to be insufficient for large IS-IS networks and does not provides much granularity for traffic engineering, especially in networks with high-bandwidth links. This therefore calls for a wider range of metrics to be also used if route leaking and MPLS traffic engineering is used.

IS-IS routers generate IS-IS link-state protocol data units (PDUs) that can include various TLV (type, length, and value) settings that specify link attributes. One such TLV is designated for an IS-IS adjacency information (Intermediate System Neighbors TLV (Code 2) **[ISO10589:2002]**), and another for IP Prefix information (IP Internal Reachability Information TLV (Code 128) **[RFC1195]**). To allow IS-IS to support traffic engineering, for example, for MPLS traffic engineering, newer TLVs with wide metrics were defined for IS-IS (e.g., Extended IP Reachability TLV (Code 135) **[RFC5305]**).

The IS-IS link-state PDUs carry the wide metric TLVs which contain link-attribute information IS-IS uses to populate its traffic engineering database. IS-IS then runs the Constrained Shortest Path First (CSPF) algorithm over this database to compute the paths that MPLS label-switched paths (LSPs) should take. Resource Reservation Protocol – Traffic Engineering (RSVP-TE) uses this path information to set up the LSPs in the network, and also make bandwidth reservations for them.

To allow wider range of metrics to be communicated in the newer IS-IS TLVs, the range of IS-IS metric values have been increased up to 16,777,215 (i.e., $2^{24} - 1$). With this, the cost metric an IS-IS router can assign to a route is an arbitrary chosen dimensionless integer that can range from 1 to 63 (for narrow metric), or from 1 to 16,777,215 if the router supports the newer wide metrics. By default, most routers will support the communication of routing updates with wide metrics. But most often, a router will allow a maximum narrow metric value of 63 and but generates both narrow and wide metric TLVs.

- **Narrow Metric**: This is often the default metric set on an interface and the default value is 10 for that interface. The narrow metric range for an interface is always set to be value from 1 to 63, and the maximum total value that can be calculated on all hops to a destination can be no more than 1023.
- **Wide Metric**: The wide metric type allows an IS-IS router to expand the metric up to a maximum value of 16,777,215 per link/interface over a route with a total path metric of 4,294,967,295 (= $2^{32} - 1$). IS-IS with wide metrics provide finer path metric granularity, and make it possible to better support applications such as traffic engineering.

2.3.2 Basic Characteristics of Link-State Routing Protocols

OSPF and IS-IS are the two most common link-state routing protocols and IGPs used in today's enterprise and service provider networks. A router running a link-state routing protocol advertises the state (link-state) and metric (link metric) for each of its connected links (including information about its directly connected networks and neighbors) to every other router in the network. OSPF sends advertisements in messages that are called link-state advertisements (LSAs), while IS-IS sends advertisements in link-state packets (LSPs). Link-state packets are also referred to as link-state PDUs **[ISO10589:2002]**. In many instances in this discussion, we use LSA to represent also LSP.

When a router receives an advertisement from a neighbor, it stores the link-state information in a local database called the link-state database (LSDB). The receiving router then advertises all the link-state information received to each of its neighbors exactly as was received. The router essentially "floods" the received link-state information unmodified, just as advertised by the originating router, throughout the network from one router to another. This method allows all the routers in the network to have a consistent, identical, and always synchronized map of the overall network.

As part of the information exchange process, each router transmits to its neighbor routers information about itself, the links that are directly connected to it, and the state of these links. This information is propagated from the source router to the neighbors, and from those routers to other routers in the network (router to router), with each router storing a copy of the information as received with no changes made to it. The flooding process results in each link-state router having a complete picture

Types of Dynamic Routing Protocols

of the entire network. When this process converges, every router will have identical information and topology map about the entire network, which then allows each router to independently compute its own set of best routes to network destinations.

Each router in the network independently runs an SPF algorithm (usually a variant of the Dijkstra's algorithm) over the complete map of the network (stored in the local LSDB) to calculate the best shortest loop-free paths to network destinations. The router then uses the resulting best paths for all the reachable network destinations from the SPF calculations to populate the local Routing Table.

The requirement of flooding link-state information, and maintaining a consistent, identical, and synchronized complete map of the network, makes link-state routing protocols require more memory, and relatively more computationally intensive than distance-vector routing protocols. The advantage, however, is link-state routing protocols make better path decisions that are less prone to routing loops.

Furthermore, link-state routing protocols have extended features and capabilities such as, opaque LSAs for OSPF, and TLVs for IS-IS, that allow the transmission of arbitrary data that these protocols were not originally designed for. These extended capabilities allow these link-state routing protocols to add extra information to LSAs or LSPs. Router can add extra information to OSPF LSAs or IS-IS LSPs to support, for example, services commonly required by service providers such as MPLS traffic engineering. Both OSPF and IS-IS support VLSM which allows both protocols to support CIDR.

The main advantage of link-state routing protocols is that, they have fast convergence and high scalability, making them more suitable for large networks. Their disadvantages, however, are they are relatively more complex to implement, and have a high resource usage (CPU processing and memory resources). These drawbacks stem mainly from the higher CPU cycles and overhead involved in processing routing updates when network changes occur, and higher memory resources that are required to store Neighbor Tables, LSDBs (containing the complete topology map), and Routing Tables. Table 2.2 summarizes the main characteristics of link-state routing protocols.

2.3.3 LINK-STATE ROUTING PROTOCOL OPERATIONS

In order to reduce traffic generated as a result of broadcasting routing updates, routers running link-state routing protocols like OSPF and IS-IS send out routing updates mainly as multicasts like RIPv2 and EIGRP. RIPv1 and IGRP routers send out routing updates as broadcasts. Similar to distance-vector routing protocols, upon system startup, a router in a routing domain running a link-state routing protocol would exchange with other routers in the domain, a complete copy of its Routing Table.

After initialization, the router would transmit routing updates in the form of multicast but only when a change in the network topology and state occurs. When this happens, the router sends out routing updates that carry only the particular change in the routing update and not the router's entire Routing Table. Routing updates reflecting the network changes are transmitted immediately (i.e., triggered by the network topology/state change), but if no changes occur, the router does not generate any routing update.

TABLE 2.2
Characteristics of Link-State Routing Protocols

Characteristic		Routing Protocol	
		OSPF	IS-IS
Route Updates	Multicasts	√	√
	Triggered updates	√	√
	After initial Routing Table exchange, updates send only network change information	√	√
Routing Databases	Neighbor (or adjacency)	√	√
	Link-state database (or topology pap)	√	√
	Routing Table	√	√
Metric		Uses cost, typically derived from bandwidth only	May use metrics of Default, Delay, Expense, or Error
VLSM		√	√
Type of Service (ToS) capable routing		√	√
Equal cost load balancing		√	√
Authentication		√	√

By allowing routers to transmit a routing update immediately upon learning of a route or network state change (i.e., a triggered update), the convergence time of the network can be significantly improved. Triggered updates can be sent as a result of events such as a link on a router failing (a router has lost connectivity to a neighbor that was previously reachable), a new link becoming available (i.e., a router has seen a new neighbor router), or the link cost to reach a neighbor router has changed. When the network changes occur and triggered updates are sent, these messages only propagate and are received by routers within the same logical area as the originating advertising router. Triggered updates all allow better utilization of network bandwidth since they are only event-driven.

Each router running a link-state routing protocols constructs and maintains three separate databases or tables (Figure 2.10): Neighbor Table (or Adjacency Database), LSDB (also known as the Topology Table), and Routing Table:

- **Neighbor Table**: This database maintains information about the link-state routing protocol neighbors of a router.
- **LSDB**: This database stores the link-state routing information learned from all neighbor routers.
- **Routing Table**: This database stores the best routes computed from the LSDB (i.e., the network topology information learned from neighbor routers).

2.3.3.1 Neighbor Discovery

As a first step, each router has to discover who its neighbors are, and establish a relationship (i.e., an adjacency) with each one of them. Each router has to determine

Types of Dynamic Routing Protocols

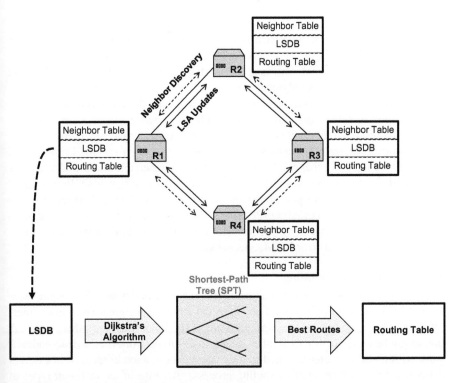

FIGURE 2.10 Link-State Routing Protocol Operations

which neighbor routers are connected to the links attached to its interfaces using a reachability protocol and its messages. The local router also sends outs these special messages periodically and separately to each of its directly connected neighbors.

The router sends these messages periodically on all interfaces in order to establish and maintain relationships with its neighbors (point-to-point links, point-to-multipoint links, multi access broadcast network segments, and virtual links). Also, a router would multicast reachability protocol messages on any of its physical networks that has broadcast or multicast capabilities to enable dynamic discovery of neighbor routers.

Link-state routing protocols achieve neighbor discovery and maintain adjacency using a Hello protocol. The Hello protocol (specific to OSPF and IS-IS) defines a Hello packet format and procedures for routers to exchange link-state routing information, and how the information in these messages are processed by receiving routers. At a minimum, a Hello message contains the originating (or Advertising) Router ID, IP address of the router's interface, network mask, Hello Interval (i.e., the time interval between sending Hello messages), and Router Dead Interval (i.e., the maximum time interval the router will wait to receive a Hello message before declaring the neighbor to be unavailable or down).

The Router Dead Interval specifies the maximum time interval a router can wait without receiving a Hello message from a particular neighbor. If the router does not receive a Hello message from a neighbor within this time interval, it modifies its

LSDB (or Topology Database) to indicate that the neighbor is unavailable or unreachable. Router running link-state routing protocols periodically monitor the status of their neighbors by sending and receiving Hello message that indicate whether each neighbor is still operational. This is to ensure that the neighbors are still sending and receiving LSA and acknowledgment messages. Link-state routing protocols transmit such messages and expects to receive the same at specified time intervals.

In addition to establishing adjacencies, Hello messages serve as keepalive messages to the routers to allow them monitor the state of the adjacency between them. If Hello messages are not received from a neighbor within a specified time period (Router Dead Interval), that neighbor is considered unreachable and the adjacency is terminated. A typical interval for the exchange of Hello messages in OSPF and IS-IS is 10 seconds (for point-to-point links and broadcast networks), and 30 seconds (for NBMA networks). The default OSPF Router Dead Interval is four times the default OSPF Hello Interval, that is, 40 seconds (for point-to-point links and broadcast networks) and 120 seconds (for NBMA networks).

2.3.3.2 Link-State Flooding

After the routers have established adjacencies with their neighbors, they begin sending out LSAs to these neighbors. Each LSA describes the originating router's interfaces or links, neighbor routers, and the state of the links (including their costs). A link might be a connection to another router, to a stub network (i.e., a dead-end network with no other router attached to it), to networks in another area, or to external networks (learned from another routing process). Because of the different types of link connectivity and link-state information available from a router, link-state routing protocols define different LSA types.

The LSAs are sent to every neighbor, who in turn, copy each received LSA, and then forward them to every other neighbor except the one that originated the LSA. The receiving router forwards LSAs almost immediately, thereby allowing link-state routing protocols to converge much faster than distance-vector routing protocols when network topology changes occur. The flooding of the LSAs throughout the routing domain allows all routers to build identical LSDBs.

2.3.3.2.1 OSPF Flooding and Parameters

The header of an OSPF LSA message contains the Link-State (LS) type, Link-State ID, and Advertising Router fields, the combination of which uniquely identifies a particular LSA **[RFC2328]**. An Autonomous System may be carrying several instances of an LSA at the same time. Therefore, routers may need to determine which LSA instance is newer or more recent. A router determines this by examining the LSA message's LS Age, LS Sequence Number, and LS Checksum fields in the 20-byte LSA header.

The LSA message header includes a field (LS Age) that specifies the age (in seconds) of the LSA since it was sent by the originating router. When a router generates an LSA, it sets this field to zero and as the message is flooded from one neighbor router to another, each one of them increments the age of the LSA accordingly. The LS Age must be incremented by an Interface Transit Delay at every router during the flooding process. The routers also increment the age of the LSAs as they are held in

Types of Dynamic Routing Protocols

the router's LSDB. Before a router propagates an LSA message out of one its interfaces, it must increment the age of the packet by the Interface Transit Delay (*InfTransDelay*) **[RFC2328]**. The Interface Transit delay sets the estimated amount of time required for the router to transmit an LSA over the interface (default Transit Delay is 1 second).

A router must never increment the age of an LSA pass the Maximum Age (*MaxAge*). LSAs with LS Age that have exceeded the *MaxAge* (e.g., 3,600 seconds or 60 minutes) are not used in the calculation of the Routing Table. However, when a router detects that an LSA's age has reached *MaxAge* for the first time, the router will reflood that LSA. The router finally flushes the LSA with *MaxAge* from its LSDB when it detects that the LSA is no longer needed to ensure database synchronization.

When a router receives two instances of the same LSA message with identical LS Sequence Numbers and LS Checksums, it examines further the LS Age field. An LSA instance with age of *MaxAge* is always accepted by the router as the most recent advertisement since this allows the router flush old LSAs quickly from its routing domain. Otherwise, if the router detects the ages of the two LSAs to differ by more than the Maximum Age Difference or *MaxAgeDiff* (e.g., 900 seconds or 15 minutes), it will accept the LSA instance with the smaller age as the most recent.

MaxAgeDiff (in seconds) represents the maximum dispersion or spread of the age of a single LSA instance as it is flooded and propagated throughout a routing domain. If the age difference between two LSAs is greater than *MaxAgeDiff*, the two LSAs are assumed to be different instances of the same LSA. This situation could take place when a router in a routing domain restarts and has lost track of the previous LS Sequence Number of a particular LSA.

Each OSPF router sends LSAs periodically (a process called "Paranoid Updates"), and when network topology changes occur. Each LSA is identified, at a minimum, by the Link-State ID, Advertising Router ID generating the LSA, LS Sequence Number (which increases every time the source router creates a new version of the LSA message), and all the other neighbor routers or networks to which the originating router is directly connected to. Each LSA is sent to all the router's neighbors, and if the LSA is determined by the receiving router to be newer (e.g., has a higher LS Sequence Number), it is saved, and a copy is propagated in turn to each of that receiving router's own neighbors.

The LSA flooding process rapidly propagates a copy of the latest version of each router's LSA to every other router in the routing domain. Networks running link-state routing protocols can also be segmented into (smaller) areas and hierarchies, to limit the scope of routing information flooding, and to allow effective routing control. The area and hierarchical routing features allow link-state routing protocols to scale and work more efficiently for larger networks.

2.3.3.3 Link-State Database

The primary objective of neighbor discovery and LSA flooding is to provide enough information to the routers for them to construct their link-state databases. Each router that runs link-state routing protocols maintains a LSDB or Topology Database that holds records of the LSAs received from all routers in the routing domain.

The LS Age, LS Sequence Number, and LS Checksum plus other related information in the LSA are mainly used to manage the LSA flooding process. On the other hand, information such as the Link-State ID, Advertising Router ID, Neighbor Router ID, Neighbor Interface ID, directly attached networks, and the cost (or metric) associated with those neighbors and networks are important for the shortest path determination process. LSAs may include information for the link type such as point-to-point connection to another router (OSPF Link Type 1), connection to a Transit Network (OSPF Link Type 2), connection to a Stub Network (OSPF Link Type 3), and Virtual Link (OSPF Link type 4).

When a router receives the complete set of LSAs from each router in the network, it creates a complete topology map of the network. The router uses an algorithm that iterates over all the LSAs received, one at a time, making links on the network map from the router which originated that LSA, to all the routers that the LSA indicates are neighbors of the originating router. Once a router builds its LSDB (or network topology map), it can construct a tree (with itself at the root) that describes the shortest path (lowest cost) to each other router and network in the routing domain by running the SPF algorithm on the LSDB.

Whenever the connectivity between a router and its neighbor changes, for example, when an interconnecting link fails, the router recalculates the LSAs that describe information about the router's neighbors, and then refloods them throughout the network. The reachability protocol (i.e., the Hello protocol and messages) which the router uses to communicate with its neighbors, is responsible for detecting any such network changes.

Note that a link is only considered to have been correctly reported when the two routers terminating it, report it. That is, if only one router reports that it is connected to the link (i.e., it is connected to a neighbor router), but the neighbor router does not report that it is also connected to the same link, then the link is not considered and is excluded from the network topology map.

2.3.3.4 Link-State Routing Process

Generally, the following processes take place before traffic can be forwarded in the network:

1. **Determining the Neighbors of Each Router**: Adjacency discovery involves all routers in the same routing domain (or simply, network) identifying themselves and who their neighbor routers are. By transmitting Hello messages, routers are able to announce their presence, identify their directly attached links, and the state of those links (up/down), as well as perform other initialization tasks. The routers use the Hello messages to establish an adjacency relationship with one another. Routers that are adjacent to each other continue to maintain this relationship by exchanging periodic Hello messages. Also, through the exchange of the Hello messages, each router in the routing domain constructs its first rudimentary Routing Table (which maps out the router itself, and its directly attached networks and neighbors).
2. **Distributing LSAs**: Once a router in the routing domain has established an adjacency relationship with one or more neighbors, it starts exchanging routing

information by flooding routing updates throughout the routing domain. A router running a distance-vector routing protocol like RIPv1 must receive routing information, process it, and wait for a periodic update timer to expire before sending out any routing updates. Even through RIPv2 sends triggered updates, it also relies on periodic updates. However, for link-state routing protocols, any router receiving information about any network change immediately floods routing updates (i.e., triggered updates) to all segments of the routing domain except the one from which the routing update was received. Furthermore, all routers receiving these updates are able to process them in parallel, thereby speeding up network convergence. The routers are virtually able to receive and process routing updates faster without waiting for update timers to expire.

3. **Creating the Network Map**: A router receiving the link-state routing updates is able to construct and maintain a local LSDB (or topology map) of the entire routing domain. Unlike a router running a distance-vector routing protocol, which does not maintain a complete map of the entire network, a router running a link-state routing protocols has a complete map of all subnets and networks within its routing domain (or area) and the different paths that can reach these subnets and networks.

4. **Calculating the Shortest Paths and Filling the Routing Table**: Each router in the routing domain then constructs a logical shortest-path tree structure with itself placed at the root and with all the subnets and networks in the domain originating from this root.

    a. Each router running the link-state routing protocol would then derive its Routing Table by running the SPF algorithm (which is based on Dijkstra's algorithm) against its local synchronized LSDB (or topology map), and then install the lowest cost paths to network destinations in its local Routing Table.

2.3.3.5 Calculating the Shortest Paths and Constructing the Routing Table

The ultimate goal of developing the LSDB is the construction of the Routing Table. Each router independently runs an SPF algorithm (which is variant of Dijkstra's algorithm) over the LSDB or network topology map it has created, to determine the shortest path from itself (the root) to every other router and destination network in the routing domain. The SPF algorithm considers the cost of each link along each path which may be based simply on link bandwidth as in many OSPF implementations. Most of the discussion here is centered around OSPF.

A router maintains the following two data structures:

- **Tree Database**: This database contains routers which have already been added to the tree under construction – the branches (or links) already assigned to the tree. The branches (or links) added to the shortest path tree being constructed are stored in this database. At the end of algorithm execution, the Tree Database describes the shortest path tree.
- **Candidate Database**: This database contains the links from which the next link to be added to the Tree Database, will be selected. Links are copied from the LSDB (or Topology Database) to the Candidate Database as candidate links to be added to the tree under construction.

The remaining links (either rejected or not considered in the tree construction) are left in the LSDB. The LSDB (or Topology Database) is still the repository of all links that have been learned in the network.

The following steps summarize the process of applying Dijkstra's algorithm to construct a shortest path tree for a network:

1. The algorithm starts with both Tree and Candidate Databases empty. Then to initialize the Tree Database (i.e., tree construction), a router adds itself as the root of the tree. This first entry in the Tree Database indicates that the router is its own neighbor with a link cost of 0.
2. The router adds to the tree, all neighbor routers which are directly connected to it, except any routers which have already been added to either the tree or the Candidate Database. All other routers are added to the Candidate Database.
3. The router compares each router in the Candidate Database to each of the routers already in the tree. The candidate router which is closest to any of the routers already added to the tree is itself moved (from the Candidate Database) into the tree and connected to the appropriate neighbor router.
    a. The router calculates the cost from the root (itself) to each router in the Candidate Database. The router in the Candidate Database with the lowest cost from the root is moved to the Tree Database. If two or more routers have equal low cost from the root, the router just chooses one of them to be added to the Tree Database.
    b. When the router moves a router from the Candidate Database to the tree, that router is removed from the Candidate Database and is not considered in subsequent iterations of the SPF algorithm.
4. The router just added to the Tree Database is further examined. With the exception of any routers already in the Tree Database, routers in the LSDB described as neighbors of the just added router are added to the Candidate Database.
5. If there are still more entries in the Candidate Database, the algorithm returns to Step 3. If there are no entries in Candidate Database, then the algorithm is terminated. Upon algorithm termination, a single router entry in the Tree Database should represent every router discovered in the network, and the shortest path tree is considered complete.

Steps 3–5 are repeated as long as there are any routers remaining in the Candidate Database. When Candidate Database is empty, all the routers in the network will have been added to the Tree Database. The algorithm terminates with the Tree Database containing all the routers in the network, with the root of the tree being the router on which the algorithm is running. The shortest path from that router (as the root) to any other router in the network, is given by the list of routers that can be traversed to get from the root of the tree, to the desired router in the tree.

2.3.3.5.1 Populating the Routing Table

As soon as the shortest path tree is constructed, the next step is for the router (root) to install best paths in its Routing Table. For any given destination router and network, the best path for that destination is the router which is the first or the next-hop from the root router, down and along the branch on the shortest-path tree leading toward the desired destination router. The next-hop for a destination in the Routing Table consists of the outgoing router interface (over which packets are to be forwarded to the destination), and the IP address of the next-hop router (if any). The next-hop can be connected to an IP unnumbered interface which is an interface that has no explicit IP address assigned to it.

To install a route in the Routing Table, what is needed is to walk down the tree, identifying the router at the head of each branch (leading to the desired destination), and installing in the Routing Table an entry for that router (i.e., the head of that branch). Whenever there is a change in the network topology (due to node or link failures or recovery, or changes in a link's routing cost), each router will have to update its LSDB (from the Hello and LSAs sent and received), recompute its shortest-path tree, and then modify the Routing Table. In most current link-state routing protocol implementations, when a new LSA is received (obviously describing a new network change), the entire shortest path tree is recomputed.

Current research focuses on how to recompute only that part of an existing shortest path tree that could have been affected by a given change in the network topology (i.e., partial recomputation of the shortest path tree) **[FRIGMARN98] [NARVSIUT00] [NARVSIUTZ01] [QUYIYANG13] [XIAOCAOSH07] [XIAOCAOZH04]**. Also, issues such as filling in the Routing Table as the shortest-path tree is recomputed, instead of handling this as a separate operation are being investigated. Most commercial routers simply delete the current shortest path tree, and construct a new one using variants of the Dijkstra's algorithm. However, such recomputation of the entire shortest path tree is inefficient and may lead to the consumption of considerable amount of CPU cycles and result in slow network convergence.

Algorithms that perform complete shortest path tree construction become very inefficient when only a small part of the tree needs to be updated for changes in the network topology and state. This is because small changes in the topology or state (even a single link-state change) still result in the recomputation of the complete tree at every router, followed by entries in the Routing Table being updated accordingly. In most cases, the new shortest path tree constructed by a router shows little or no difference when compared with the old tree particularly when a small change occurs in a large network.

The traditional method of constructing the complete shortest path tree even when a small network change occurs incurs a lot of unnecessary computation and Routing Table updates. Furthermore, the LSA flooding and the resulting routing protocol message exchange that take place as a result of that small change (because of the need to reconstruct the complete new shortest path tree), may create some undesirable traffic load and fluctuations in the network. This has created the need for algorithms that can dynamically update shortest path trees efficiently when network changes occur. Dynamic shortest path tree update algorithms tend to perform much better since they utilize the information available from the existing tree in the new tree computation.

The neighbor discovery and the LSA flooding process has to ensure that all the routers in the network develop exactly the same network topology map and are working from an identical (or synchronized) LSDB, if not, routing loops can form in the network. If any two or more routers end up with different topology maps (unidentical LSDBs), then it is possible to have scenarios in which routing loops can be created (similar to those in networks running distance-vector routing protocols).

2.3.3.6 Areas

We have discussed above that, in a network running a link-state routing protocol, each link is associated with a cost, and routers exchange link-state information to allow each one of them to have a complete and consistent picture of the network topology. The basic idea is to have every router construct (in the form of a graph) a complete map of the router and link connectivity in the network. This map shows which routers are connected to which other networks and routers in the network.

Using the link cost and other routing information in the LSDB, each router independently computes the shortest path tree (consisting of the best logical paths) to every possible destination in the network with the router itself as the root. Then, the resulting shortest path tree is used to construct a Routing Table, which stores the routes with the least cost from the router (root) to each destination in the network. The collection of best paths to each destination forms each router's Routing Table.

Implementing a link-state routing protocol in a large network exposes some serious issues particularly regarding the processing and memory requirements in the routers. Therefore, partitioning a large network into areas is a way to address the three concerns commonly expressed about link-state routing protocols, namely, the memory requirements for storing the LSDBs, CPU cycles (time) required to process the relatively more complex link-state routing algorithms, and the effects of LSA flooding on the available bandwidth in the network, particularly in unstable networks. An area is simply a partition of the network that contains a subset of the routers that make up the entire network (Figure 2.11).

When a (large) network is segmented into areas, the routers within each area flood LSAs only within that area, and maintain a LSDB (topology map) only for that area. Furthermore, routers in an area require a smaller LSDB, which implies less memory in each router, and smaller CPU cycles to run the SPF algorithm (on the smaller LSDB). In the event that network topology changes occur in an area, the routers only have to flood LSAs only within that area. Furthermore, if there are any routing instabilities (e.g., route flaps) in the area, they will be confined to that area only.

By utilizing areas with link-state routing protocols, network managers seek to make routing in network more manageable, and at the same time, save on system memory and router computing resources. Implementing areas can also be used as a way of introducing a hierarchy to a network architecture, where an extra layer can be added to a network hierarchical structure, by grouping (smaller) areas into larger areas.

As discussed earlier on, an Autonomous System is an interconnection of networks (or routing prefixes) that are under a common administration, and share a common routing strategy. The links and routers that make up the Autonomous System are typically divided and organized in logical groups called areas (Figure 2.11). An Autonomous System must define at least one area. Each area in a network running OSPF or IS-IS is identified by a uniquely assigned (area) number.

Types of Dynamic Routing Protocols

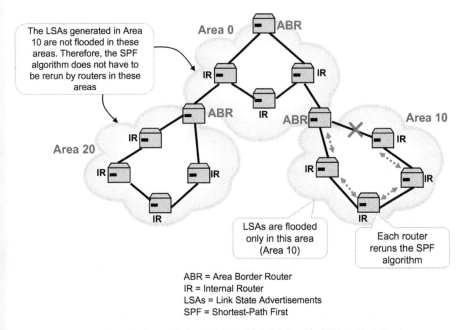

FIGURE 2.11 Use of Areas to Optimize Routing and Minimize Router Resource Usage

When an Autonomous System is partitioned into multiple areas, a router referred to as an area border router (ABR), in OSPF, or Level 2 router in IS-IS, is used to interconnect the various areas. An ABR supports different OSPF interfaces that connect the router to the multiple different OSPF areas to allow it operate and handle the communication of routing information between the areas. The ABR learns link-state routing information for each area it is attached to, and also maintains a copy of the same LSDB for each of those attached areas.

2.4 PATH-VECTOR ROUTING PROTOCOLS

Typically, a path-vector routing protocol is used for exchanging routing information between Autonomous Systems. A path-vector routing protocol carries as part of its routing information, the Autonomous System Numbers (ASNs) the routing information has passed through for a set of network destination address prefixes that share the same path attributes. The routing information (listing the Autonomous Systems traversed) is used by the various routers along the path to the network destination for preventing routing loops. We discussed in Chapter 1 that BGP **[RFC4271]**, including its protocol enhancements and extensions, is presently, the only path-vector routing protocol in use today. We discuss in this section, the main features of a path-vector routing protocol, and how it is used for exchanging routing information between Autonomous Systems.

2.4.1 Why an IGP Is Not Recommended for Routing between Routing Domains or Autonomous Systems

IGPs provide network reachability and path information, but there are a number of reasons why an IGP is used for routing within a single routing domain or Autonomous System (interior or internal routing), but not between different routing domains or Autonomous Systems (the later referred to as exterior or external routing). This is because a routing protocol may not only be required to provide network reachability information, but also provide or work with routing policy information regarding network administration and control.

Most IGPs have extensions (e.g., OSPF extensions in **[RFC3630] [RFC7471]** and IS-IS extensions in **[RFC5305] [RFC8570]**) that allow them to carry traffic engineering and class-of-service (CoS) (or quality-of-service [QoS]) information for use within a single routing domain or Autonomous System. However, they are not designed to route traffic between routing domains or Autonomous Systems, subject to administrative policy decisions and constraints. An EGP such as BGP is designed with these concerns in mind and provides the following advantages **[CISCWHMCDA04]**:

- **Preventing Network Changes in One Routing Domain from Propagating to Other Routing Domains**: During the design of the Internet, it became apparent that separating the routing information learned from within a routing domain or Autonomous System, from the routing information learned from outside that domain or system is very important and beneficial. One important reason is to limit the effects of network changes and failures occurring within a routing domain and not allowing them to propagate and impact other or external routing domains or Autonomous Systems.

    Figure 2.12 illustrates how network changes in one routing domain could propagate to another routing domain and have a serious negative impact on the operation of that domain **[CISCWHMCDA04]**. Failures in one routing domain can have unintentional consequences in other routing domains when the domains share routing information via a protocol that does not limit or control the propagation (or cascading) of such failures. BGP was designed with this important function in mind.

    Considering Figure 2.12, BGP and a routing policy (see Chapter 7), can be used to prevent a routing domain from accepting routing information from outside the domain that would interfere with the internal routing information. This provides a more efficient and effective solution without having to manually configure a list of route filters on a regular basis.

- **Hiding Information about a Routing Domain and Not Making It Visible to Other Routing Domains**: BGP also provides the capability of hiding specific routing information about a particular routing domain or Autonomous System from external domains or systems. Using simple IGP route redistribution (see Chapter 7) in Figure 2.12 to share routing information between the routing domains does not entirely solve the problem described. Using an IGP, the information about one partner's internal network infrastructure will still be

Types of Dynamic Routing Protocols

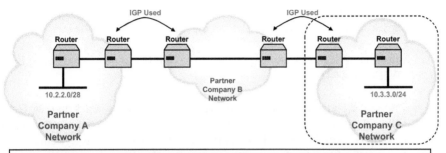

- Let us assume partner companies A and B have decided to use an IGP to share routing information between them, including specific information about how hosts and servers within each other's networks can be reached.
- Let us also assume that 10.2.2.0/28 is one of the subnets within Company A that the two partners need to share routing information about. So, Company A and Company B decide to set up an IGP with route redistribution between their networks to allow this information be to leaked between the two routing domains.
- After some time of partnering, Company B also decides to partner with Company C, and again uses IGP route redistribution to share information about reachable networks in their two routing domains.
- However, in this scenario, the routing information passed by Company C into Company B's routing domain will also be leaked into Company A's routing domain, creating possibly overlaps (or conflicts) with the internal routing information in Company A's routing domain.
- This scenario can result in some destinations within Company A's network becoming unreachable to even sources within Company A's network because, for example, some actions of the network administrators in Company B's routing domain has caused a fault in Company A's network.
- Problems such as these are not only difficult to identify, but are also difficult to troubleshoot and fix, since this may involve actions from the network administrators from, possibly, all three routing domains.

FIGURE 2.12 Illustrating the Impact of Routing Changes in one Routing Domain on Another When Using IGP Instead of BGP

passed on to the other partners. Instead, BGP can be used to define policy-based rules to prevent information about one routing domain/autonomous system from leaking into other domains or systems.

Using BGP between routing domains in Figure 2.12, Company C can mark the routing information it advertises to Company B so that this information will be filtered and not be passed on to Company A. This prevents information from being leaked inappropriately without having to maintain manual access lists by Company B. BGP when implemented between the routing domains, would mark routes so that they are not advertised beyond the adjacent routing domain (Company B).

- **Implementing Routing Policies between Routing Domains or Autonomous Systems**: Propagating routing policies within a routing domain or autonomous system is generally not important or a priority when routing with a single system. This is because the routers within the Autonomous System or routing domain are under a single administrative control, and management and administrative policies can be implemented on all the routers (normally through manual configuration or via a Dynamic Host Configuration Protocol [DHCP] server). Thus, the IGPs do not need to propagate this kind of information.

IGPs such as EIGRP, OSPF, and IS-IS are expressly designed for routing protocols within a routing domain or Autonomous System, and consider network speed of convergence as one of the most important design attributes. These routing protocols focus on collecting and propagating accurate information about a network topology as quickly and efficiently as possible. BGP is designed to allow a wide range of routing policies to be implemented to prevent problems like those described in Figure 2.12 from occurring.

The following types of routing policies that can be implemented with BGP [CISCWHMCDA04]:

- **Prevent a Network from Accepting Certain Routing Information from Other Routing Domains**: A network may configure routing policies that reject or suppress certain route advertisements by other Autonomous Systems.
- **Prevent Traffic from Traversing Certain Routing Domains or Autonomous Systems**: A network may use BGP and policy-based routing to force traffic to go over certain paths and avoid certain networks on their way to the destination.
- **Allow Traffic to Take the Cheapest Exit Point**: A network may have several exit links with different pricing that is dependent on the amount of traffic sent on a particular link or set of links. In this case, the network may want to route traffic to external destinations based on the cheapest exit point, rather than the closest point.
- **Specify the Closest Exit Point for Traffic to Always Take**: A routing policy can be set to allow traffic to always take the closest exit point out of a network, rather than use the best path toward the destination. This form of routing is sometimes referred to as hot potato or closest-exit routing. A network uses this form of routing when there is the need to allow traffic from other networks to traverse the local network, but at the same time minimize the amount of bandwidth needed to provision the transit path to be used.
- **Specify the Closest Exit Point to the Final Destination for Traffic to Take**: In some cases, a network may want user traffic to always take the best, or shortest, path to the final destination, rather than the shortest path out of the network (best-exit routing). This is done in order to provide better service to user traffic from another Autonomous System passing through the local network.

BGP allows several non-conflicting routing policies to be combined with the end goal of achieving the desired routing behaviors in an internetwork. Chapter 7 discusses in greater detail, the path controls tools used with IGPs and BGP.

2.4.2 Using an EGP between Routing Domains: Path-Vector Routing Protocol

In path-vector routing, it is assumed that there is one node in each Autonomous System, called the Speaker node or router, that communicates routing information on behalf of the entire Autonomous System. The Speaker node in an Autonomous System learns routes or paths to network destinations, and advertises them to the Speaker nodes in the neighboring Autonomous Systems. A Speaker node advertises the path information (and not the link metrics) that can reach an IP prefix (i.e., the destination address) in its Autonomous System or other Autonomous Systems.

Types of Dynamic Routing Protocols

A path-vector routing protocol has a few similarities to a distance-vector routing protocol, except it does not rely on the distance (or hop count) to a destination network to determine or guarantee a loop-free path. Instead, a path-vector routing protocol analyzes the different paths available in the network using a routing policy it imposes on the network, in order to arrive at the most suitable or optimum path. A path-vector routing protocol is typically deployed in network environments (inter-autonomous system routing) where different routing policies and metrics are generally used, and it is difficult to ensure or guarantee a consistent routing metric across the different routing domain.

The main functions of the Speaker node are the following:

- **Prevention of Routing Loops**: When a Speaker router receives a routing message, it checks the message to see if its Autonomous System is in the path list to the specified destination network. If yes, then the router assumes looping is involved and the routing message is ignored. The path information is gathered at each router, and carried in each routing advertisement, so that any router receiving the routing advertisement can validate the loop-free path before propagating the routing information.
- **Implementing Routing Policy**: When a Speaker router receives a routing messages it can check the path indicated to the specified network destination to see if any one of the Autonomous Systems listed in the path does not meet its routing policy. If the routing policy is not met, the router can ignore this path and not update its Routing Table with this path information, or it may choose to not advertise this routing message to its neighbors.
- **Determination of Optimum Path**: The Speaker router (based on its routing policy) determines the path to an IP prefix or destination address that is the best for the organization or administrative entity that runs the Autonomous System.

A path-vector routing protocol such as BGP determines the best loop-free path to a destination by checking and considering a number of BGP Path Attributes. It also analyzes any given path to a destination to determine if the path is loop free or not as illustrated in Figure 2.13. IGPs advertise a list of network address prefixes and the routing metrics to reach each address prefix. In contrast, routers in a path-vector routing domain (e.g., BGP) exchange network reachability information, called path-vectors, made up of path attributes (Figure 2.13). The path-vector information includes [RFC4271]:

- A list of the full path ASNs (listed hop-by-hop) necessary to reach a network destination.
- Other path attributes including the interface IP address needed to get to the next Autonomous System (the BGP Next-Hop Attribute), and how the network address prefixes at the end of the path were introduced into the path-vector routing protocol (the BGP Origin Attribute).

In path-vector routing, the Speaker router determines the best route from a source using a routing policy that is defined for the network without assigning costs to the

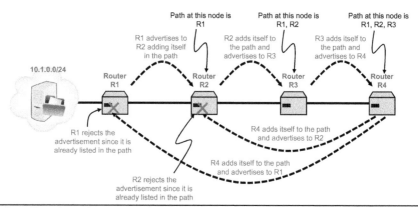

FIGURE 2.13 How a Path-Vector Routing Protocol Works

links and routes as done in distance-vector or link-state routing. The route from a source to all network destinations is determined by the best spanning tree that satisfies a predefined set of criteria that is based a routing policy. The best path selection algorithm a Speaker router uses is based on a predefined set of criteria. This algorithm is used to determine the most efficient or best routes through the internetwork. In the event that no specific or predefined set of criteria is created, the Speaker router will route traffic over the shortest path that spans the lowest number of intervening Autonomous Systems hops. Network administrators generally can change from time to time the path selection criteria to route traffic according to their network needs.

When a Speaker router starts up, it creates a path-vector based on the routing information it obtains from its Autonomous System. After the creation of the initial path-vector, each Speaker router, sends this to all its immediate neighbor Speaker routers. When a Speaker router receives a path-vector from a neighbor, it updates its path-vector by applying its own routing policy instead of using an algorithm that is based on link-state or least cost metric of routes.

2.4.3 BGP: A Path-Vector Routing Protocol

Although BGP is an EGP, the networks within many organizations have become so complex that BGP is used to simplify these networks. Speakers that exchange routing messages with each other are referred to as BGP peers. A group of peers that share the same set of policies are called a peer group.

Types of Dynamic Routing Protocols

2.4.3.1 Internal and External BGP Peering

Using internal BGP (iBGP) peering sessions, BGP peers within the same organization are able to exchange routing information. BGP peers that are in different Autonomous Systems are referred to as external BGP (eBGP) peers while BGP peers that are in the same Autonomous System are iBGP peers.

Typically, eBGP peers are adjacent and share a common IP subnet, while iBGP peers can be located anywhere in the same BGP Autonomous System. The eBGP peers are generally directly connected, while iBGP peers do not necessarily have to be. As long as an IGP exist within an Autonomous System to allow any two iBGP neighbors to communicate with each other, iBGP peers do not have to be directly connected. However, there are situations where two eBGP neighbors cannot be physically connected but rather, logically connected (referred to as multihop eBGP peering).

A network that resides within a particular Autonomous System is said to originate from that system. To inform other Autonomous Systems about its internal networks, the eBGP routers connected to an Autonomous System advertises these networks to eBGP routers of the other Autonomous Systems.

2.4.3.2 Basic Characteristics of BGP Routes

Although BGP is an EGP and is a path-vector routing protocol used for exchanging routing information between Autonomous Systems, it assumes that an IGP (like RIP, EIGRP, OSPF, and IS-IS) is used within an Autonomous System for routing within that system (intra-autonomous system routing). BGP also does not impose any restrictions on the type of network technology used in the underlying internetworks. To route information, BGP Speakers (or routers) exchange routing information, and each constructs a *graph* (a *tree*) that depicts the Autonomous Systems a piece of routing information (update) has passed through since it left its originator.

BGP treats each Autonomous System on the path to any given destination as a single point (Figure 2.13). An entire Autonomous System is treated as a single hop in the Autonomous System path information (AS-Path) to hide the topological details of the Autonomous System. One Autonomous System does not know what the paths through another Autonomous System look like, but only has to know that the destination is reachable through that Autonomous System. One interesting drawback of treating each Autonomous System as a single entity is that, without addition mechanisms or rules, loop free paths inside an Autonomous System cannot be guaranteed. BGP can only use the Autonomous System path-vector to detect loops between Autonomous Systems (see Figures 2.14 and 2.15).

Each Autonomous System in this directed graph is represented by its unique ASN, and any connection between two Autonomous Systems represents a *path* to a potential network destination (and has associated with it some path information). A collection of paths in this graph to a specific network destination (along with their path information) is referred to as a *route* (see Figures 2.14 and 2.15) [CISCHALABS00] [CISCWHMCDA04].

A BGP route is a unit of routing information that pairs a set of destination IP addresses with a list of path attributes to those destinations. BGP refers to the set of

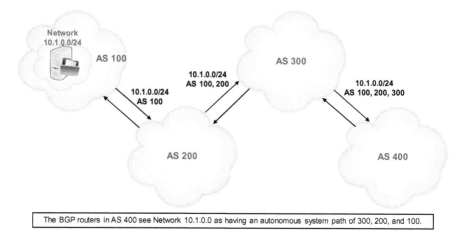

FIGURE 2.14 BGP AS-Path Advertisement

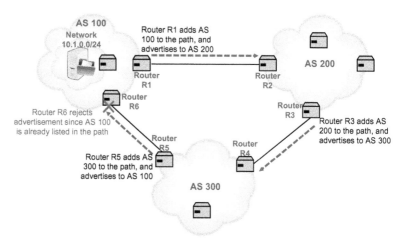

- Network 10.1.0.0/24 is advertised as a *prefix* with the list of autonomous systems the routing update has passed through. The list of autonomous systems in the routing update is called the *AS Path*.
- AS 100 originates the prefix 10.1.0.0/24, adding itself to the AS Path, and advertises it to AS 200.
- AS 200 is added to the AS Path, and the prefix is advertised to AS 300.
- When AS 300 advertises the prefix 10.1.0.0/24 to AS 100, the prefix is rejected, since the AS 100 is already listed in the AS Path, and accepting the advertisement would result in a routing information loop.

FIGURE 2.15 Path-Vector Routing over a set of Autonomous Systems

destinations as the Network Layer Reachability Information (NLRI) **[RFC4271]**. An NLRI is a set of networks whose IP addresses are represented by one IP network address prefix. In general, a route is expressed as the n-tuple <destination IP address prefix, next-hop node, set of Autonomous Systems traversed, set of path attributes…> **[RFC4098]**. A path-vector routing protocol such as BGP also uses the path information to prevent routing information loops (Figure 2.15).

2.4.3.3 BGP Autonomous System Path Advertisement

BGP peers exchange network reachability information using BGP routing updates (BGP UPDATE messages). BGP uses UPDATE messages to advertise new routes, withdraw routes that were previously advertised, or both. When BGP sends new routing information, it includes one or more network prefixes and a set of BGP attributes associated with those routes. While BGP can advertise multiple routes with a common set of attributes in a single BGP UPDATE message, it advertises new routes with different attributes in separate BGP UPDATE messages. Placing multiple network prefixes sharing a common set of path attributes in a single BGP UPDATE message is referred to as *update packing*.

A routing update information contains the network IP address prefixes, a list of ASNs that a route has traversed to reach a destination network, and other BGP path-specific attributes **[RFC4271]** (Figures 2.14 and 2.15). The ASNs are written into the BGP AS-Path attribute (AS_PATH), which in turn is carried in BGP UPDATE messages that the BGP peers exchange. Whenever a BGP routing update passes through an Autonomous System, BGP prepends its ASN to the routing update. The AS_PATH attribute as it is propagated from one Autonomous System to another, captures the list of ASNs that the routing update has traversed in order to reach a network destination (see "BGP and Path Attributes" discussion below).

To prevent routing loops from occurring, a BGP router rejects any routing update that contains its local ASN. A received routing update that contains the local ASN indicates that the routing update has already passed through that Autonomous System, and a routing loop would therefore be created if the update is accepted (Figures 2.14 and 2.15). BGP uses the path-vector routing algorithm and the AS-PATH routing loop detection mechanism to allow communication across internetworks (e.g., the Internet) consisting of many Autonomous Systems.

2.4.3.4 Loop-Free Paths within an Autonomous System

Although BGP relies on the AS-Path information to prevent routing loops from forming when routing between Autonomous Systems, it cannot provide loop free routing within an Autonomous System and has to rely on other mechanisms and a number of rules. As explained in Figure 2.16, an Autonomous System must ensure that every router within it makes the same routing decision as to which exit point to use when forwarding packets to a given destination. Also, all router must use the same constrained set of route advertisement rules within the Autonomous System **[CISCWHMCDA04]**. BGP relies on the IGP running within the Autonomous System to determine the best path to each of the Autonomous System's exit points. To prevent routing loops within the Autonomous System, the best path chosen throughout the system must be consistent.

2.4.3.5 Manually Configured BGP Connections over TCP

A BGP router does not discover other BGP routers automatically. Instead, a BGP connection is manually configured over TCP between any two BGP routers so that they establish a BGP relationships. A BGP peer is a BGP router that has an active TCP connection to another BGP router.

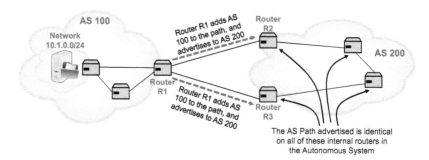

FIGURE 2.16 BGP Path-Vector Routing within an Autonomous Systems

Two BGP routers set up a TCP connection between themselves at the beginning of a BGP peering session, and this TCP connection is maintained throughout the session. By using TCP, BGP delegates all reliable transport, data sequencing, retransmission, error control, and connection keepalive issues to TCP. This allows BGP to focus solely on the processing of the routing information exchanged with its peers.

After two BGP routers have established a TCP connection, each BGP peer initially sends all of its BGP Routing Table to the other. After this initial routing information exchange, each router only sends incremental routing updates when there has been a network topology change, or when a routing policy has been modified or a new one implemented. During periods of inactivity between routing updates, the BGP peers exchange BGP KEEPALIVE messages which are special messages used to maintain BGP session connectivity between the peers.

2.4.4 BGP and Path Attributes

BGP Path Attributes contain the characteristics of a BGP Route that a BGP router advertises to its BGP peers. BGP uses BGP Path Attributes as pieces of information to describe the different network prefixes that are included in BGP UPDATE messages. BGP routers also use BGP Attributes to communicate the routing information needed for implementing BGP routing policies. BGP Path Attributes belong to one of the two categories **[RFC4271]**:

- **Well-Known Path Attributes**: These Path Attributes must be supported and recognized (i.e., understood) by all BGP implementations. The well-known Path Attributes further have two sub-categories:
    o *Well-Known Mandatory*: These Path Attributes must always be carried/exist in all BGP UPDATE messages sent to BGP peers. A BGP peer has to

recognize and accept this path attribute, and also advertise it to its own peers. Examples attributes are: AS_PATH, ORIGIN and NEXT_HOP.
    o *Well-Known Discretionary*: These Path Attributes must be recognized by a BGP implementation but may or may not be written and carried in a specific BGP UPDATE message. Their inclusion or not in a BGP UPDATE message sent to BGP peers is at the discretion of the BGP implementation. Examples attributes are: Local Preference (LOCAL_PREF) and ATOMIC_ AGGREGATE.
- **Optional Path Attributes**: A BGP implementation may decide to support or not support these path attributes. Optional BGP Path attributes also have two sub-categories:
    o *Optional Transitive*: If this path attribute is sent in a BGP UPDATE message, but not recognized by the BGP peer that receives it, the BGP peer should passed it on to the next Autonomous System. The BGP peer has to accept the path attribute in which it is included, and should advertise it on to other peers even if the peer does not support these attributes. If the BGP peer does not recognize any of these attributes, it checks if the transitive flag in the attribute has been set. If it is set, then the peer should accept and pass the attribute on to its other peers. Examples attributes are: The Aggregator of the Route (AGGREGATOR) and Community String (COMMUNITY)
    o *Optional Non-transitive*: These path attributes may not be supported in a BGP implementation, and a BGP peer will not pass them on when received. The BGP peer is not required to pass them on, and will simply ignore such optional attributes. The transitive flag is not set in these attributes and the BGP peer can quietly ignore them – it does not have to accept and advertise them to its other peers. Examples attributes are: Multi-Exit Discriminator (MULTI_EXIT_DISC or simply, MED), Route Originator ID (ORIGINATOR_ID), Route Cluster List (CLUSTER_LIST), Multiprotocol Reachable NLRI, and Multiprotocol Unreachable NLRI.

Each BGP route advertised, carries well-known mandatory, well-known discretionary, and possibly, optional transitive BGP path attributes that BGP routers use in their BGP best path selection process (see Volume 2 of this two-part book). The various path attributes are used by every BGP router along a path to make a comparison among the different network paths available, and to select the best paths to be installed in the IP Routing Table.

Each BGP router uses the best path selection algorithm to analyze the path attributes in order to determine the best routes to be installed in the IP Routing Tables. When a single path exists to a given destination, BGP selects that single path (by default) as the best path to that destination. When a BGP router receives routing updates from multiple BGP peers that describe different paths to the same network destination, it must select a single best path for reaching that destination. Once selected, the BGP router advertises that best path to its neighbors. The BGP path selection process can be influenced through standard BGP routing policy configuration, and also by altering some of the path attributes carried in routing updates as discussed in Chapter 7.

As discussed earlier, the AS_PATH attribute provides to a receiving BGP router, the list of Autonomous Systems through which a particular routing advertisement has traversed. The BGP AS_PATH attribute captures the inter-autonomous system path taken to get to a destination. The list of ASNs taken to reach a destination are written in this path attribute. A BGP router guarantees loop-free paths by ensuring that its local ASN is not among the ASNs that are contained in a routing advertisement.

Any time a BGP router receives a routing advertisement which already lists its ASN as part of the AS_PATH attribute, that routing advertisement is rejected. This is because accepting the AS_PATH would definitely result in a routing loop. Thus, when a BGP router receives a route with the AS_PATH attribute listing its own local ASN, that route is considered invalid and will be rejected to avoid creating an Autonomous System path routing loop.

2.5 THE IP ROUTING TABLE AND SELECTION OF BEST PATHS

A number of parameters, at various levels of the IP routing and forwarding process, play a key role in the selection of best paths for packet forwarding. These parameters are discussed in this section. We also discuss how they relate to the construction of the IP Routing Table.

2.5.1 PATH METRICS AND ROUTING PROTOCOLS

As discussed in Chapter 1, a routing metric is one of many variables specified in a Routing Table, and is a value or measure used by a routing protocol in the calculation of best paths to network destinations. When a routing protocol learns about multiple different routes to the same network destination, it uses a routing metric to decide which of these routes will be presented to the route selection process (algorithm), for possible installation in the IP Routing Table. A routing metric allows a routing protocol to determine which specific route to choose when multiple routes exist to a given destination.

Each routing protocol (RIP, EIGRP, OSPF, etc.) specifies its own set of metrics, mainly used for best path calculations. As discussed in Chapter 1, a routing metric can be formed from any number of variables that the routing protocol uses to determine the best path among multiple paths to a destination. Typically, a routing metric is based on information such as hop count, bandwidth, traffic load, delay, path cost, path length, reliability, Maximum Transmission Unit (MTU), and communications cost.

RIP uses a simple hop count, while OSPF uses a "cost" parameter as its metric which can be a function of other variables (but typically based on bandwidth). EIGRP typically uses bandwidth and delay to compute a composite metric. Even though, the primary use of a routing metric is for best path calculation, metrics of different routing protocols are not equivalent and cannot be directly compared.

However, when multiple routing protocols present routes to the same network destination to a router, it has to choose which of those routes to install in the Routing Table. The router decides which route to install in the Routing Table based on the Administrative Distance (also called Route Preference) of the source routing protocol. The route selection process selects routes to be entered in the Routing Table based on the routing protocol's Administrative Distance. The best routes installed in the Routing Table are those from the routing protocols with the lowest Administrative Distance.

If a particular routing protocol learns multiple routes to the same network destination, then these routes would have the same Administrative Distance, and the best route is chosen based on the routing protocol's metrics. The routing protocol associates metrics with specific routes it has learned, allowing the protocol to rank routes from most preferred to the least preferred routes.

2.5.1.1 Equal-Cost Multipath (ECMP) Routing

The parameters (e.g., hop count, bandwidth, traffic load, delay, path cost, path length, reliability) a routing protocol uses to calculate its metrics, or directly as its metrics, are specific to that protocol, and can differ from one routing protocol to another. The routing protocol selects the route with the lowest metric as the optimal (best) path and as a candidate route to be installed in the Routing Table. If there are multiple routes to the same network destination all with equal metrics, these can be installed as equal cost paths, and load balancing of traffic can be performed on these paths.

In load balancing, traffic is distributed across multiple paths to the same network destination with the aim of maximizing network throughput, optimizing network resource use, minimizing network delay, and avoiding the overload of any single path. Using multiple paths with load balancing, instead of forwarding traffic over a single path, has the added benefit of increasing reliability and availability (because of path redundancy).

If a routing protocol determines multiple paths to a single destination and all are best paths, the router can install these paths into the Routing Table as a bundle, or as a single logical path. If the router supports multiple path entries, it will install the maximum number of paths allowed per destination in the Routing Table. This type of routing is known as ECMP routing, and also allows the router to load balance/share traffic across all the multiple paths.

For example, if a router running OSPF identifies two equal-cost paths to the network prefix 10.2.2.0/24, it will install these two paths in the Routing Table as ECMP. Although each path will have its own outgoing interface and next-hop IP node, the two will be treated as one logical path mapped to the destination network 10.2.2.0/24.

2.5.2 Administrative Distance and Route Selection

The Administrative Distance (AD) (called the Route Preference value in Juniper Network terminology) is a decimal number used to rate the trustworthiness of the routing information source (whether static mechanism or dynamic routing protocol). This provides a mechanism for selecting (best routes to be installed in its Routing Table when multiple sources supply routes to the same destination. If different routing protocols including static routing provide different routes to the same destination, the Administrative Distance is used to decide which of these routes should be installed in the Routing Table (Figure 2.17). The route selection process gives preference to the route with the lowest Administrative Distance.

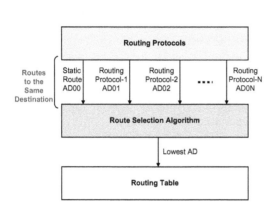

Routing Method	Default Administrative Distances
Directly connected interface	0
Static route	1
EIGRP summary route	5
eBGP	20
Internal EIGRP	90
OSPF	110
IS-IS	115
RIP	120
External EIGRP	170
iBGP	200
Unknown (Others)	255

FIGURE 2.17 Routing Protocol Cisco Default Administrative Distances

The reason for using an Administrative Distance is that, different routing protocols use different metrics and algorithms for best path computations that are not similar or compatible with each other. In a router running multiple routing protocols, the situation always arises that the routing function needs to select the best path for packet forwarding locally, or exchange route information across the multiple protocols (via a process called route redistribution as discussed in Chapter 7).

The Administrative Distance is a mechanism that the routing function can use to select the best path when there are two or more routing protocols providing best routes to the same destination. The Administrative Distance defines the level of believability, trustworthiness, or reliability of a routing protocol. Each routing protocol is assigned a number that prioritized its level of trustworthiness (in order of most to least). This number (which most often is configurable) is referred to as the Administrative Distance value of that particular routing information source.

Figure 2.17 lists the default Administrative Distances for the routing protocols used in Cisco routing devices. A directly connected interface (or network) is assigned a default Administrative Distance of 0 while a static route has Administrative Distance of 1. The smaller the Administrative Distance, the more the trustworthiness of the protocol. For example, OSPF (default Administrative Distance of 110) is considered more believable/reliable than RIP (default Administrative Distance of 120). This means the routing function will install routes from OSPF into the Routing Table over routes from RIP.

2.5.2.1 Administrative Distance Use Case Example

Each routing protocol in a routing device receives routing information and updates from its neighbors, and then calculates and selects the best path to any given network

Types of Dynamic Routing Protocols 71

destination. Each routing protocol then presents its best path to be installed in the Routing Table. The Administrative Distance is a measure of the trustworthiness of the supplier or source of the competing best path information. *It has only local significance for route selection in a routing device and is not information that is advertised to neighboring routers in routing updates.*

For example, if RIP determines a best path leading to network 10.1.2.0/24, it first checks the Routing Table to see whether an entry exists for this network destination. If no entry exists, this path (or route) is installed in the Routing Table. If an entry already exists in the Routing Table for this destination, the routing function determines whether to install the new route presented by RIP based on the Administrative Distance of RIP and the Administrative Distance of the routing protocol that installed the existing route in the Routing Table. If RIP has the lowest Administrative Distance to the destination when compared to the routing protocol that installed the existing route in the Routing Table, then the new route is installed in the Routing Table. If RIP is not the routing protocol with the lowest Administrative Distance, the new route is rejected.

Now let us assume three routing protocols EIGRP (internal), OSPF, and IS-IS in a router have each determined a best path with a different metric to the same destination network 10.1.5.0/24. Each of these three routing protocols will then present their best paths to 10.1.5.0/24 to be installed in the Routing Table. Since the destination is the same, the route to be chosen will be based on the Administrative Distance of these protocols. The routing protocol with the lowest Administrative Distance will have its route installed in the in the Routing Table. EIGRP internal has the lowest Administrative Distance (90) and so, will have its route installed in the Routing Table.

The routing protocol that did not get its route installed in the Routing Table (OSPF and IS-IS in this example) will keep that route so that it can be used as a backup route when the accepted route entered in the Routing Table fails. The route selection algorithm will request for these backup routes if the best path previously installed fails, so that they can be examined once again if they qualify to be reinstalled in the Routing Table.

For example, if the best route learned by EIGRP to destination 10.1.5.0/24 (the one installed in the Routing Table) fails for some reason, the route selection algorithm takes the backup routes kept by OSPF and IS-IS, and examines them to see which one can be installed in the Routing Table to replace the failed EIGRP route. In this case, the preferred route is selected based once again on the Administrative Distance of OSPF and IS-IS, which results in the route from OSPF being selected because of its lower Administrative Distance (of 110).

2.5.3 PREFIX LENGTH AND LONGEST PREFIX MATCHING LOOKUP

A routing device identifies the path (i.e., next-hop IP node and egress port/interface) a packet should take by examining the destination address of the packet, and performing a lookup in the IP Forwarding Table for the network with the longest matching prefix. The network prefix length is the key parameter in the lookup process.

The prefix length represents the number of leading (most significant) contiguous 1 bits in the IP address, and is determined from the network mask. The prefix length is determined starting at the most significant bit of the first byte or octet (i.e., the left-hand-side of the binary equivalent) of the network mask.

To calculate the prefix length (or network mask length or netmask), we first convert the dotted-decimal representation of the network mask to a binary equivalent. Then, we count the number of leading contiguous 1 bits in the binary number. For example,

255.255.248.0 in binary is **11111111 11111111 11111**000 00000000

The number of contiguous 1 bits is 21, resulting in a prefix length of /21. This means the prefix length of the IPv4 network address 128.42.6.5 with a network mask 255.255.248.0 is /21. This IPv4 network address can be written, equivalently, as 128.42.6.5/21

Finally, to calculate the (exact) network address (or prefix), we perform the logical AND of the corresponding bits in the binary representation of the IP address and network mask. We then convert the individual four octets of the result back to dotted-decimal representation as follows:

128.42.6.5 in binary:	10000000 00101010 00000110 00000101
255.255.248.0 in binary:	**11111111 11111111 11111**000 00000000
Logical AND:	**10000000 00101010 00000**000 00000000
Dotted-decimal representation:	128.42.0.0

This shows that the IPv4 network address (prefix) of 128.42.6.5/21 is 128.42.0.0.

To explain the longest prefix matching lookup, let us assume that an IPv4 Forwarding Table has the following routes with different prefix lengths: 10.1.2.0/28, 10.1.2.0/26, and 10.1.2.0/24 (Figure 2.18). These prefixes or routes have different prefix lengths (or network masks), and are installed in the Forwarding Table because they are considered as different destinations. We consider the situation where the IP forwarding engine has to select a route (next-hop node and egress interface) when the packet's destination address is within the network prefix range for multiple entries in the Forwarding Table.

When a packet is to be forwarded, the forwarding engine selects its route based on the longest matching prefix entry. For example, the prefix length /28 is given preference over /26, and /26 has preference over /24. An arriving packet with destination address 10.1.2.14, would match all three entries, but 10.1.2.0/28 gives the longest prefix match, meaning the packet would be forwarded to next-hop address 10.2.2.1 through egress interface Gig 1/1.

A packet with destination address 10.1.2.42 would match network prefix 10.1.2.0/24 and 10.1.2.0/26, but because 10.1.2.0/26 gives the longest prefix match, the packet would be forwarded to next-hop address 10.3.3.1 through egress interface Gig 2/2. A packet with destination address 10.1.2.100 matches only 10.1.2.0/24, so the packet is forwarded to next-hop address 10.4.4.1 through egress interface Gig 3/3.

Let us assume that a router is running EIGRP, OSPF, and RIP routing protocols and each protocol has learned the following best routes each with a different prefix length:

- EIGRP (internal): 192.168.30.0/26
- OSPF: 192.168.30.0/20
- RIP: 192.168.30.0/24

Types of Dynamic Routing Protocols

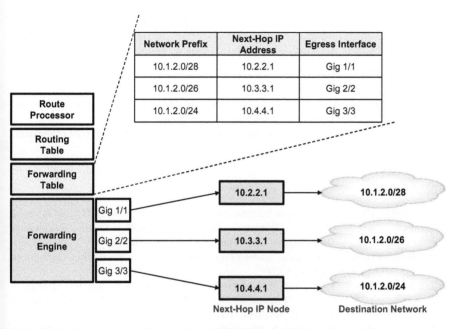

FIGURE 2.18 Entries in the Forwarding Table with Different Prefix Lengths

The router has to decide which of these best routes will be installed in its Routing Table. One might assume that since EIGRP internal route has the best Administrative Distance, this route will be installed in the Routing Table. However, since all of these three protocol-dependent best routes have different prefix lengths (of /26, /20, and /24), the router will consider them as different network destinations, and will therefore, install all of them in the Routing Table.

REVIEW QUESTIONS

1. Why does RIPv2 send routing updates to a multicast address instead of a broadcast address?
2. Explain the meaning of count-to-infinity in RIP.
3. Explain briefly how Poison Reverse works in Distance-Vector Routing Protocols.
4. Explain briefly how Split Horizon works in Distance-Vector Routing Protocols.
5. What is the use of the Holddown Timer in Distance-Vector Routing Protocols?
6. Explain the differences between the Update, Invalid, and Flush Timers in Distance-Vector Routing Protocols.
7. Explain the main difference between the Link-State Database and the Routing Table in Link-State Routing Protocols.
8. Explain briefly the benefits of creating Areas in networks running Link-State Routing Protocols.
9. Why is an EGP (like a Path-Vector Routing Protocol like BGP) preferred over an IGP for routing between Autonomous Systems?
10. What that main functions of a Speaker in a Path-Vector Routing Protocol?

11. What is the difference between an internal BGP peer and an exterior BGP peer?
12. How does a Path-Vector Routing Protocol (like BGP) detect and prevent routing loops?
13. What are BGP Path Attributes?
14. Explain the main difference between a Routing Metric (or Cost) and Administrative Distance (also called Route Preference).
15. What are directly connected interfaces/networks preferred over all other routing information sources?

REFERENCES

[CCIEDOYDEH]. Jeff Doyle and Jennifer De Haven Carroll, *Routing TCP/IP Volume 1*, Chapter "Dynamic Routing Protocol", CCIE Professional Development, Cisco Press, October 19, 2005.

[CISCHALABS00]. Sam Halabi, *Internet Routing Architectures*, 2nd Edition, Cisco Press, August 23, 2000.

[CISCWHMCDA04]. Russ White, Danny McPherson, and Srihari Sangli, *Practical BGP*, Chapter "Introduction to the Border Gateway Protocol", Addison-Wesley Professional, July 6, 2004.

[FRIGMARN98]. D. Frigioni, A. Marchetti-Spaccamela, and U. Nanni, "Fully Dynamic Output Bounded Single Source Shortest Path Problem," *Proceedings of the 7th Annual ACM-SIAM Symposium on Discrete Algorithms*, Atlanta, GA, pp. 212–221, 1998.

[ISO10589:2002]. ISO/IEC 10589:2002 – Information technology – Telecommunications and Information Exchange between Systems – Intermediate System to Intermediate System Intra-Domain Routing Information Exchange Protocol for use in Conjunction with the Protocol for Providing the Connectionless-Mode Network Service (ISO 8473)", International Organization for Standardization (ISO). November 2002.

[NARVSIUT00]. P. Narvaez, K.-Y. Siu, and H.-Y. Tzeng, "New Dynamic Algorithms for Shortest Path Tree Computation," *IEEE/ACM Transactions on Networking*, vol. 8, pp. 734–746, December 2000.

[NARVSIUTZ01]. P. Narvaez, K.-Y. Siu, and H.-Y. Tzeng, "New Dynamic SPT Algorithm based on a Ball-and-String Model," *IEEE/ACM Transactions on Networking*, vol. 9, pp. 706–718, December 2001.

[QUYIYANG13]. H. Qu, Z. Yi, and S. X. Yang, "Efficient Shortest-Path-Tree Computation in Network Routing Based on Pulse-Coupled Neural Networks", *IEEE Transaction on Systems, Man, and Cybernetics: Part B, Cybernetics*, Vol. 43, No. 3, pp. 995–1010, June 2013.

[RFC1195]. R. Callon, "Use of OSI IS-IS for Routing in TCP/IP and Dual Environments", IETF RFC 1195, December 1990.

[RFC1517]. R. Hinden, Ed., "Applicability Statement for the Implementation of Classless Inter-Domain Routing (CIDR)", IETF RFC 1517, September 1993.

[RFC1518]. Y. Rekhter and T. Li, "An Architecture for IP Address Allocation with CIDR", IETF RFC 1518, September 1993.

[RFC1519]. V. Fuller, T. Li, J. Yu, and K. Varadhan, "Classless Inter-Domain Routing (CIDR): An Address Assignment and Aggregation Strategy", IETF RFC 1519, September 1993.

[RFC1878]. T. Pummill and B. Manning, "Variable Length Subnet Table For IPv4", IETF RFC 1878, December 1995.

[RFC2328]. J. Moy, "OSPF Version 2", IETF RFC 2328, April 1998.

[RFC2453]. G. Malkin, "RIP Version 2", IETF RFC 2453, November 1998

[RFC3630]. D. Katz, K. Kompella, and D. Yeung, "Traffic Engineering (TE) Extensions to OSPF Version 2", IETF RFC 3630, September 2003.

[RFC4098]. H. Berkowitz, E. Davies, Ed., S. Hares, P. Krishnaswamy, and M. Lepp, "Terminology for Benchmarking BGP Device Convergence in the Control Plane", IETF RFC 4098, June 2005.

[RFC4271]. Y. Rekhter, Ed., T. Li, Ed., and S. Hares, Ed., "A Border Gateway Protocol 4 (BGP-4)", IETF RFC 4271, January 2006.

[RFC4632]. V. Fuller, T. Li, "Classless Inter-Domain Routing (CIDR): The Internet Address Assignment and Aggregation Plan", IETF RFC 4632, August 2006.

[RFC4822]. R. Atkinson and M. Fanto, "RIPv2 Cryptographic Authentication", IETF RFC 4822, February 2007.

[RFC5305]. T. Li and H. Smit, "IS-IS Extensions for Traffic Engineering", IETF RFC 5305, October 2008.

[RFC7471]. S. Giacalone, D. Ward, J. Drake, A. Atlas, and S. Previdi, "OSPF Traffic Engineering (TE) Metric Extensions", IETF RFC 7471, March 2015.

[RFC7868]. D. Savage, J. Ng, S. Moore, D. Slice, P. Paluch, and R. White, "Cisco's Enhanced Interior Gateway Routing Protocol (EIGRP)", IETF RFC 7868, May 2016.

[RFC8570]. L. Ginsberg, Ed., S. Previdi, Ed., S. Giacalone, D. Ward, J. Drake, and Q. Wu, "IS-IS Traffic Engineering (TE) Metric Extensions", IETF RFC 8570, March 2019.

[XIAOCAOSH07]. B. Xiao, J. Cao, Z. Shao, and E. Hsing-Mean Sha, "An Efficient Algorithm for Dynamic Shortest Path Tree Update in Network Routing", *Journal of Communications and Networks*, Vol. 9, No. 4, pp. 499–510, 2007.

[XIAOCAOZH04]. Bin Xiao, J. Cao, Q. Zhuge, Z. Shao, and E.H.M. Sha, "Dynamic Update of Shortest Path Tree in OSPF", *7th International Symposium on Parallel Architectures, Algorithms and Networks*, Hong Kong, China, 10–12 May 2004.

3 Routing and Forwarding Tables in Routing Devices

3.1 INTRODUCTION

A router maintains a number of databases which hold information about its neighbor routers and adjacencies, and the routing information it has learned from other routers in the network. This chapter discusses the main components of a router, and the databases relevant to IP routing and packet forwarding. The main components discussed are the control engine (also called the route processor), and the forwarding engine. The databases described are the IP Routing Table, IP Forwarding Table, and the router Adjacency Table. The chapter also discusses the high-level architectures of routers and their advantages and limitations.

3.2 FUNCTIONAL COMPONENTS OF AN IP ROUTER

Like many other computing devices, a router has one or more processors, Operating System (OS), random-access memory (RAM), read-only memory (ROM), non-volatile random-access memory (NVRAM), erasable programmable read-only memory (EPROM), Flash memory, and interfaces (for networking and device management). However, from a routing and packet forwarding perspective, the control engine (or route processor) and the forwarding engine, which can be entities residing in or implemented as part of the router components cited above, are the most important components worth a deeper understanding. In this section, we discuss these specialized components in greater detail, since they play the most important roles in the operations of a router.

3.2.1 IP Control Engine (or Route Processor)

IP routers determine best/optimal paths to network destinations by communicating with neighboring routers to share information about their directly attached networks, and network state. The control engine (Figure 3.1; also called the routing engine, but many times called the route processor), is a network module that is outfitted with a processor and memory and is responsible for running the routing protocols, best path computations, and other router housekeeping, and network and device management functions [AWEYA1BK18] [AWEYA2000] [AWEYA2001] [AWEYA2BK19]. We use interchangeably, the terms control engine, routing engine, and route processor in our discussion.

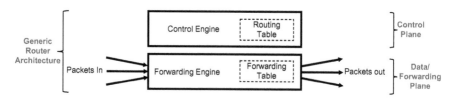

FIGURE 3.1 Functional Components of an IP Router

The route processor in its simplest form, typically, consists of the following: CPU; RAM for storage of the Routing Tables, Forwarding Tables, and data and instructions for other processes; a primary storage (a flash drive) for configuration files, software images, and microcode; secondary storage (hard disk); other storage for software upgrades; and Ethernet interfaces for out-of-band device management access. It should be emphasized that the control engine (and equally, the forwarding engine) is not necessarily a physical module in the router but could be a logical software component running in a processor in the router. In the architectures discussed below, the control engine and forwarding engine are viewed more as logical entities than physical ones.

The route processor is responsible for the control plane functions in the router. The route processor, in effect, holds the "intelligence" behind the router, and is dedicated to communicating with neighboring routers, plus gathering all the routing information required for forwarding packets to their destinations. This communication enables the route processor to build a network topology map of some sort (depending on the routing protocol used), and a comprehensive Routing Database (or Routing Table), that enables the forwarding engine to determine the router interfaces on which to send packets toward their network destinations.

In most of today's router architectures, route processing plus device management is a centralized function on a centralized computing module in the router. This centralized architecture is based on the reasoning that, having a single repository for all Routing Table information for the router, significantly reduces system complexity. Running the routing protocol, computing best paths, plus device management, are all activities that are do not required the kind of real-time urgency, as data plane functions such as receiving user packets, performing Forwarding Table lookups, and forwarding them to their next hops.

The route processor does not perform functions on a packet-by-packet basis, and as a result can be a centralized module on the router. More so, running routing protocols, best path computations, plus device management, are more compute-intensive functions that are more suitable to conventional CPUs, unlike the data plane functions that are more simplistic and repetitive. Routing protocol updates are sent at a much more-slower rate than the rate of user data packet transmission, essentially, making the control plane operations and control data reception and transmission, independent of the data plane packet forwarding process.

This in essence, makes it unnecessary to scale route processing resources in a router to be directly proportional to the number line cards or their speeds, in order to maintain system throughput. Roughly, the route processing resource scale in proportion to the size of the network, since mapping out the network topology and determining optimal routes, is after all the main objective of route processing.

3.2.2 IP Forwarding Engine

We explained above that the route processor runs the routing protocols, which enables the router to exchange network status information with its neighboring routers. The forwarding engine receives IP packets, performs IP Forwarding Table lookups, and then forwards the packets to their correct outgoing interfaces. The relationship between the IP Routing Table and Forwarding Table are explained below. Typically, in high-performance, high-end routers, the forwarding engine is implemented in ASIC. This is to ensure efficient movement of packets through the router without loss of data when handling high-speed interfaces.

The main functions of the forwarding engine can be summarized as follows:

- Receive packets from router interfaces and perform packet verification to see if they are valid for IP forwarding. This process, depending on the type of receive interface and router architecture, may include decapsulation and reassembly of packet entities (e.g., ATM cells), managing the buffering of cells in memory and priority queueing, etc.
- Perform IP Forwarding Table lookups to determine if a received IP packet is destined to the router itself, or is in transit to another network destination. If it is a transit packet, the outcome of the lookup determines the next hop node and outgoing interface for the packet.
- Perform a number of packet header updates (this will be explained in detail below).
- Forward the packet out the correct outgoing interface to the next hop. This process, depending on the type of outbound interface and router architecture, may include segmentation and encapsulation of packet entities (e.g., ATM cells).

As discussed above, the main function of the control engine is to communicate with other routers and populate the IP Routing Table, which in turn then gets distilled into the IP Forwarding Table (as explained below). On the other hand, the function of the forwarding engine is to receive IP packets, perform Forwarding Table lookups, and forward them out the correct interface to their destinations. This means, the operations of the control engine and the forwarding engine can be decoupled, making them operationally independently, with the Forwarding Table being the only coupling entity **[AWEYA1BK18] [AWEYA2000] [AWEYA2001] [AWEYA2BK19]**.

Most high-performance, high-end routers adopt this architecture, where the routing engine and the forwarding engine perform their primary tasks independently, although the Forwarding Table is continuously updated any time the control engine makes changes to the Routing Table. This architecture eliminates processing and traffic bottlenecks on the data plane, and streamlines routing and forwarding, allowing routers and networks to scale to high speeds.

3.3 HIGH-LEVEL ROUTER ARCHITECTURES

With advances in technology leading to the growth of the Internet, new services and applications continue to emerge. The network devices that drive residential networks, enterprise networks, service provider networks, and the Internet have evolved considerably and architecturally over the years. The capacities and performance of switches and routers, in particular, are still evolving to keep up with user traffic and new service requirements. The continuous improvements in wireless technologies and user mobility is creating more demand for network bandwidth. It is not surprising that the introduction of new generation of services and applications is placing tremendous stress on network devices, calling for more performance-driven devices.

In this section, we give a brief review of the main architectures found in routers. Low-end, low-capacity routers (typically used in residential and small business networks), tend to have a single centralized forwarding engine. High-end, high-performance routers (typically used in aggregation and core networks) tend to adopt architectures with distributed forwarding engines in the line cards.

3.3.1 ROUTER ARCHITECTURES WITH CENTRALIZED FORWARDING ENGINE

In a router architecture with centralized forwarding engines (Figure 3.2), a single processor with an IP forwarding function is used to make all the packet forwarding decisions in the router. In this architecture, the single processor runs both the routing engine and forwarding engine functions. Even in architectures where the control engine and forwarding engine reside on separate processor, a centralized forwarding engine can still be used to handle packet forwarding for the whole system **[AWEYA1BK18] [AWEYA2BK19]**. The simplest form of the centralized architecture is using a single general-purpose CPU to run all the control engine and forwarding engine functions. These architectures have traditionally been based on

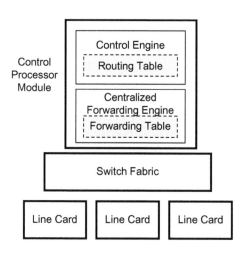

FIGURE 3.2 Router Architectures with Centralized Forwarding Engine

shared-memory and shared bus switch fabrics, where packets cross the shared-bus twice (to and from the centralized forwarding engine) before leaving the system to their destinations.

In the centralized forwarding engine router architecture, all packets received on any incoming interface, are forwarded to the centralized forwarding engine. The forwarding engine parses the IP destination address of the received packet, and performs a Forwarding Table lookup to determine the outgoing interface on which it should be sent out. The forwarding engine, while competing with other system tasks for the shared processing and memory resources, eventually completes all its processing, and then forwards the packet to the outgoing interface to be sent on to the next hop.

The first-generation router architectures implemented the control engine and forwarding engine functions on the centralized processor, resulting in poor packet forwarding performance and, consequently, poor network performance as network traffic grow. In these architectures, all processing functions, including device configurations and management, had to contend for the single centralized and finite pool of processing and memory resources, regardless of the network load. These architectures are still suited for residential and small business networks, because such networks tend to be stable with little growth over relatively long periods of use.

3.3.2 Router Architecture with Multiple Centralized Forwarding Engines

The centralized forwarding engine architectures are not able to deliver the desired performance when deployed in aggregation and core networks, where high-speed links are predominant, and data rates are very high. Aggregation and core networks tend to have high interface speeds, and are required to deliver very high throughputs. This in turn drives the need for higher capacity forwarding architectures that can replace the traditional centralized forwarding architectures, where all system line cards have to contend for a single, centralized shared processing and memory resources.

One of the ways to improve the performance and capacity of the centralized forwarding engine architectures is to use multiple parallel forwarding engines (Figure 3.3). Typically, the system implements some form of load sharing on the multiple forwarding engines as the traffic load increases. With the pool of multiple forwarding engines and load sharing, the system has the advantage of handling, efficiently, router interfaces with different speeds and utilization levels.

The forwarding engines can be implemented in NPUs (Network Processing Units), FPGAs, or specialized ASICs. Using any of these approaches, each forwarding engine can be designed with specialized and optimized forwarding architectures to handle high-speed Forwarding Table lookups, as well as, QoS classification, priority queuing, and security access control lists (ACLs) filtering. The system can be optimized such that these extra forwarding functions can be processed at the same time the Forwarding Table lookups are performed. ASIC implementations, in particular, allow the forwarding functions to be custom built to achieve extremely high packet forwarding rates.

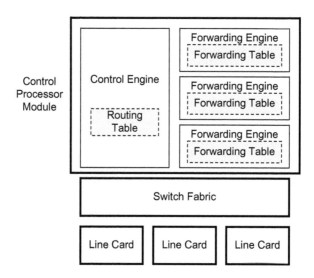

FIGURE 3.3 Router Architecture with Multiple Centralized Forwarding Engines

3.3.3 ROUTER ARCHITECTURE WITH DISTRIBUTED FORWARDING ENGINES

In the router architecture with distributed forwarding engines (Figures 3.4 and 3.5), the forwarding engines are distributed to the line cards to enable them to make packet forwarding decisions locally. The control engine is not consulted any time a packet is to be forwarded – the forwarding engine in the line performs all forwarding

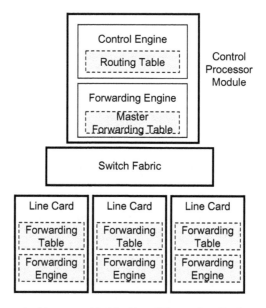

FIGURE 3.4 Router Architecture with Distributed Forwarding Engines

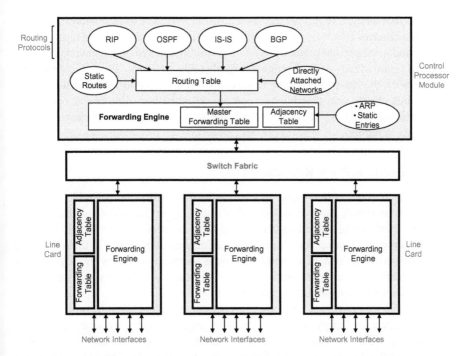

FIGURE 3.5 IP Router with Distributed Processing and Forwarding

operations. The control engine is only responsible for building the master Forwarding Table from its Routing Table, and then distributing copies of that master table information to the line cards.

The control engine is also responsible for synchronizing the contents of the distributed Forwarding Tables (in the line cards) with the master Forwarding Table, whenever the routing protocols make changes to the Routing Table. Any updates to the Routing Table are reflected in the master Forwarding Table. Route and network topology changes trigger routing updates which are captured by the routing protocols, and reflected in the Routing Table.

In the distributed forwarding engine architecture, a received packet is processed directly by the incoming line card – the packet is processed by the local forwarding engine on the card. A Forwarding Table lookup is performed by the local forwarding engine to determine if the outgoing interface is local, or on another line card in the router. If the outgoing interface is local to the line card, the forwarding engine forwards the packet out that local interface. If the outgoing interface is determined to be on another line card, the local forwarding sends the packet across the switch fabric directly to the outgoing line card (which then forwards it on the correct outgoing interface). The line card forwarding engine bypasses the control engine all this time. A packet being forwarded, crosses the switch fabric only once (for intercard forwarding), leading to lower forwarding delays and better switch fabric utilization.

Even in the distributed forwarding architecture, the control engine is most often still a centralized function (which also provides the added benefit of supporting control redundancy, if required [see discussion below]). As explained above, the control plane requires relatively more complex operations (running routing protocols, IP control and management protocols, system configuration and management, etc.). So, using a centralized control engine significantly reduces system complexity. Furthermore, the functions running in the control engine tend to have a network-wide impact and, change very slowly when compared to the forwarding engine operations (e.g., routing updates, IP control and network management packets, system management inputs).

3.3.4 Control Plane Redundancy

To optimize overall network and system availability, it is often desirable to design high-end routers with route processor redundancy (Figure 3.6) – control plane redundancy by configuring redundant centralized route processors. This mostly involves adding a second, identical route processor module into another slot in the router, and then configuring the system to switchover to the secondary (standby) route processor in the event the primary route processor fails. Aggregation and core routers in general demand such high-availability architectures.

Most high-end routers or carrier-grade routers support a special feature called hitless switchover (or failover). In such a feature, switchover of the active controller to the standby route processor takes place without reloading or resetting the routing protocols in the standby processor, and without any noticeable packet loss of services and protocols that are supported. The network remains stable during the switchover from active to standby, and after the switchover, the router, using the standby processor, continues to learn network routes by communicating with adjacent routers. All this while during the switchover, the data plane (via the forwarding engine) continues to operate, and packet forwarding remains uninterrupted, achieving nonstop forwarding (NSF).

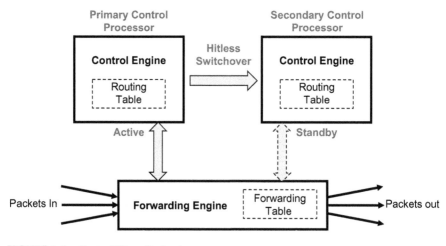

FIGURE 3.6 Control Plane Redundancy

A router may support dual route processor modules to provide 1+1 redundancy for the control and management plane **[CISCNEXHA16]**. The dual route processor configuration operates in an such that only one of the route processor modules is active at any given time, while the other serves as a backup or standby route processor. The state and configuration of the active processor is constantly synchronized with the standby processor module to provide stateful switchover in the event the active processor module fails.

Typically, an online diagnostics subsystem (plus additional monitoring processes) on the active route processor will trigger a stateful failover to the redundant processor when the diagnostics subsystem monitoring processes detect hardware failures, unrecoverable kernel errors, service restartability errors, or any critical failures. If an unrecoverable failure occurs in the active processor module, it triggers a switchover. The standby processor then assumes the new active role, and uses its synchronized state and configuration, for router operations while the failed previously active module is reloaded.

If the failed router processor module is able to reload, and successfully goes through self-diagnostics and pass, it initializes, and becomes the new standby route processor module. It then synchronizes its operating state and configuration with the newly active route processor module. This feature is stateful (and nondisruptive), and allows control traffic to flow unaffected. There is no disruption to data traffic because the forwarding engine modules are not affected, and there is no need to reset them.

3.4 IP ROUTING AND FORWARDING TABLES

To better appreciate the details of Layer 3 unicast and multicast forwarding in switch/routers and routers, we provide first, an overview of the most common Routing and Forwarding Tables maintained in routing devices.

A routing device, in general, maintains two key databases that hold the routing information required for forwarding packets in a network.

3.4.1 ROUTING TABLE

The Routing Table contains a list of the routes to particular IP network destinations (or IP address prefixes) in the network. These routes are discovered by all the dynamic routing protocols, as well as, the directly connected interfaces (also called interface routes or direct routes), and the static routes configured by the network administrator and (Figure 3.7). Each entry in the Routing Table describes one or multiple (in the case multipath unequal/equal cost routing) best paths to a particular network destination. The Routing Table is sometimes referred to as the Routing Information Base (RIB). Generally, the Routing Table stores and keeps track of routes to network destinations, the source of the routing information, and the metrics associated with those routes.

3.4.1.1 Routing Table Entries

Different routing protocols (RIPv4, EIGRP, OSPFv2, IS-IS, BGPv4) have protocol-dependent parameters for each entry in the Routing Table, in addition to some

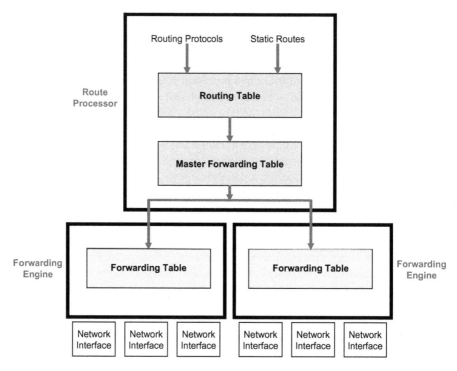

FIGURE 3.7 Populating the Routing Table and Forwarding Tables

standard parameters. Each entry in the Routing Table contains at least the following parameters:

- **Network ID**: This is the destination network address corresponding to a route.
- **Network Mask**: This is the network mask associated with the network ID.
- **Next Hop**: This is the IP address of the network entity to which the packet is to be forwarded.
- **Outgoing Interface**: This is the router interface to which the packet should be sent to reach the next hop and/or its destination network.
- **Metric**: This is the routing metric or cost of the path to the destination (network ID).

The additional parameters a RIPv2 router adds to its Routing Table are: the Address Family; and Route Tag (see Chapter 5). An EIGRP router, for example, will contain additional parameters such as: Successor (best route); Feasible Distance; and Advertised Distance (see Chapter 6). An OSPFv2 router will have additional parameters such as: Destination Type (i.e., Network, Area Border Router [ABR], or Autonomous System Boundary Router [ASBR]); Optional Capabilities; Area; Path Type (i.e., Intra-area, Interarea, Type 1 External, or Type 2 External); Type 2 Cost; Link-State Origin; and Advertising Router (see Chapter 7).

The entries of a BGP router has the following parameters (see Chapter 9): Network Prefix; Next Hop (BGP NEXT_HOP Attribute); Metric (BGP Multiple Exit Discriminator [MED] Attribute); BGP Local Preference Attribute; Weight (for Cisco routers); and BGP AS_PATH Attribute. Essentially, each routing protocol has a number of protocol-specific parameters that it maintains in its Routing Table.

The entries of the Routing Table listed above can be constructed to store the following types of routes:

- Directly attached network IDs (i.e., route to a network or host that is directly attached to a router interface).
- Remote network IDs (i.e., route to networks that can be reached through one or more next hops).
- Default routes (see Chapter 1). Default routes are used by a router as the conditional default gateway or route of last resort when no specific contributing route is found. Packet with destinations which do not match the more specific routes in the Routing Table will be forwarded using the default route. A default route matches any destination address. The IPv4 default route is designated in CIDR notation as the zero-address 0.0.0.0/0 (network mask is always 255.255.255.255), and for IPv6, it is 0:0:0:0:0:0:0:0 or ::/0.

If a router receives a packet and does not have a more specific route for the packet's destination address, it forwards the packet to the next hop associated with the default route entry, if one is configured. Typically, network core routers or top tier routers in a network have default-free Routing Tables – have no default routes configured. These routers forward every packet to a specific next hop using the longest prefix matching (LPM) of the packet's destination IP address and the network address prefixes in the Forwarding Table.

If the destination is not a directly attached network/host, the next hop is the address of the next best router on the outgoing interface that can best route the packet to the destination. The entries of the Routing Table can be maintained using any of the following methods:

- Routes to directly attached networks/hosts (see Chapter 1). When a router powers up and starts its routing processes, it automatically adds all of its directly attached networks to its Routing Table (interfaces that are in the up state).
- Routes that have been statically configured (static routes, see Chapter 4).
- Routes that have been learned by dynamic routing protocols (see Chapters 5 and 6).

Most Routing Tables include the following parameters to each Routing Table entry to describe the source of the route:

- **Route Source**: This identifies the routing method that provided the route, that is, how the route was learned. Cisco routers identify routing sources using the following route source codes **[CISCINTROUT14]**:

- Directly connected interfaces are assigned two route source codes.
  - **C**: Identifies a directly connected network. The router creates a directly connected networks automatically whenever any one of its interfaces is configured with an IP address and is activated.
  - **L**: Identifies that the interface leads to a local route. The router creates a local route automatically whenever any one of its interfaces is configured with an IP address and is activated.
  - **S**: Identifies that the route is a static route to a specific network.
  - **D**: Identifies that the route was learned dynamically from another router using EIGRP.
  - **O**: Identifies that the route was learned dynamically from another router using OSPF.
  - **R**: Identifies that the route was learned dynamically from another router using the RIP.
- **Administrative Distance**: This is used to select the best route when multiple routing information sources provide routes to the same network destination. The route with the lowest Administrative Distance is considered the best route and is placed in the IP Routing Table. The Administrative Distance of a route is not included in the IP Forwarding Table.

For purely IP packet forwarding purposes, the most important parameters from the Routing Table are the three information fields: network ID, next hop, and outgoing interface. The IP network mask and network ID are not considered as separate parameters since together they produce the IP network prefix.

It is important to understand the difference between the local Routing Database in which a given routing protocol stores all routes to all known network destinations, and the global IP Routing Table in which a router stores the best or optimal routes to all known network destinations. The former is more like a topology database and contains all non-optimal and optimal routes, plus directly connected routes and static routes. The non-optimal and optimal routes are routing information the router has learned from other routers running the same routing protocol. The best routes from this table are installed in the IP Routing Table.

The discussion in this chapter refers to the global IP Routing Table that contains the best routes. The exact implementation of these databases is vendor-specific and will not be discussed further here. This book simply views these two databases (local Routing Database and global IP Routing Table) as logically different and separate.

3.4.1.2 Routing Tables in a Router with Multiple Protocols

The Internet Operating System (IOS) powering a routing device may maintain, to some extent, a unified (or integrated) Routing Table or separate tables for IPv4 unicast routes, IPv6 unicast routes, IPv4 and IPv6 multicast routes, and Multiprotocol Label Switching (MPLS) routes. Depending on the IOS and protocols supported, a number of Routing Tables can be configured in the device. Typically, an IOS maintains separate Routing Tables for IPv4, IPv6, and MPLS routes. An IOS may also maintain separate Routing Tables for unicast routes, multicast routes, and MPLS routes. Generally, a router may support additional Routing Tables where there is the need to separate and maintain together a particular group of routes (sharing a specific routing and forwarding characteristics), or where there is the need for greater flexibility in managing or manipulating routing information.

Supporting additional Routing Tables may allow several specific uses such as, importing routes from another routing protocol into more than one Routing Table, applying different routing policies when routes are exported to different peer routers, and providing greater flexibility in managing routing information when dealing with multicast topologies or multitopologies that are not the same or not compatible. In general, the use of additional Routing Tables can allow more complex manipulations and operations to be performed on routing information.

The IOS installs all optimal or best routes (those with the lowest Administrative Distance) into the Routing Table which are then copied into the Forwarding Table. The optimal routes, indexed by their destination network addresses or prefixes, along with their next-hop IP addresses and outgoing interfaces, are installed in the Forwarding Table, to be used for forwarding packets to their destinations. The IOS may maintain a master Forwarding Table, copies of which are distributed to other forwarding entities in the system (Figure 3.7). The IOS supports mechanisms for making copies of the master Forwarding Table for the forwarding engines, which are the actual components in the system responsible for forwarding packets. These mechanisms are also responsible for synchronizing the distributed Forwarding Tables with the master Forwarding Table whenever changes are made to the Routing Table.

The routing protocol route selection algorithm generally determines the best route for the Routing Table by selecting the route with the lowest Administrative Distance (or Route Preference Value). A router uses the Administrative Distance (see Chapter 2) to select the best path when there are multiple routes to the same destination learned by different routing methods. An Administrative Distance provides a router with ranking information when selecting which routing source (directly connected, static, or dynamic routing) to consider when more than one method is available for supplying routes to the Routing Table. The router prioritizes each routing source in the order of most to least trustworthy using the Administrative Distance of the routing information source. The IOS typically uses some sort of tiebreaker resolution mechanisms when a number of alternate routes or route options are available. Routing protocols such as BGP and MPLS are designed to use additional mechanisms alternate route and tiebreaker preferences when multiple routes exist.

Typically, an IOS would create and maintain several Routing Tables concurrently, with each Routing Table holding routing information for a specific routing protocol or forwarding method (unicast, multicast, Virtual Private Network [VPN], etc.). The Forwarding Table could be partitioned based on Routing Table type, with each Routing Table populating its corresponding portion of the Forwarding Table. This mechanism allows a network administrator to define a specific forwarding behavior for each Routing Table. For example, when creating VPNs, where each VPN has its own Routing Table (VPN-based outing tables or Virtual Routing and Forwarding [VRF]), each VPN would be assigned its own VPN-specific partition in the (integrated) Forwarding Table.

3.4.1.3 Types of Unicast Routing Tables

It is not uncommon for the IOS to maintain unicast routes and multicast routes in separate Routing Tables. In general, an IOS may maintain the following types of Routing Tables:

- **IPv4 Unicast Routes**: This table maintains dynamically learned IPv4 routes, static routes, direct routes, and interface local routes.
- **IPv6 Unicast Routes**: This table maintains dynamically learned IPv6 routes, static routes, direct routes, and interface local routes.
- **IPv4 MPLS Routes**: This table is used only at a device that is the ingress node (i.e., Label Edge Router [LER]) to a Label-Switched Path (LSP). An LSP is an MPLS virtual circuit or path between two LERs. The table maintains the egress address of an LSP, the name of the LSP, and the name of the outgoing interface on the LER.
- **MPLS Labels (for MPLS Network Label Switching Operations)**: This table is used at a device that is a transit router in an MPLS network (i.e., Label Switch Router [LSR]). The table maintains the short path labels (not the long network addresses) that identify the virtual paths between an LER and LSR, LSR and LSR, or LSR and LER.
- **Layer 2 VPN Routes Learned from BGP**: This table maintains Layer 2 VPN routes learned via BGP from other Provider Edge (PE) routers. The Layer 2 routes are copied and maintained in Layer 2 VPN VRF (Virtual Routing and Forwarding) instances based on BGP Route Target Communities **[RFC4360]**. A Route Target Community, in the MPLS VPN context, identifies a set of routers that are part of a VPN (i.e., members of the community) that need to know about the routes within the VPN (i.e., receive a particular address prefix). A BGP Community, in general, defines a group of network nodes that share a common understood property. The Community information identifies members of the BGP Community and is carried as a path attribute in BGP UPDATE messages. The Community information allows a network to perform specific actions on the group of Community members without having to provide details on each member.
- **Layer 3 VPN Routes Learned from BGP**: This table maintains Layer 3 VPN routes learned via BGP from other PE routers **[RFC4364]**. The Layer 3 routes in this table are copied and maintained in Layer 3 VRFs that have matching route tables.
- **IS-IS Routes**: Intermediate System-to-Intermediate System (IS-IS) **[ISO10589:2002]** is a link-state Interior Gateway Protocol (IGP) that allows a network (i.e., routing domain) to be organized into a group of flooding subdomains similar to OSPF **[RFC2328]**. The subdomains are referred to as areas and each is assigned an Area Address. A routing device in IS-IS is referred to as an Intermediate System (IS), and is identified by a Network Entity Title (NET). Specifically, the NET identifies an IS-IS routing protocol instance (router) running on an IS. A NET contains (defines) an Area Address and the

System ID of the router (or IS). When IS-IS is used to support IP routing, this IS-IS table maintains the NETs. Level 1 routing refers to routing within an area, while Level 2 routing refers to routing between Level 1 areas.

3.4.1.4 Aggregate or Summary Routes in the Routing Table

Route aggregation (or summarization) allows a router to combine or consolidate a number of routes with close addresses into a single entry (i.e., aggregate destination address) in the Routing Table. Each contributing route is relatively more specific within the aggregate route. Only when an aggregate route has at least one contributing route, then can it become active. The routing device can advertise aggregate route in a single route advertisement. An aggregate route is also referred to as a summary route.

Route aggregation can be used to decrease the size of the Routing Table (number of entries maintained) as well as the number of route advertisements the router sends out. An aggregate route is not used by a routing device for actual packets forwarding – only for route aggregation and advertisement (e.g., in BGP route advertisement). The router can advertise aggregate routes, but it relies on the more specific routes in the aggregate for packet forwarding. Packet with destinations matching the more specific routes in the aggregate route (and not the aggregate route itself) will be forwarded according to those more specific routes.

3.4.2 Forwarding Table

The Forwarding Table (Figure 3.7) contains all the active routes along with the most relevant forwarding information contained in the Routing Table directly used or relevant to data plane operations. It contains the information used for making actual packet forwarding decisions (network address prefixes, next-hop IP addresses, outgoing interfaces, next-hop MAC addresses [adjacency information]). The Forwarding Table is sometimes referred to as the Forwarding Information Base (FIB). Using the destination IP address of an IP packet, the forwarding engine consults the Forwarding Table to determine the appropriate outgoing interface and next-hop router, and forwards the packet accordingly.

An active route is a route that is still retained in the Routing Table and has not been marked for removal from the table (e.g., in RIP, the RIP route flush timer has not expired). Note that when the RIP invalid timer expires (as discussed in Chapter 2), the route is simply marked as inaccessible and advertised as unreachable, but the route may still be used to forward packets. The router copies the active routes from its Routing Table to the Forwarding Table.

When routing or topology changes occur in the network, the router updates the IP Routing Table and those changes then get reflected in the Forwarding Table. The synchronization of the routing information in the Routing Table and Forwarding Table is handled by a routing protocol process in the router's operating system. Any time a network change occurs, the routing protocol process determines all the routes in the Routing Table and installs these in the Forwarding Table.

3.5 A NOTE ON LAYER 2 ADJACENCY TABLE

To complete the forwarding of an IP packet over the outgoing interface, for example, an Ethernet interface, a rewrite of the destination and source MAC addresses is required. The adjacency information, that is, the Ethernet address of the next-hop's receiving interface (typically obtained through ARP (Address Resolution Protocol) or configured manually), specifies the destination MAC address needed for the Ethernet MAC address rewrites.

The Adjacency Table is a database of outbound interfaces and Layer 2 next-hop addresses for all Routing Table entries. An IP router uses ARP to learn the Ethernet MAC address that corresponds to the IP address of any host or router on networks that are directly connected to it. When a router does not have a direct connection to a packet's final destination, it will forward the packet to the next-hop router (the neighbor router adjacent to it) that is closer to that destination. This process repeats until the packet reaches its final destination.

The adjacency information used by the IP forwarding engine can be integrated with the IP Forwarding Table or maintained as a separate Adjacency Table (Figure 3.5) **[AWEYA1BK18] [AWEYA2BK19]**. Either way, the Ethernet MAC address entries of the next-hop can be populated using any one of the following methods:

- Obtained from ARP requests sent to neighbor devices (routers and hosts)
- Gleaned from ARP request sent by neighbor devices
- Configured manually by the network administrator by considering the devices that are directly connected to the router, that is, connected by a Layer 2 network (point-to-point, VLAN/subnet).

The Forwarding Table is updated when one of the following occurs:

- When an address prefix in the Routing Table entry is removed, or a new prefix is added.
- When the next-hop IP address for a Routing Table entry changes or is removed.
- The ARP cache entry for a next hop IP changes, removed, or times out.

3.6 IP FORWARDING OPERATIONS

As discussed above, the process of forwarding an IP packet across internetworks can be divided into two distinct parts or two processing planes. The control plane in a router is responsible for communicating with peer routing devices to construct and maintain the IP Routing Table, which defines where and how IP packets should be forwarded to, based upon the IP destination addresses of the packets. The forwarding information is defined in terms of a next hop IP address and the outgoing router interface, in addition to any other relevant information, that can be used to forward the packet on its way to the destination. The IP Routing Table is generally distilled and maintained in a simplified table referred to as the IP Forwarding Table or FIB.

The data plane process is responsible for using the routing information maintained in the Forwarding Table to actually forward IP packets to their respective destinations. The two planes work together to get packets to their correct destinations. The control plane defines the routes on which IP packets should be forwarded on, while the data plane executes exactly how the IP packet should be routed. The forwarding information includes the underlying Layer 2 addresses required in an outgoing packet (so that it they can reach the next hop IP destination), as well as other operations required for forwarding the packet (Figure 3.8). The data plane performs operations such as decrementing the Time-To-Live (TTL) field and recomputing the IP header checksum.

The following steps summarized the main processing at the router when an IP packet carried in an Ethernet frame is received **[AWEYA1BK18] [AWEYA2BK19]**:

- **Bit/Symbol Reception**: Interface receives bits and Ethernet symbols from the transmission medium and constructs an Ethernet frame.
- **Ethernet Frame Verification**: Interface performs verification of Ethernet frame length, Ethernet checksum (or Frame Check Sequence [FCS]), destination MAC address, etc.
- **Encapsulated Protocol Demultiplexing**: Interface demultiplexes the encapsulated packet according to its Ethertype or protocol number (IPv4 (= 0x0800), IPv6 (= 0x86DD), ARP (= 0x0806), etc.).
- **IP Packet Validation**: IP forwarding engine validates the IP(v4) packet by verifying the total data length passed by the Data Link Layer, IP checksum, IP version, IP header length, IP packet total length, etc.

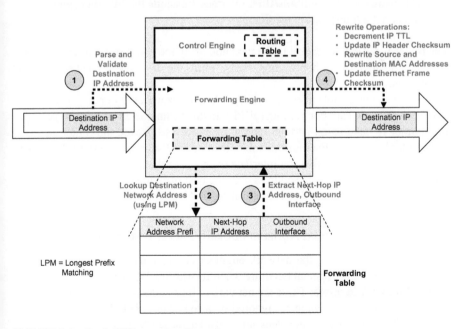

FIGURE 3.8 Basic IP Forwarding Operations

- **Local or Remote Packet Delivery Decision**: IP forwarding engine decides if received IP packet is for local delivery, or is to be forwarded to another external node (a next-hop node).
- **IP Forwarding Table Lookup and Packet Forwarding Decision**: IP forwarding engine performs a LPM search in its IP Forwarding Table to determine the next-hop node and outbound interface for the IP packet. IP Layer also decrements the IP TTL, and updates the IP header checksum.
- **Ethernet Frame Parameter Mapping**: IP forwarding engine determines the Ethernet frame parameters to be used in encapsulating the IP packet (e.g., source and destination Ethernet MAC addresses, VLAN mappings, Class-of-Service [CoS] mappings, etc.).
- **Ethernet Frame Construction and Frame Rewrites**: IP packet is encapsulated in in an Ethernet frame with appropriate source and destination Ethernet MAC addresses, and updates all relevant fields in the frame such as VLAN and CoS fields, and then updates the Ethernet checksum.
  o Destination MAC Address: The MAC address of the next hop is written into this field in the Ethernet frame. The MAC address information is built from the ARP process or configured manually [AWEYA1BK18][AWEYA2BK19].
  o Source MAC Address: The MAC address of the outgoing interface of the router is written into this field in the Ethernet frame.
  o Ethernet Frame Checksum: This value is recomputed as the Source and Destination MAC addresses have changed.
- **Mapping of Ethernet Frame into Symbols**: Ethernet Physical Layer receives the Ethernet frame and maps it into corresponding Ethernet symbols.
- **Transmission of Symbols/Bits**: Interface transmits the Ethernet symbols and bits on the transmission medium.

The router forwards the packet to the next hop using the above process, and this process is repeated on a hop-by-hop basis, until the IP packet reaches its final destination.

The Forwarding Table also plays a key role in certain security operations, such as in unicast Reverse Path Forwarding (uRPF). In this technique, in addition to performing normal IP destination address lookup of a received packet, the router also performs a look up in the Forwarding Table, using the IP source address of the packet. If the router determines that the interface over which the packet was received does not lead back to the IP source address over the best path, the packet is assumed to be involved in a network attack or is malformed, and is dropped.

3.6.1 Handling Special Addresses during Packet Forwarding

In some routing devices, the network engineer can identify, for example, Martian addresses, that require special handling when forwarding IP packets:
- **Martian Addresses**: These are network or host addresses whose routing information must be completely ignored by a router (see RFC 1812, Section 5.3.7 [RFC1812]). The router does not consider these routes/addresses when discovering the network topology. These routes are generally propagated by

devices that are improperly configured on the network. Packets carrying Martian addresses obviously have invalid destination addresses, or have source or destination IP addresses reserved for only special-use IP address prefixes or ranges. In IPv4 and IPv6, an example of a default Martian address is the loopback address. Martian addresses are commonly used in IP address spoofing, for example, in denial-of-service attacks, but can also be generated by host and network nodes that are malfunctioning or misconfigured.

3.7 REDISTRIBUTING ROUTING INFORMATION AND ROUTING METRIC TRANSLATION

Most modern routers used in enterprise and service provider networks are designed to allow multiple routing protocols to be run on a single router connecting different networks, or connecting routing domains using different routing protocols. For example, a router can connect one routing domain running RIP to another domain running EIGRP, and allow these two domains to exchange routing information in a controlled fashion.

Routing protocols have different characteristics, and each protocol collects and exchanges different types of routing information. Each routing protocol also reacts to network topology changes in a different way. For example, RIP uses a hop-count routing metric, while EIGRP uses a composite routing metric (see Chapter 1). When there is the need to exchange routing information between different networks that use different routing protocols, the differences in the routing protocol characteristics have to be taken into consideration as explained in this section (and also discussed in greater detail in Chapter 7).

3.7.1 THE NEED FOR ROUTE REDISTRIBUTION

Route redistribution is the process by which a routing protocol advertises routes that are learned by another routing information source (such as static routes, directly connected routes, or routes from another routing protocol). A router that runs multiple routing protocols simultaneously can redistribute routing information from one routing protocol to another. Essentially, it is possible to redistribute routes learned by any routing protocol into any other routing protocol. For example, taking redistribution between IGPs, OSPF can import routes learned by RIP, EIGRP, or IS-IS. Routes learned by OSPF can also be exported into RIP, EIGRP, or IS-IS. For redistribution between and IGP and EGP, any IGP can import routes learned by BGP. IGP routes can also be exported into BGP.

3.7.2 FILTERING INBOUND AND OUTBOUND ROUTING INFORMATION

A network administrator also can conditionally control how routes are redistributed between routing domains by employing route filters such as route maps, distribute lists, prefix lists, and so on (see Chapter 7). In BGP, route maps, distribute lists, or prefix list can be defined on a per BGP neighbor basis to filter routes and modify the BGP path attributes associated with the routes.

Route filters can be applied to either inbound or outbound routing information at a router, to allow only the routes that pass the route filters' conditions to be accepted or sent to other neighbor routers. For instance, in the inbound direction, a BGP router can be configured to match routes on BGP Autonomous System path and community information. In the outbound direction, the router can be configured to match based on IP address prefixes, Autonomous System path, and community information.

A network administrator can filter routing protocol information by performing the following tasks:

- *Suppress the Sending of Routing Updates on a Particular Router Interface*: This is done to prevent neighbor routers on a network segment attached to a router interface from learning about routes (see "Passive Interfaces" in Chapter 7).
- *Suppress Routes from Being Advertised in Routing Updates*: This is done to prevent neighbor routers from learning one or more routes.
- *Suppress Routes Listed in Routing Updates from Being Accepted and Acted Upon by a Routing Process*: This is done to prevent a router from accepting and using certain routes.
- *Filter Routes Based on the Routing Information Source*: This is done to prioritize routing information from different routing information sources, because some routes (to a given network destination) may be more accurate than others. In a large network, some routing information sources may provide more reliable routes than others. By specifying different Administrative Distance values to rank different routing information sources, a router can intelligently discriminate between different routes to a destination. The router always prefers the route whose routing source has the lowest Administrative Distance.
- *Apply an Offset to Routing Metrics*: This is done to increase the value of the routing metrics of a redistributed route. An offset list (discussed in Chapter 7) is a mechanism for increasing the routing metrics of incoming and outgoing routes learned via distance-vector routing protocols (RIP and EIGRP). Optionally, the routing metric of a route can be increased using a route map.

3.7.3 The Need for Routing Metric Translation

When a routing protocol determines that multiple routes exist to a given network destination, it compares the routing metrics of these routes to select the best route to that destination. The router computes the routing metric of each route by assigning a characteristic or set of characteristics (i.e., numeric parameters) to each network link that forms part of the route to the destination. The routing metric for a route is an aggregation of the parameters of each link that lies on the route. It is discussed in Chapter 1 that different routing protocols use different techniques for assigning parameters to individual links in a network. Further, each routing protocol aggregates the set of parameters on a route in a different way.

Differences in routing protocol characteristics, such as classful or classless routing capabilities, routing metrics, and Administrative Distances, can affect route redistribution. As discussed in Chapter 1, different routing protocols use different routing metrics, and the routing metrics of one routing protocol (RIPv2, EIGRP,

OSPF, or IS-IS) do not necessarily translate seamlessly into the routing metrics of another routing protocol. This means serious consideration must be given to these differences when carrying out route redistribution between different routing protocols. Careless manipulation of routing metrics and Administrative Distances, when redistributing routes between different routing protocols, can create routing loops, thereby resulting in serious degradation of network performance.

For example, RIP's routing metric is based on hop count, while that of EIGRP is based on a composite metric that uses bandwidth, delay, load, Maximum Transmission Unit (MTU), and reliability, where bandwidth and delay are the only default parameters. Therefore, when routes are redistributed from one routing protocol to another, a routing metric must be defined for the redistributed route that is understood by the receiving protocol. During redistribution, the redistributed route is assigned an artificial seed or starter routing metric, that is compatible with the receiving routing protocol (see Chapter 7).

As discussed in Chapter 2, the Administrative Distances (or Route Preferences) are used in route selection when different routing protocols provide routes to the same network destination. The Administrative Distance is used to rate the trustworthiness of a routing information source when multiple sources provide routes to a particular network destination. A router uses the Administrative Distances to select which route to install in its Routing Table when several routing information sources provide routes to the same network destination. The route with the lowest Administrative Distance is preferred. The discussion in Chapter 7 shows that when not properly used, the Administrative Distance setting of the routing protocols can cause problems during route redistribution. These problems can be in the form of inefficient routing, convergence problems, or routing loops.

REVIEW QUESTIONS

1. What are the main functions of the control engine (also called the routing engine or route processor) in an IP router?
2. What are the main functions of the forwarding engine in an IP router?
3. What is the main difference between the IP Routing Table and the Forwarding Table.
4. What are the advantages of distributed forwarding architectures over centralized forwarding architectures?
5. What are the advantages of control plane (or route processor) redundancy?
6. Why is the IP checksum recomputed when a packet is being forwarded?
7. Why is the Ethernet checksum recomputed when a packet is being forwarded?
8. When forwarding an IP in an Ethernet frame to the next hop, what gets written into the destination MAC and source MAC address fields of the Ethernet frame?
9. How is the Layer 2 Adjacency Table of an IP router populated?

REFERENCES

[AWEYA1BK18]. James Aweya, Switch/Router Architectures: Shared-Bus and Shared-Memory Based Systems, Wiley-IEEE Press, ISBN 9781119486152, 2018.

[AWEYA2000]. James Aweya, "On the Design of IP Routers. Part 1: Router Architectures," *Journal of Systems Architecture (Elsevier Science)*, Vol. 46, April 2000, pp. 483–511.

[AWEYA2001]. James Aweya, "IP Router Architectures: An Overview," *International Journal of Communication Systems (John Wiley & Sons, Ltd.)*, Vol. 14, Issue 5, June 2001, pp. 447–475.

[AWEYA2BK19]. James Aweya, *Switch/Router Architectures: Systems with Crossbar Switch Fabrics*, CRC Press, Taylor & Francis Group, ISBN 9780367407858, 2019.

[CISCINTROUT14]. Cisco Networking Academy, Routing Protocols Companion Guide, Chapter "Cisco Networking Academy's Introduction to Routing Dynamically", Cisco Press, February 24, 2014.

[CISCNEXHA16]. Cisco Nexus 9000 Series NX-OS High Availability and Redundancy Guide, Release 6.x, Chapter "System-Level High Availability", May 13, 2016.

[ISO10589:2002]. ISO/IEC 10589:2002 – Information technology – Telecommunications and Information Exchange between Systems–Intermediate System to Intermediate System Intra-Domain Routing Information Exchange Protocol for use in Conjunction with the Protocol for Providing the Connectionless-Mode Network Service (ISO 8473)", International Organization for Standardization (ISO). November 2002.

[RFC1812]. F. Baker, Ed., "Requirements for IP Version 4 Routers", IETF RFC 1812, June 1995.

[RFC2328]. J. Moy, "OSPF Version 2", IETF RFC 2328, April 1998.

[RFC4360]. S. Sangli, D. Tappan, and Y. Rekhter, "BGP Extended Communities Attribute", IETF RFC 4360, February 2006.

[RFC4364]. E. Rosen and Y. Rekhter, "BGP/MPLS IP Virtual Private Networks (VPNs)", IETF RFC 4364, February 2006.

4 Static Routes in the Routing Table

4.1 INTRODUCTION

Many of today's networks of all sizes use a combination of static and dynamic routing. Static routing is very appealing and widely used, because it does not require the same amount of processing and memory resources, and routing information messaging overhead, as in dynamic routing protocols. In this chapter, we contrast static routing with the widely implemented dynamic routing protocols. We discuss, in addition, the different methods used for configuring static routes in Routing Tables.

4.2 BENEFITS OF DYNAMIC ROUTING PROTOCOLS

Routers can learn routes to remote networks via static configuration (static routes) and dynamic routing protocols (dynamic routes). However, in large networks with many VLANs or subnets, a great deal of administrative and operational overhead can be incurred when configuring and maintaining static routes between the different VLANs or subnets. The initial configuration and maintenance of static routes can be time-consuming. It becomes especially cumbersome, and requires significant operational overhead to configure static routes when network changes occur, such as when a link goes down, or fails, or when a new VLAN or subnet is added to the network.

The use of dynamic routing protocols can reduce significantly the burden of configuring and maintaining routes in a dynamic network, allowing greater network design flexibility and scalability. Dynamic routing protocols allow routers to share routing information in a network without manual intervention, and the best discovered routes are installed in the Routing Tables. Thus, dynamic routing protocols play a major role in a network, and are generally implemented in all of today's networks because of the many benefits they provide.

The primary benefit of dynamic routing protocols is that, they allow routers to exchange routing information automatically when changes in the network topology or state occur. The automatic exchange of routing information allows routers to discover new remote networks, and also to determine alternate paths when existing paths to network destinations become unavailable. Dynamic routing protocols are suitable for basically all network topologies where multiple routers are used. Using dynamic routing, the best paths to network destinations always depend on the current topology or state of the network (Figure 4.1).

Generally, most of today's dynamic routing protocols like EIGRP, OSPF, IS-IS, and BGP can operate independent of network size, and allow a network to automatically reconfigure routing to reroute traffic, and adapt to changes in topology and state. Even though dynamic routing protocols can be relatively more complex to

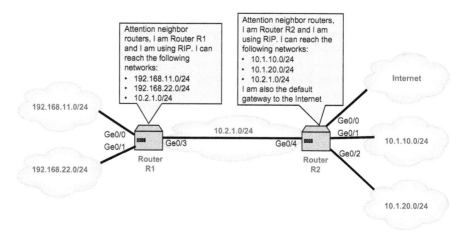

FIGURE 4.1 Illustrating Dynamic Routing

implement initially, they enable network managers circumvent the tedious and often time-consuming process of configuring and maintaining only static routes in a network. However, in real-world networks, dynamic routing protocols are not used alone, but are used in combination with static routing.

4.3 BENEFITS OF STATIC ROUTING

Even though, dynamic routing protocols require significantly lower administrative overhead to configure and maintain routes, this comes at the expense of using more of a router's computing and memory resources for protocol operation. The protocol messaging used by dynamic routing protocols also consumes more network bandwidth. Dynamic routing protocols are also relatively less secure when used natively without additional security mechanisms, due to the broadcast and/or multicast mechanisms they use to propagate routing updates.

The interfaces used by the dynamic routing protocol to send out routing updates, require additional mechanisms for the authentication of routing protocol exchanges. The network administrator may configure additional features such as passive interfaces for added routing security. The need for configuring passive interfaces during dynamic routing often arises in large complex networks. In most dynamic routing protocols, the configuration of passive interface prevents or restricts routing updates from being sent over some interfaces, while permitting routing updates to be advertised normally over other interfaces.

By default, a router running RIP forwards routing updates out all its RIP-enabled interfaces. However, in reality, the router only needs to forward RIP updates out interfaces that are connected to other RIP-enabled routers. To address the problems of wasted bandwidth, processing and memory resources, as well as, reduce security risks, a passive interface can be configured to prevent a router from sending routing updates on those interfaces. A passive interface prevents the transmission of routing updates through it but will still allow the network connected to it to be advertised to

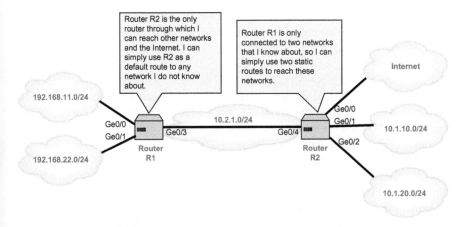

FIGURE 4.2 Illustrating Static and Default Route

other routers. The network connected to the passive interface can still be advertised in routing updates sent out other router interfaces.

Even with the many benefits of dynamic routing, making it a better choice for routing, there are some networking scenarios where static routing is more suitable (Figure 4.2). Many of today's networks, especially those with moderate levels of complexity, employ both dynamic and static routing. It is not uncommon today to see large networks that employ a combination of dynamic and static routing. In such scenarios, the routers will maintain Routing Tables that contain directly attached networks/interfaces, routes learned dynamically via routing protocols, and manually configured static routes.

Static routing is most suitable for the following networking scenarios [CISCNETAIRD14] [CISCNETAISR14]:

- **Ease of Configuring and Maintaining Routes in Small Networks**: In networks that are small, simple, and have less complex topologies, and are not expected to have significant growth or change over time, static routing provides an easy way of configuring and maintaining routes in the Routing Table. Static routes are not that difficult to implement in small networks where the route to a traffic destination is known, does not change, and always predictable (the flow of traffic is very predictable).
- **Access to a Stub Network**: A stub network has only one path to the outside world, so, using static routing to and from it, for simplicity, makes sense. A stub network is a network that has only one default path in and out to non-local devices. A stub network is reachable via a single route, and its access router has only one neighbor router. A stub networks may have multiple internal connections, but is essentially a dead-end network that has only one logical path to and back from the outside network.
- **Access to a Default Route**: Static routing can be used as a means of accessing a single default route. A default route is an entry in the Routing Table that

specifies where to forward packets when no matching route can be determined for a given destination address. Packets destined to network destinations that are not specified in the Routing Table can be forwarded via the default route.

Static routing is more applicable to the above networking scenarios because the routes stay the same, and do not change, which also makes network connectivity problems fairly easy to troubleshoot. Another benefit is, static routes do not require routing update messages to maintain them, and require very little administrative and operational overhead as mentioned above.

Unlike dynamic routing protocols, static routing does not require routing update algorithms and mechanisms, and route advertisement over the network. As a result, static routing uses less network bandwidth (as routers do not exchange routing information), has better security, does not use extra router processing and memory resources (to calculate and communicate routing information) as in dynamic routing (Figure 4.3).

However, it should be recognized that static routing is not easy to implement in a large network, where configuring and managing a large number of static routes can be time consuming, and a complete knowledge of the entire network is required for proper implementation. In such networks, if there are link failures, a static route that passes through such links cannot reroute traffic. Whenever link failures occur, such events require manual intervention to reroute traffic in the network.

Thus, static routing does not scale well in networks that are growing and undergoing changes because configuration and maintenance of routes can become overwhelming and cumbersome. Manual intervention is always required to reconfigure and manage changing routing conditions, and the configuration of routes can be error-prone, especially in large and/or complex networks.

Despite the importance and benefits of implementing dynamic routing in a network, network design practices recognize that static routing still has a role play in networking. As discussed above, static routing is suitable for simple topologies such as routing to a stub network, as a default static route, or to a smaller network with only one path to the outside world. However, network engineers have come to

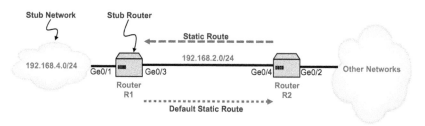

- From the figure, network 192.168.4.0/24 attached to Router R1 has only one path to reach other networks attached to R2 or beyond.
- This makes network 192.168.4.0/24 a stub network and Router R1 a stub router.
- Therefore, running a routing protocol between Routers R1 and R2 would be unnecessary and a waste of resources.
- With this, a static route can be configured on Router R2 to reach the R1.
- Furthermore, because Router R1 has only one path to send out traffic to other destinations outside network 192.168.4.0/24, a default static route can be configured on R1 pointing to R2 as the next hop for all other networks.

FIGURE 4.3 Illustrating a Stub Router and Stub Network

recognize that configuring and managing static routes in large networks can become cumbersome and time consuming and the complexity of configuration increases dramatically as the network grows.

Static routes are installed permanently in the Routing (and Forwarding) Tables until manually removed or changed. These routes often contain only one or very few hops to a destination prefix (e.g., a well-known server), and are configured manually by the network manager and do not change. Through the use appropriate configuration commands (depending on the routing platform), a static route can be created in the Routing Table to a network destination. A network prefix can be added to the Routing Table, by defining at minimum, the route as static, and associating with it, an outgoing interface and a next-hop IP address.

The router then installs the static route in the Routing Table when the associated next-hop IP address is reachable, and then forwards all traffic destined for the static route to the next-hop IP address for delivery to the destination. As discussed in Chapter 2, static route, by default, has a lower Administrative Distance (or Route Preference) than any dynamic routing protocol to the same destination. When multiple paths to a network destination exists including dynamically learned routes and a static route, the static route will take precedence over all the routes discovered via the dynamic routing protocols since the default Administrative Distance of a static route is 1.

4.4 CONFIGURING DYNAMIC ROUTING VERSUS STATIC ROUTING

When a router running a dynamic routing protocol first powers up, it has to configure each one of its interfaces with information that includes an IP address and network mask [RFC1878]. The configuration information is typically saved in a configuration file that is stored in the router's nonvolatile RAM (NVRAM). Before power up and normal operation, the routers do not even know that there are devices at the other end of its attached links, and also nothing about the overall network topology.

After successfully powering up, and applying the saved configuration to configure the IP addresses of its interfaces correctly, the router will then initially discover its own directly connected hosts and networks. The router adds any discovered directly connected networks (i.e., their IP addresses along with their network masks [RFC1878], or equivalently, network prefix) to its Routing Table. The router updates the Routing Table with all directly connected networks and the interfaces on which these networks are attached.

After entering this initial information about all directly attached networks, the router then proceeds to discover additional routing information to remote networks for its Routing Table. For the particular routing protocol configured, the router begins to exchange routing updates with other routers in the network to learn about any remote networks and the routes to these networks.

The router sends a routing update message out all of its configured interfaces with each update containing information about all the current directly connected networks in its Routing Table. At the same time, other routers in the network send similar routing update messages which the local router also receives and processes. When the router receives a routing update, it checks the update for new routing information, and where there are any remote networks and routes that are not currently installed in its Routing Table, it will add these to that table.

The routers in the network continue the exchange of routing information until full knowledge and/or a converged view of the network takes place (depending on the routing protocol in use). At some point in time, each router in the network will have knowledge about its own directly connected networks, in addition to the routes to the connected networks of its neighbor routers. The routers continue to exchange periodic and/or triggered updates as a process of bringing the network toward convergence. Each time a router receives an update, it checks it for new network information.

However, with static routing, the network administrator manually configures any required static route to a specific remote network in the route table. The static routes are not automatically or dynamically reconfigured, but instead have to be manually reconfigured and reentered into the Routing Table whenever changes in the network topology or state occur. This means a static route that becomes unavailable, still remains in the Routing Table and does not change, until the network administrator manually removes it, and reconfigures another route to that remote network.

Static routes can be redistributed into dynamic routing protocols, but obviously, routing information learned by dynamic routing protocols cannot be redistributed into a static Routing Table. We discuss below the types of static routes that can be configured in routing platforms **[CISCNETAISR14]**.

4.5 STANDARD STATIC ROUTE

A network administrator can configure a static route to connect to basically any type of remote network. To manually configure a standard static route in Routing Table, the following minimum configuration information is needed:

> `ip route` [*destination_network*]_[*network_mask*] [*outgoing_interface*] [*next_hop*] [*admin_distance*]

- **Command to Add a Static Route**: The standard command used in IP routers to define a route as static is `ip route`.
- **Outgoing Interface**: The parameter *outgoing_interface* specifies the exit interface on the router that leads to the next-hop router. In most routers, specifying a [*next_hop*] also implies it is associated with a given outgoing interface, so this is generally not a required parameter. However, this parameter can be used when a physical interface is being configured as the next hop to a destination network.
- **Destination Network IP Address**: The parameter *destination_network* specifies the IP address of the destination network to installed in the Routing Table.
- **Network Mask**: The parameter *network_mask* specifies the network mask of the destination network IP address **[RFC1878]**. Equivalently, the destination IP address and prefix length can be used above in place of the destination IP address and network mask (e.g., 10.0.0.0/24 instead of 10.0.0.0 255.0.0.0).
- **Next-Hop IP Address**: The parameter *next_hop* specifies the IP address of the next-hop router to which packets should be forward to reach the destination network.

Static Routes in the Routing Table

- **Administrative Distance**: The optional parameter *admin_distance* specifies the Administrative Distance value of the static route being configured. When not specified, the default Administrative Distance for directly connected routes (or networks) is 0, while that for static routes is 1. Directly connected routes take precedence over static routes, which in turn are preferred over routes discovered by dynamic routing protocols.

For example, when configuring a static route over a physical interface that leads to the next hop, the IP address and prefix length (or enter the IP address and network mask), can be used, followed by the outgoing interface number to be used as the next hop:

```
ip route 10.128.2.65/24 Eth 1/2
```

This example configures an IP static route to the destination network address of 10.128.2.65/24 (i.e., 10.128.2.65 with network mask of 255.255.255.0), and Ethernet port Eth 1/4 as the outgoing interface to the next hop (Figure 4.4).

4.5.1 Concept of Qualified Next Hop

A destination address can have several static routes configured to it, each route with a different next hop assigned to it as discussed in reference **[JUNIPROINGUI20]**. In this set up, the destination address has multiple next hops associated with it. In this case, the source router will insert multiple static routes with different next hops into it Routing Table, and will use a route selection method to determine which route to

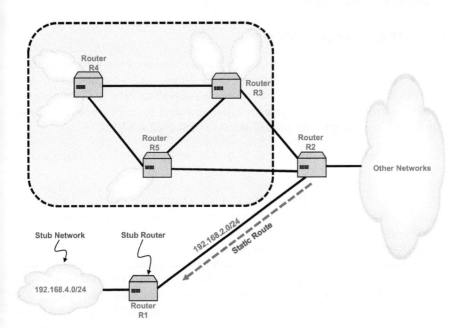

FIGURE 4.4 Using a Standard Static Route to Connect to a Stub Network

use. The primary criterion for route selection is the Administrative Distance (or Route Preference), which allows the router to control which of the routes to use as the primary route for the destination.

Route selection can be influenced by setting the Administrative Distance (or Route Preference) associated with each next hop to a specified value (to allow the ranking of static routes) **[JUNIPROINGUI20]**. The static routes with a lower Administrative Distance are always preferred and used for traffic forwarding. When a route preference is not set, the source router can select in a random fashion one of the next-hop addresses to install in its IP Forwarding Table to be used for traffic forwarding.

In general, when multiple static routes with different next-hop addresses are configured for a given destination, they are all treated as standard static routes (they have the default static routes properties). However, if there is the need to configure two different next-hop addresses for a particular destination and have them treated differently, one of them can be defined as a *qualified next hop* **[JUNIPROINGUI20]** (see also the concept of *floating routes* below).

Using qualified next hops, the network administrator can associate one or more properties with any one of the next-hop addresses. In Juniper Network Operating System (JUNOS), an overall preference can be set for a particular static route, and then a different preference specified for the qualified next hop **[JUNIPROINGUI20]**. For example, let us assume two next-hop addresses (10.10.20.20 and 10.10.20.17) are associated with the static route 192.168.50.6/32. The network administrator can assign a general preference to the entire static route (192.168.50.6/32), and then a different preference to only the qualified next-hop address 10.10.20.17. The qualified next hop 10.10.20.17 can be assigned the preference 7, while the next-hop 10.10.20.20 is assigned the preference 6.

4.6 DEFAULT STATIC ROUTE

A default static route can be used to forward packets destined to network destinations that are not explicitly listed in a Routing Table. This is used when no other address entries in the Routing Table match a packet's destination address. In Cisco routers, a default static route can be configured by defining a static route (in the Routing Table) with 0.0.0.0/0 as the IPv4 destination address **[CISCNETAISR14]**:

```
ip route 0.0.0.0 0.0.0.0 [outgoing_interface] [next_hop] [admin_distance]
```

This means, entries in the Routing Table that identify a specific destination with a larger network mask (or network prefix) for a packet's destination address will take precedence over the default static route.

The default static route identifies a gateway (or next-hop router) to which packets with destination addresses that do not match specific dynamically learned or static route, are forwarded. Configuring a static route to network 0.0.0.0/0 is a way of adding a gateway of last resort on a router. At a minimum, the configuration information required is the outgoing interface, destination IP address (0.0.0.0), network mask (0.0.0.0), IP address of the IP next-hop router (or gateway), and an optional Administrative Distance value (Figures 4.5 and 4.6).

Static Routes in the Routing Table

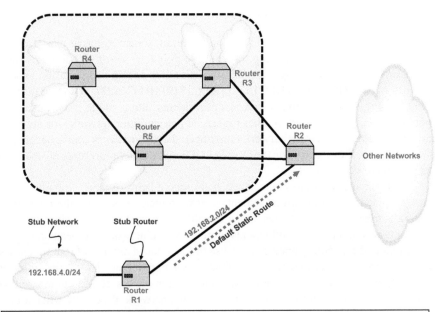

Router R1 only needs to know about its directly connected networks and then use a default static route pointing to R2 to send traffic to all other networks.

FIGURE 4.5 Using a Default Static Route to Connect to a Stub Router

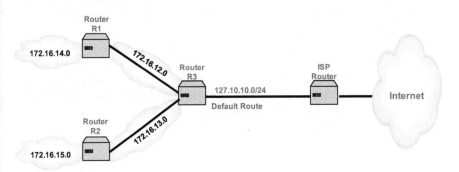

- Let us assume the Routers R1, R2, and R3 in this network are running RIP.
- Router R3 is connected to the ISP's router and has a static default route that points to it.
- It is possible to use RIP to advertise that default route to other routers in the local network, that is, Routers R1 and R2.
- On Router R3, we simply need to enter the *default-information originate* command in the RIP configuration mode:

    ```
    R3(config)#ip route 0.0.0.0 0.0.0.0 127.10.10.1
    R3(config)#router rip
    R3(config-router)#default-information originate
    ```

- Routers R1 and R2 do not need any additional configuration since they learn the default route just like any other RIP route

FIGURE 4.6 Advertising a Default Route Using RIP

4.7 SUMMARY STATIC ROUTE

Route summarization (also called route aggregation, supernetting, or prefix aggregation) is the process of grouping together a contiguous set of IP addresses (networks or subnets) into a single IP address with a less-specific address and shorter network mask **[RFC1517] [RFC1518] [RFC1519] [RFC4632]**. With summarization, a contiguous set of routes can be consolidated into a single route advertisement that can be advertised to other routers. When multiple networks (or subnets) are aggregated into a larger network (supernet), the new network prefix becomes a single Routing Table entry for the aggregated network, and represents all the constituent networks.

The advantages of route summarization are that, it helps to reduce the number of entries carried in routing advertisements sent by routers, as well as, the number of routing entries in Routing Tables. Other benefits are that, it helps to reduce the processing load and memory requirements on routers, and network overhead and bandwidth required for routing updates (because of the reduced number of routing entries in the routing protocol updates). The smaller Routing Tables also results in faster Forwarding Table lookups and processing latencies in routers. Route summarization can also reduce the overall end-to-end latency in a network, especially when a path consists of many routers.

Route summarization also hides or isolates routing instability in the networks behind the summary route (which remains valid), even if any network contained in the summary route is unavailable. Summarization can improve network stability by eliminating or reducing unnecessary propagation of routing updates (or routing churn), after any part of the networks behind the route summary experiences a change in topology or route flapping. Thus, using route summarization, a router can isolate network topology changes from other routers, thereby limiting the propagation of routing updates after routes become unavailable. This helps to improve the stability of the overall network (Figure 4.7).

By advertising, only a single summary route to neighbor routers, a router does not need to advertise any changes to specific constituent networks within the summarized address. In addition to significantly reducing any unnecessary routing updates, and allowing for a more stable network following a topology change, route summarization can increase the speed of network convergence.

The following steps can be used to calculate the summary route (address and mask) for a given set of IPv4 networks or subnets:

Step 1: Convert all the IPv4 addresses from dotted decimal notation to binary format, and list and align each address in a row.

Step 2: Starting from the far-left bit, locate the bit position where the common pattern or matching binary bits in each address ends.

Step 3: The network mask and address prefix for the summary route occupies the bit positions where all the leftmost bits match consecutively, and ends before the column of bits that do not match, which is, the summary boundary.

Static Routes in the Routing Table

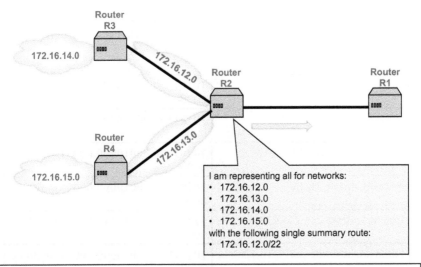

- Router R1 would require four separate static routes to reach the 172.16.12.0/24 to 172.16.15.0/24 networks.
- Instead, Router R1 can be configured with one summary static route and still provide connectivity to these networks.

FIGURE 4.7 Using a Summary Static Route

Step 4: Count the number of leftmost matching bits (common pattern) found above to determine the network address prefix length or, equivalently, the network mask for the summary route.

Step 5: The IP address of the summary route is represented by the matching bits (common pattern) followed by an all 0 bits up to the rightmost end to make the 32 bits of the IPv4 address. The summary route address can be expressed as the common pattern (in dotted decimal notation), followed by a slash (/), then followed by the prefix length (or number of common pattern bits) determined above.

A single static summary route can be configured in the Routing Table to replace a set of static routes with contiguous IP addresses using the steps described above. This helps reduce the number of (static) Routing Table entries and the number of routes advertised, in addition to the other benefits described above. The steps for configuring a summary static route are similar to that of a standard static route.

For example, instead of configuring four separate static routes to reach the networks 172.20.0.0, 172.21.0.0, 172.22.0.0, and 172.23.0.0, one summary static route 172.20.0.0/14 (with network mask 255.252.0.0) can be configured, and still provide connectivity to all of these four networks. If these four static routes already exist in the Routing Table, they can be removed and then replaced with the new summary static route.

If multiple static routes use the same outgoing router interface or next-hop IP address to reach destination networks that have contiguous IP addresses, then they can be

Example 4.1: Using these steps, the summary route for the networks 172.16.12.0, 172.16.13.0, 172.16.14.0, and 172.16.15.0 can be determined as follows:

Network 1: 172.16.12.0	**10101100.00010000.000011**00.00000000
Network 2: 172.16.13.0	**10101100.00010000.000011**01.00000000
Network 3: 172.16.14.0	**10101100.00010000.000011**10.00000000
Network 4: 172.16.15.0	**10101100.00010000.000011**11.00000000
Network Mask: 255.255.252.0	**11111111.11111111.111111**00.00000000
Summarized Address: 172.16.12.0/22	**10101100.00010000.000011**00.00000000

The four networks have been summarized into the single network address and prefix 172.16.12.0/22 (with network mask 255.255.252.0). After the summary route has been created, a router can replace all the component routes with the single summary route in its Routing Table.

Example 4.2:

Network 1: 192.168.98.0	**11000000.10101000.0110**0010.00000000
Network 2: 192.168.99.0	**11000000.10101000.0110**0011.00000000
Network 3: 192.168.100.0	**11000000.10101000.0110**0100.00000000
Network 4: 192.168.101.0	**11000000.10101000.0110**0101.00000000
Network 5: 192.168.102.0	**11000000.10101000.0110**0110.00000000
Network 6: 192.168.105.0	**11000000.10101000.0110**1001.00000000
Network Mask: 255.255.240.0	**11111111.11111111.1111**0000.00000000
Summarized Address: 192.168.96.0/20	**11000000.10101000.0110**0000.00000000

Static Routes in the Routing Table

The summary route is obtained as 192.168.96.0/20 with network mask 255.255.240.0. It should be noted that the summary route covers networks that are outside Networks 1–6 above which are:
- 192.168.9x.0 – 192.168.96.0, 192.168.97.0
- 192.168.1xx.0 – 192.168.103.0, 192.168.104.0, 192.168.106.0, 192.168.107.0, 192.168.108.0, 192.168.109.0, 192.168.110.0, 192.168.111.0

However, to use the summary route 192.168.96.0/20, the network administrator must ensure that these missing network addresses do not exist elsewhere in the network outside this summary route.

Example 4.3:

Network 1: 10.1.32.0	00001010.00000001.00100000.00000000
Network 2: 10.2.45.0	00001010.00000010.00101101.00000000
Network 3: 10.3.0.0	00001010.00000011.00000000.00000000
Network Mask: 255.252.0.0	11111111.11111100.00000000.00000000
Summarized Address: 10.0.0.0/14	00001010.00000000.00000000.00000000

In this example, summary route is 10.0.0.0/14 with network mask 255.252.0.0. Similarly, the networks, 10.1.1.0/24, 10.1.2.0/24, and 10.1.3.0/24 can be replaced with the summary route 10.1.0.0/16.

aggregated into a single summary static route. Configuring summary static routes minimizes the administrative and operational overhead of managing a large number of static routes in a network, and also makes their configuration less prone to errors.

4.8 FLOATING STATIC ROUTE

A floating static route defines a static route that is used as a backup route to a dynamic route, or backup to another static route that serves as a primary route. The floating static route is kept on standby, and is only used when the primary dynamic or static route becomes unavailable, for example, in the event of a link failure along the path.

To place the floating static route in a backup mode, it must be configured with a higher Administrative Distance than the primary dynamic or static route. A static route has a lower default Administrative Distance than a dynamic route. So, to use a floating static route as a backup to a dynamic route, it must be assigned an Administrative Distance value higher than that of the dynamic route.

As discussed above, the Administrative Distance defines the relative trustworthiness or reliability of a route. When multiple routes exist to a network destination, the router chooses the route with the lowest Administrative Distance. A floating static route configured as a backup to a primary static route will be assigned a higher Administrative Distance to give precedence to the primary route which has a default value of 1.

For example, if a floating static route is configured to serve as a backup to an OSPF learned route, the floating static route must be configured with a higher administrative Distance than OSPF which has a default Administrative Distance of 110. If the floating static route is configured with an Administrative Distance of 112, then the router will prefer the OSPF learned route over the floating static route. In the event the OSPF learned route becomes unavailable, the floating static route will be used in its place. If the OSPF learned route fails, it will disappear from the Routing Table, and the router will select the floating static route as the next best route to reach the destination.

The Administrative Distance of a floating static route can be set to a higher value to make the primary static route or a route learned through a dynamic routing protocol more preferable. The floating static route is configured and enabled, but is not used to forward traffic when the primary (static or dynamic) route which has a better Administrative Distance, is active. Only when the preferred primary route becomes unavailable that the floating static route can take over, and traffic can be forwarded over this backup route (Figure 4.8).

The use of a floating static route is independent of the type of encapsulation used on an interface **[CISCNETAISR14]**. This mean a floating static route can be created on any interface to forward traffic irrespective of its encapsulation type. A floating static route can also be created to serve as a backup route to multiple interfaces on a router. As a backup route, a floating static route is also affected by route flapping and

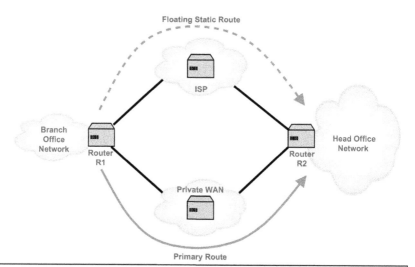

FIGURE 4.8 Using a Floating Static Route as a Backup Route

network convergence time. A primary route that experiences flapping (i.e., being continuously dropped and re-established) can cause the floating static route and its backup interface to be unnecessarily activated and deactivated.

REVIEW QUESTIONS

1. What are the benefits of dynamic routing?
2. What the benefits of static routing?
3. What are the limitations of static routing?
4. What is a stub network?
5. What is a default static route?
6. What are the benefits of route summarization?
7. What is a summary static route?
8. What is a floating static route?
9. Explain briefly how the Administrative Distance (also called the Route Preference) is used when employing floating static routes.

REFERENCES

[CISCNETAIRD14]. Cisco Networking Academy's Introduction to Routing Dynamically, March 24, 2014, Chapter from Cisco Networking Academy, Cisco Press.

[CISCNETAISR14]. Cisco Networking Academy's Introduction to Static Routing, March 27, 2014, Chapter from Cisco Networking Academy, Cisco Press.

[JUNIPROINGUI20]. Juniper Networks, *Protocol-Independent Routing Properties User Guide*, March 26, 2020.

[RFC1517]. R. Hinden, Ed., "Applicability Statement for the Implementation of Classless Inter-Domain Routing (CIDR)", IETF RFC 1517, September 1993.

[RFC1518]. Y. Rekhter and T. Li, "An Architecture for IP Address Allocation with CIDR", IETF RFC 1518, September 1993.

[RFC1519]. V. Fuller, T. Li, J. Yu, and K. Varadhan, "Classless Inter-Domain Routing (CIDR): An Address Assignment and Aggregation Strategy", IETF RFC 1519, September 1993.

[RFC1878]. T. Pummill and B. Manning, "Variable Length Subnet Table for IPv4", IETF RFC 1878, December 1995.

[RFC4632]. V. Fuller, T. Li, "Classless Inter-Domain Routing (CIDR): The Internet Address Assignment and Aggregation Plan", IETF RFC 4632, August 2006.

5 Routing Information Protocol (RIP)

5.1 INTRODUCTION

This chapter and the next describe, respectively, RIP and EIGRP, which are two examples of the most common routing protocols used in today's networks. We discuss the main features of each routing protocol, protocol packet formats, authentication mechanisms, and their high-level router architectures, processes, and databases. The discussion covers some relevant topics related to RIP and EIGRP operations, and the way these protocols generate routing information for use in IP packet forwarding. Each routing protocol maintains a number of databases which hold information about neighbor routers, routing information learned from other routers in the network, and information used for protocol-specific operations. We discuss these databases as well.

The router architecture discussions include the protocol-specific processes, routing databases and their contents, and the various router control plane and data plane processes required for forwarding IP packet in an IGP routing domain. The discussion covers in particular, how the routing information is used in the control and data planes operations in a router. The router architecture discussions are limited to forwarding of IP packet since forwarding at Layer 2 is relatively simple. The Layer 3 forwarding components are much more complex, and require the cooperation of many more network devices to realize IP packet routing.

5.2 ROUTING PROTOCOLS AND THEIR DATABASES

A router uses a routing protocol to determine the best paths to network destinations, shares this routing information with immediate neighbor routers and other routers in the network, and adapts to changes in the network topology and state. Thus, all dynamic routing protocols support procedures for a router to advertise as well as receive reachability information about network destinations including directly connected networks.

A router has to be able to receive and process routing information from other routers, and to pass on the routing information it has learned to other routers yet to be informed about the information. Whenever multiple paths exist to a particular network destination, a routing protocol defines a metric by which the best path to that destination is determined. All these procedures and processes are implemented with the assistance of databases that are specific to the routing protocol, and are tailored for its requirements and operations.

Different routing protocols use different databases that have different contents and serve different purposes. The procedures and algorithms that a routing protocol uses to learn about remote network destinations, and to dynamically detect and adapt to changes in the network topology and state, depend very much on the specific procedures and algorithms the protocol uses and its operational characteristics.

In this chapter and the next, we discuss RIP and EIGRP, respectively, with each chapter describing the most identifying characteristics, operations, and databases of each protocol. We discuss the main features of each routing protocol, which include:

- **Routing Protocol Messages**: These are the messages the routing protocol sends and receives in order to discover neighbor routers, exchange routing information, and maintain an up-to-date picture about the state of the network. Each router in the network transmits and receives routing protocol messages from neighbor routers reachable on its interfaces. The routers send routing messages to indicate network reachability along with the cost associated with routes to those networks. Exchanging routing information enables routers to discover remote networks, detect network topology changes, and allow the routers to advertise such changes or events to other routers in the network. Routing updates generally are sent at regular intervals, and/or after a change in network topology occurs, depending on the routing protocol type.
- **Data Structures**: These are the databases used to store all the routing information the routing protocol needs for its operations.
- **Best Path Computation Algorithm**: These are the algorithms the routing protocols use to derive a picture of the network topology from which the best paths to network destinations are determined.

Out of the different IPv4 IGPs available for distributing routing information within an Autonomous System, the most widely deployed protocols today are RIPv2, EIGRP, OSPFv2, and IS-IS. EIGRP is a proprietary IGP protocol commonly implemented in Cisco routing devices **[CISCID16406] [RFC7868]**. As discussed in Chapter 2, both RIP and EIGRP are distance-vector routing protocols, while OSPF and IS-IS are link-state routing protocols.

As discussed in Chapter 2, a router running a distance-vector routing protocol does not have a complete map of the entire network topology, and has no knowledge of the entire path it takes to reach any particular remote network destination. The router only knows the distance metric (e.g., hop count in RIP, or bandwidth and delay in EIGRP) to reach that destination, and the outgoing interface (i.e., next hop) that can be used to get there. RIPv2 uses the Bellman–Ford algorithm, while EIGRP uses the Diffusing Update Algorithm (DUAL) to determine the best routes to each network destination. These best routes are then installed in the router's Routing Table.

The Routing Table generally contains the directly connected networks of the local router, routes learned dynamically, and configured statically. As discussed in Chapter 4, a static route defines an explicit path that is configured manually by the network administrator between two routers (a source router and a next-hop). It cannot be

Routing Information Protocol (RIP)

updated automatically when network topology changes occur but instead, must be manually reconfigured.

5.3 RIP OVERVIEW

RIPv1 is a first-generation IPv4 routing protocol specified in RFC 1058 **[RFC1058]**. It was designed to be simple and easy to configure, and was very suitable for small networks. RIPv1 had several limitations and had to be improved to a classless routing protocol version, RIPv2 **[RFC2453]**. RIPv2 introduced a number of improvements which included the following: support of VLSM and CIDR (it includes the network mask in routing updates); transmission of routing updates to multicast address 224.0.0.9 for increased network efficiency, instead of the broadcast address 255.255.255.255; support of manual route summarization on any router interface (to reduced Routing Table and update entries); and support of authentication mechanisms to secure the exchange of routing updates between neighbor routers.

A RIPv2 router will advertise to neighbors its directly connected routes/networks as well as the routes it has learned from other routers and installed in its Routing Table. RIPv2 routers do not form neighbor relationships (adjacencies) with other RIPv2 routers in the network, and do not use any form of Hello protocol to track established adjacencies. They simply send routing updates message to the well-known multicast address 224.0.0.9.

RIPv2 support the authentication of RIPv2 routing update messages using plaintext passwords **[RFC1388] [RFC1723] [RFC2453]**, or cryptographic authentication **[RFC4822]** (e.g., Message Digest 5 [MD5] algorithm authentication **[RFC1321]**, see appropriate section below). Authentication helps to ensure that the routing updates a router receives originate from authorized routing information sources. An IPv6 version of RIP was also developed RIPng (RIP Next Generation) **[RFC2080]** based on RIPv2 which still has a 15-hop limit, and a default Administrative Distance of 120 (as RIPv2).

As discussed in Chapter 2, RIP is distance-vector routing protocol and an IGP used for distributing routing information within an Autonomous System. RIP determines the best path to destinations within an Autonomous System by considering only the number of hops it takes to get to the destinations. This technique does not consider the differences in link speed on the available paths nor their traffic loads and utilization, and all other metrics (as in EIGRP and OSPF), many of which constitute important factors in determining the best path to a destination. This chapter describes the main characteristics of RIPv2 and the processes and databases involved in the creation of the Routing Table. RIP has already been described in greater detail in Chapter 2; so, most of that material will not be repeated here.

5.4 RIPV2 MESSAGE FORMAT AND OTHER CHARACTERISTICS

Both RIPv1 and RIPv2 messages are encapsulated in IPv4 packets using UDP as the transport protocol (see Figure 5.1). Both RIPv1 and RIPv2 use the UDP well-known port number 520 (for sending and receiving).

| IPv4 Header (20 Bytes) | UDP Header (8 Bytes) | RIP Message |

FIGURE 5.1 RIP Message in an IP4/UDP Packet

5.4.1 RIPv2 Message Format

Figure 5.2 shows the RIPv2 message format along with the meanings of the various fields in the message. The Command field is used to describe the purpose of the RIPv2 message. The RIPv2 Request and Response messages are two of the most common messages, but others such as for use over demand circuits have been defined in **[RFC1582]** and **[RFC2091**:

- **Command Field Value Equal to 1 Indicates the RIP Message Is a Request Message**: This message is sent asking a receiving RIP router to send all or part of its Routing Table.
- **Command Field Value Equal to 2 Indicates the RIP Message Is a Response Message**: This message contains all or part of a router's Routing Table. A RIP router sends this message in response to a RIP Request message, or as an

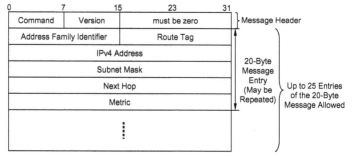

Field	Meaning
Command (8 Bits)	Indicates the type of message being sent (e.g., request or response message). A request message is sent for the responding router to send all or part of its routing table. A response message may be sent in response to a request message, or it may be an unsolicited routing update transmitted by a router. Other command field type shave been defined for RIP running on demand circuits (RFC1582, RFC2091).
Version (8 Bits)	Specifies version of RIP used by the originating router
Address Family Identifier (AFI) (16 Bits)	Specifies address family used by the originating router (The AFI for IP is 2). Each entry has an AFI to indicate the type of addresss pecified.The AFI is set to 0xFFFF for the first entry to indicate that the remainder of the entry contains authentication information.
Route Tag (16 Bits)	Provides a method for distinguishing between internal routes (learned by RIP within the routing domain) and external routes (learned and imported from other routing protocols). This field can also be used for other purposes, including routing policy control.
IPv4 Address (32 Bits)	Specifies the destination IPv4 address (IPv4 address of the route being advertised)
Subnet Mask (32 Bits)	Mask associated with the destination IPv4 address (if this field is 0,no sub net mask has been specified for the address)
Next Hop (32 Bits)	IPv4 address of the next-hop router to which packets are forwarded to reach the destination. Specifying a value of 0.0.0.0indicates that routing should be done via the router that originated the RIP advertisement.
Metric (32 Bits)	Value of the metric advertised for the destination IPv4 address (metricis between 1 and 15 for a valid route or 16 for an unreachable route)

FIGURE 5.2 RIPv2 Message Format

Routing Information Protocol (RIP)

unsolicited routing update sent to other routers. RIP sends Response messages periodically, and also when a Request message is received. Periodic RIP Response messages are referred to as routing update messages.

Each RIPv2 message contains a 4-byte header containing the Command and Version fields, with the remaining portion of the message consisting of routing entries with each entry being of 20 bytes in length. A single RIP message can carry at most 25 routing entries, with each routing entry containing the AFI identifying the type of network-layer protocol address used in the routing entry. A RIPv2 message has the following properties:

- Each route entry in a RIPv2 message has five 32-bit (4-byte) words and these are repeated for each entry (AFI [16 bits]; unused [16 bits]; Route tag [16 bits]; IPv4 address; Subnet mask [32 bits]; Next-hop [32 bits]; Metric [32 bits]).
- Each route entry takes five 8-byes words (20 bytes) and each RIP message can carry up to 25 route entries.
- The maximum RIPv2 message size is limited to 512 bytes, excluding the IPv4 header (which has a fixed header size of 20 bytes).
  o Multiplying 25 routes by 20 bytes (per route), plus the RIPv2 header (4 bytes), plus, the UDP header (8 bytes), gives a maximum RIPv2 message size of 512 bytes.
- If the RIPv2 carries authentication fields (i.e., AFI specifies an authenticated message), the RIPv2 message can only carry only 24 route entries.

5.4.2 Interpreting the Address Family Identifier (AFI) Field in RIPv2

The design of RIP allows routing information from different protocols to be carried. The AFI specifies the address family (i.e., the type of address) used by the router originating the RIP message. As shown in Figure 5.2, each message carries an AFI to indicate the type of address being used. RIP can be used for non-IPv4 protocols and allows an AFI field to be specified for each route entry, so that the protocol to which the route belongs can be easily identified. The AFI field is also used for other purposes other than identifying the routing protocol of a route entry:

- In the case of IP (IPv4 and IPv6) route entries, the AFI is 2 (the IPv4 Address Family Identifier is referred to as AF_INET).
- When a RIPv2 carries authentication information, the first route entry space in the RIPv2 routing update message is used to carry the authentication information, and the AFI value in this route entry is set to 0xFFFF in hexadecimal (or 65535 in decimal to indicate authentication present).
- A RIP Request message that is sent asking for a neighbor router's full Routing Table update contains just one route entry, with the AFI for this single route entry set to zero (0x0000).
  o Anytime a RIP router processes an incoming Request message, it does so entry by entry. Many times, a router will receive a Request message that has

a single entry in it with a routing metric (hop count) of 16, and an AFI field that consists of all zeros. A Request message containing such information is interpreted by the receiver as "Send me your entire Routing Table".

A router may receive a Request message that contains one or more entries for specific routes. The receiving router in this case will consult its Routing Table for each of the destinations listed. If the router finds a listed destination, it will send the corresponding information about that route back to the requesting router in a Response message via unicast. If the router finds no corresponding requested destination or route, it will send by unicast back to the requesting router, a Response message containing a metric field for the route set to infinity (16).

5.4.3 RIPv2 Routing Table

Each RIP router maintains a Routing Table which contains one entry for every destination that is reachable in the network. The RIPv2 protocol maintains the following information in the Routing Table for each destination:

- **IPv4 Address**: This is the IPv4 address of the destination network or host
- **Next-Hop Router**: This specifies the next router on the path to the destination
- **Outbound Interface**: This is the exit interface on which packets should be forwarded to reach the destination
- **Metric**: This indicates the number of hops that would be traversed to get the destination
- **Timers**: These specifies various periodic time intervals for maintaining routes, and include a timer that indicates the elapse time since a routing entry was last updated

The Routing Table is initialized with networks are directly connected to the router, and the contents are updated with routing information received in routing update messages from neighboring routers. Each router in the routing domain sends routing update messages that describe its current Routing Table.

5.4.4 RIPv2 Timers

RIPv2 uses a number of timers which have been described in Chapter 2 with the following default settings: Update timer (30 seconds); Invalid timer (six updates periods, that is, $30 \times 6 = 180$ seconds); Flush timer (60 seconds longer than the default setting of the invalid timer and is 240 seconds); Holddown timer (180 seconds). RIPv2 also supports a number of loop-prevention strategies such as counting to infinity, poison reverse, and split horizon.

5.4.5 RIPv2 Request Message

The two main message types exchanged by RIP routers are Request and Response messages. Each message type performs a specific function. A RIP router may receive

… Request messages from its neighbors asking for all or some part of the local current Routing Table contents. Also, when a router first boots up, or restarts its RIP routing process, it will send Request messages asking for the Routing Tables of all active neighbor routers. The router will send a Request message to each neighbor seeking routing information that can be used to populate its local Routing Table as quickly as possible.

5.4.6 RIPv2 Response Messages

A Response message can be sent as a response to a specific RIP query (RIP Request), as an unsolicited response (i.e., a regular RIP update sent upon expiration of the RIP Update timer), or as triggered update sent as a result of a network topology change. A RIP router that has just been powered up can multicast (for RIPv2 only) a RIP request on all of its interfaces. Neighbor RIP routers connected to these interfaces will receive the RIP Request messages and then respond by sending RIP Response messages immediately to the originating router.

Generally, RIPv2 routers process all Response messages as follows:

- A router upon receiving a Response message will validate the message by checking certain fields. For example, the router will check if the Response message is from a valid RIP neighbor, if the message is from one of the router's own interfaces, and/or if the source address of the message is on a directly connected network. The validity checks performed by the receiving router will determine if the message will be accepted or rejected.
- Next, the router will process the listed route entries one by one by first checking the validity of each entry. The receiving router will check if the destination address of the message is valid, if the routing metric (or hop count) is between 1 and 16. Once the router has validated an individual route entry, it will update the metric for that route by adding to it the metric of the router interface on which the message was received. If the final metric determined by the local router is now greater than 16 (or infinity), then the router will advertise the route with the infinity metric of 16 as the final metric.
- The router receiving the message will then consult its local Routing Table to determine if there is already an explicit route to the destination listed in the Response message. If the router determines that no such entry exists, then it will add the received route to the local Routing Table, except when the route's metric is currently set to infinity.
- If the receiving router determines that the route does currently exist in its Routing Table, it will check the next-hop address of the route. If it determines that the next hop of the received route is equal to the next-hop address in its Routing Table, it will reinitialize the update timer and compare the metrics of the two routes.
- If the receiving router determines that the metric values are the same, or the metric of the received route is higher, it will retain the existing route in the table, and will stop processing the received route entry. If the router determines the metric of the received route is lower than the current metric, then it will install a new copy of the route with the lower metric in its Routing Table, and

then sends a triggered update. If the router determines the metric of the received route to be equal to the infinity metric (16), it will start the process of deleting the route from its Routing Table.

Prior to sending a Response message, a router will examine each route in its current Routing Table. If the router determines that it should include a route in the Response message due to local administrative controls, then it will add the destination address and metric of the route to the message. Due to limitations imposed on the maximum size of a RIP message, a Response message can contain no more than 25 routing entries, or routes. When more than 25 routes need to be sent, the router will send multiple Response packets covering the routes.

5.4.7 SENDING AND RECEIVING RIPv2 REQUEST AND RESPONSE MESSAGES

A RIP Response message contains the Routing Table information that the receiving router requires to construct and maintain its local Routing Table. In the absence of RIP Request messages, all RIP routers in the routing domain are required to multicast a RIP Response message every 30 seconds (the default update period) on all of their interfaces. The periodic RIP multicast is the primary method by which network routing information is propagated throughout the network.

Once a RIP router learns about a particular route to a network destination and installs the route in its Routing Table, it will start a timer (the invalid timer). Every time the router receives a new RIP Response message with information about that route (i.e., a route refresh update), the router will reset the timer to zero. However, if the router receives no routing update for that particular route for 180 seconds (default of the invalid timer), the route will be marked as unreachable (invalid).

In addition to the periodic transmission of routing updates every 30 seconds to neighbor routers, when a RIP router detects that one of its interfaces is unavailable or a new neighbor has come online, it will generate a triggered update. The (RIPv2) router will immediately multicast new routing information on all of its interfaces, and this change will be reflected in all subsequent RIP Response messages sent by routers in the network.

When a RIP router receives a set of route updates on a particular interface, it determines through a technique known as split horizon that these updates do not need to be retransmitted out the same interface (see Chapter 2). Split horizon helps to limit the amount of routing traffic propagated throughout the network by eliminating routing information that other neighbor routers on that interface have already learned.

5.5 RIPV2 AUTHENTICATION

RIP routers exchange routing information through RIP Request and Response messages. RIPv2 supports authentication using plaintext passwords **[RFC1388]** **[RFC1723]** **[RFC2453]**, in addition to cryptographic authentication **[RFC4822]** (e.g., using the MD5 algorithm **[RFC1321]**). RIPv2 does not set aside a specific field in the message header for carrying authentication information. Instead, RIPv2 carries authentication information in one of the fields used for a single route entry. When

Routing Information Protocol (RIP)

RIPv2 sends routing updates on an interface that has authentication enabled, the authentication information is placed in the space normally reserved for the first route entry in each routing update message.

RIPv2 authentication is on a per message basis. Since a RIPv2 message header has only one 2-byte field available (the 2-byte unused field must be set to zero), and since any good authentication scheme will take more than two bytes, RIPv2 authentication uses the space occupied by an entire RIP route entry.

5.5.1 Plaintext Authentication

RIPv2 supports plaintext authentication by placing information in the fields that would normally carry the first route entry of the RIPv2 message. RIPv2 identifies the field as an authentication field, rather than a normal route entry, by specifying 0xFFFF in the first two bytes, that is, the AFI field. The remainder of the first route entry space contains the RIPv2 authentication information. RIPv2 has the following Authentication Type values:

- Type = 0 indicates no authentication used
- Type = 1 indicates Authentication Trailer and only used with cryptographic authentication (see **[RFC4822]**)
- Type = 2 indicates plaintext password authentication (see **[RFC1388]**)
- Type = 3 indicates cryptographic authentication which requires an Authentication Trailer

With the first route entry field used for carrying authentication information, a RIPv2 routing update message sent out an interface enabled for authentication can carry a maximum of 24 routes. Figure 5.3 shows the format of a RIPv2 message carrying plaintext authentication.

As shown in Figure 5.3, a RIPv2 message indicates the presence of authentication information by setting the AFI field to 0xFFFF (all 1s). The Authentication Type for simple plaintext authentication is 0x0002 (two). The remaining 16 bytes in the RIPv2 authentication header (Figure 5.3) can carry a simple plaintext password of up to 16 characters. The password is written in the 16-byte field left-justified, and if the password occupies less than 16 bytes, the unused bits of the 16-byte field are set to zero.

This plaintext password is transmitted along with the RIPv2 message to neighbor routers who have a security association with the sender. RIPv2 messages are sent to the IPv4 multicast address 224.0.0.9. The receiving router, knowing the same password, compares its own password to that in the transmitted RIPv2 message. If the local password matches the password transmitted with the message, the message is accepted, if not, it is rejected.

Plaintext authentication does not provide strong security because a router includes the unencrypted password (authentication key) in every RIPv2 message it sends. This method of authentication is very vulnerable to attacks, thereby, making cryptographic authentication preferable. The growth of internetwork has created the need for stronger authentication of routing information. Plaintext authentication is not really useful from a security perspective because by simply sniffing information on the

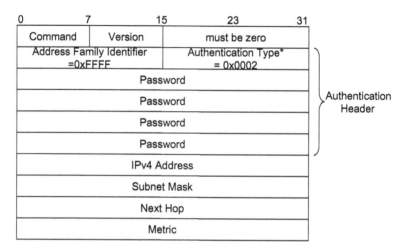

FIGURE 5.3 RIPv2 Plaintext Authentication Message

transmission medium, an attacker can read the plaintext RIPv2 password, and use that knowledge to inject malicious routing information into the RIPv2 Routing Table.

5.5.2 Cryptographic Authentication

With both plaintext and cryptographic authentication, the maximum number of route entries in a single RIPv2 routing update is reduced to 24. RIPv2 supports the keyed-MD5 cryptographic authentication mechanism, and even stronger authentication based on the SHA family of hash algorithms (HMAC-SHA-1, HMAC-SHA-256, HMAC-SHA-384, and HMAC-SHA-512) **[RFC4822]**. The RIPv2 cryptographic authentication mechanism is algorithm-independent, allowing for different types of algorithms to be used as needed.

RIPv2 cryptographic authentication is intended to reduce the risk of attacks on the exchange of RIPv2 routing information. Such a mechanism greatly reduces the vulnerability of the routing information from malicious attacks. When cryptographic authentication is used, instead of transmitting a plaintext password directly in the RIPv2 message, the router will include the output of a keyed cryptographic one-way function in the RIPv2 message's authentication field. Only authorized routers involved in the RIPv2 routing information exchange know the RIPv2 authentication key.

The RIPv2 authentication key is never transmitted over the network to the receiver. Also, RIPv2 authentication does not provide routing information confidentiality, because RIPv2 messages are transmitted in the cleartext format. This is because the

Routing Information Protocol (RIP)

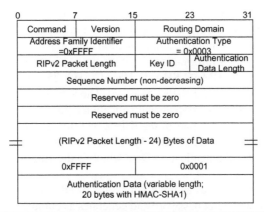

Field	Meaning
RIPv2 Packet Length (16 Bits)	Specifies an offset from the start of the RIPv2 header to the end of the regular RIPv2 message excluding the authentication trailer.
Key ID (8 Bits)	Contains a value to identify the RIPv2 Security Association in use for this message.
Authentication Data Length (8 Bits)	Contains the length in bytes of the trailing Authentication Data field
Sequence Number (32 Bits)	Contains a non-decreasing number sent in all messages from a given source router with a given Key ID value (initial sequence number is a random value).
Authentication Data	Contains the cryptographic Authentication Data used to validate this message. The length of the data is stored in the Authentication Data Length above

FIGURE 5.4 RIPv2 Message Format using Cryptographic Authentication

objective of a routing information exchange is to advertise the routing topology and routes to network destinations, thereby making data confidentiality less important.

Figure 5.4 shows the format a RIPv2 message when cryptographic authentication is used. The basic RIPv2 message has an 8-byte header plus an array of 20-byte blocks that hold RIPv2 data content (Figure 5.2). When RIPv2 cryptographic authentication is used, the same 8-byte header and blocks are used as in the case with the RIPv2 message with plaintext authentication (Figure 5.3), however, the 16-byte authentication password field is now used to carry a RIPv2 packet length, Key Identifier (ID), Authentication Data Length, and Sequence Number (which is non-decreasing).

A RIPv2 security association specifies a set of shared configuration parameters required by the sending and receiving routers for correct authentication operations. The 8-bit Key-ID when combined with the router's interface, identifies the security association in use for a particular RIPv2 message. The receiver of a RIPv2 authentication message uses both the router interface on which the RIPv2 message was received and the Key-ID (carried in the message) to uniquely identify the correct security association.

It should be noted that the RIPv2 security association is always associated with a router interface, and not with the router itself. The security association includes the authentication key that the router uses to create the authentication data carried in the

RIPv2 message. Each RIPv2 security association established in a router also has a lifetime (start time [valid time] and stop time [invalid]) specified for it.

If the router supports multiple authentication algorithms, then the RIPv2 security association also includes information about the type of authentication algorithm and mode (KEYED-MD5, HMAC-SHA-1, HMAC-SHA-256, HMAC-SHA-384, and HMAC-SHA-512) used for a RIPv2 message. The actual authentication key that is used with the selected cryptographic authentication algorithm (and is never carried in any RIPv2 message), is also part of the RIPv2 security association. The authentication trailer (with AFI = 0xFFFF and Authentication Type = 0x0001) as shown in Figure 5.4 carries the Authentication Data which is the output of the keyed cryptographic hash function.

The MD5 algorithm takes as input a message of arbitrary length (RIPv2 routing update), and a known secret password and generates as output, a 128-bit (or 16-byte) one-way message digest (hash value). This makes it a more secure authentication method than using plaintext passwords. The message digest is transmitted along with the RIPv2 message (Figure 5.4). The receiving RIPv2 router, knowing the same secret password (not transmitted in any RIPv2 message), computes its own message digest (or hash value). If the contents of the received RIPv2 message has not been modified or tampered with, the receiver's computed hash value should match the hash value transmitted with the RIPv2 message.

5.5.2.1 RIPv2 Authentication Message Generation

This section describes the general steps involved in generating RIPv2 authentication messages using, for example, the Keyed-MD5 and HMAC-SHA1 algorithms. The router creates a RIPv2 message the normal way but with the following exceptions:

1. The router should calculate the standard UDP checksum (for the UDP datagram carrying the RIPv2 message), but may choose to set it to zero because the use of any cryptographic authentication mechanism will provide stronger data integrity check than the standard UDP checksum.
2. The router sets the Authentication Type field to 0x0003 to indicate cryptographic authentication.
3. The 16-byte Authentication password field (i.e., 16 bytes after the Authentication Type field) is used to carry a RIPv2 Packet Length (or offset), Key-ID, Authentication Data Length, and Sequence Number.

The router uses the following process to create the RIPv2 Packet:

1. The size of the main body of the RIPv2 message is written into the Packet Length field of the RIPv2 header.
2. The router selects an appropriate RIPv2 security association for the RIPv2 message being created packet, based on the message's outbound interface. The router can use any valid RIPv2 security association for that outbound interface. The router then fills in the Packet Length (offset), Key ID, and Authentication Data Length fields appropriately.

3. The router now performs cryptographic algorithm-dependent processing using, for example, the Keyed-MD5 algorithm or the HMAC-SHA algorithm family.

The router writes the output of the authentication algorithm which is the Authentication Data value into the Authentication Data field. The router will not transmit any trailing pad, as it is entirely predictable from the Authentication Algorithm used and the message length.

When the Keyed-MD5 algorithm is used, the router will append the RIPv2 Authentication Key (which is always 16 bytes when Keyed-MD5 is used) to the RIPv2 message in router memory. The router will also add the Trailing Pad for the MD5 algorithm and message length fields in memory as illustrated in Figure 5.5 (which shows how these fields appear in memory). The router will then calculate the Authentication Data to be carried in the RIPv2 message according to the MD5 algorithm defined in **[RFC1321]**.

5.5.2.2 RIPv2 Authentication Message Reception

When the receiver gets the RIPv2 authentication message, the process used above is reversed:

1. The receiver extracts the Authentication Data in the received RIPv2 message and stores it for processing later.
2. The receiver determines the appropriate RIPv2 security association from the Key ID field value and the interface on which the RIPv2 message was received. If the router finds no valid RIPv2 security association on that interface for the received Key ID, then the router will cease all processing for the incoming RIPv2 message, and log a security event for the message as described in **[RFC4822]**.
3. The router then performs cryptographic algorithm-dependent processing, using the algorithm specified by the RIPv2 security association for the RIPv2 message. This calculation produces Authentication Data based on the information carried in the received RIPv2 message, and the RIPv2 security association information for the message.

Authentication Key (16 Bytes Long)
Zeros or More Pad Bytes (Defined by RFC 1321)
64-bit Message Length (Most Significant Word)
64-bit Message Length (Least Significant Word)

Field	Meaning
Authentication Key (16 Bits)	Specifies value of the cryptographic authentication key used with the chosen Authentication Algorithm. The authentication key must never be sent over the network in cleartext via any protocol. The length of this key depends on the Authentication Algorithm used.

FIGURE 5.5 Creating Trailing Pad for Keyed MD5 and Message Length Fields in Sender's Memory

4. The router next compares the resulting Authentication Data with the received Authentication Data.
5. If the calculated and received authentication data do not match, then the RIPv2 message must be discarded and a security event must be logged for the message.
6. If the receiver has heard from the neighbor recently and long enough to have viable routes in its Routing Table, and the Sequence Number in the received RIPv2 message is less than the last sequence number received, then the router must discard the message without further processing, and must log a security event.

When a router loses connectivity with the neighbor, it should accept from that neighbor, either a RIPv2 message with a Sequence Number of 0, or a message with a sequence number higher than the Sequence Number of the last received RIPv2 message.

7. When the receiver accepts a RIPv2 message, it will truncate the received message to the standard RIPv2 message (discarding the authentication trailer), and then process this part normally (in accordance with the RIPv2 standard in [RFC2453]). The receiver will then update the last received Sequence Number for that sender and RIPv2 security association.

5.6 HIGH-LEVEL RIP ROUTER ARCHITECTURE, PROCESSES, AND DATABASES

Figure 5.6 shows a high-level block diagram of the processes, interfaces, and databases in a typical RIP implementation. These elements are linked by well-defined asynchronous interfaces as shown in the figure. We describe these elements in greater detail below.

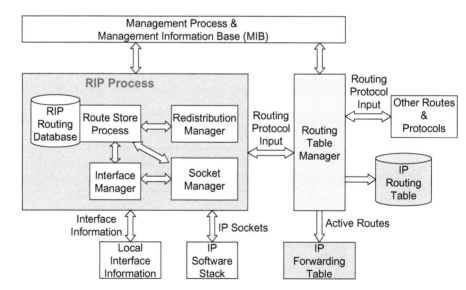

FIGURE 5.6 RIP Processes and Databases

5.6.1 THE RIP PROCESS

The RIP Process is a process within the control engine of the router, which controls the various components within the protocol that run on the router. Its functionality includes sending and receiving routing information, maintaining the Routing Table, and implementing routing policies in the network among other discussed below. The RIP Process starts all configured features of the protocol and handles the receipt and transmission of all routing messages. The RIP Process maintains the local RIP Routing Database and supplies routes for consolidation in the (combined) IP Routing Table that holds the routing information learned from all other routing protocols. *The IP Routing Table and the RIP Routing Database are simply logical components and a RIP router may choose to implement these as one single component.*

From the (combined) IP Routing Table, the RIP Process determines the active routes to all reachable directly attached and remote network destinations, and programs these routes into the router's Forwarding Table. The RIP Process also supports components that implement routing policies in the network, which allows network administrators to control the routing information that is exchanged between the routing protocols and the Routing Table. A network administrator can use a routing policy to filter and limit the routing information that can be exchanged, and the properties associated with specific routes.

Other processes not discussed here, include the router control processes that control the router's interfaces, management process that support tools that can be used to manage and configure the router as well as control user access to the router, chassis process that control attributes such as temperature and other properties within the router, and processes that control the communication between the router's route processor and forwarding engine(s). The route processor also supports specialized additional functionality and processes such as Virtual Router Redundancy Protocol (VRRP), Simple Network Management Protocol (SNMP), Quality of Service (QoS), and Class of Service (CoS) mechanisms.

The router's operating system (sometimes called an Internetwork Operating System [IOS]) is responsible for running multiple processes that perform the actual functions in the router. Modern router operating systems (e.g., Cisco IOS) are designed such that each process operates in its own protected memory space, while the operating system still controls the communication among all the processes. The design provides separation and isolation between the processes, as well as resiliency in the event of any process failure. Such operating systems have become very essential in routers operating in core or backbone networks because failure of a single process does not interrupt operation of the entire router.

5.6.2 THE MANAGEMENT PROCESS

As a network device used for interconnecting multiple networks and subnets, a router has visibility that extends to more of an entire network than most other devices like Layer 2 switches and end-hosts. With today's networks being more complex, and handling diverse traffic and network topologies, it has become critical to equip routers with advanced features to enable them monitor and debug the processes that run in their environments. This allows routers to detect and/or solve many networking problems using information that routinely passes through them.

The Management Process in Figure 6 is responsible for gathering and displaying system and network statistics. This allows a network administrator to display specific router statistics such as traffic statistics (e.g., forwarded packets, dropped packets, queue occupancies, and link utilization) and the contents of router databases (e.g., Routing Tables, Forwarding Tables, Route/Flow Caches, and Adjacency Tables). For example, the traffic statistics can be used to determine system and network resource utilization, and also to troubleshoot and solve network problems. A network administrator may also use various management tools to gather statistics that provide information about the reachability of other nodes and discover the routes that packets take in the network.

The Management Process typically supports tools and commands that can be used to display the following:

- The current state of the Routing Table, including the network prefixes, next hop IP addresses, outgoing router interfaces, routing protocol that derived each route, Administrative Distance of the routing information source, and whether the destination network is remote or directly connected, and possibly, any routing metrics.
- The current state of the active routing protocol process, including the settings of its routing update timers, Administrative Distance, and active networks for which the routing protocol process is the routing information source.
- The current state of the active accounting and statistics counters/databases, including the number of packets and bytes received/forwarded over interfaces, exchanged between particular sources and destinations.
- The contents of the Adjacency Table (or ARP cache), including the destination or next hop IP address, the interface through which that destination is reachable, the encapsulation method used, and the Layer 2 address of that destination.
- The current state of the router interfaces, including whether an interface and its Physical Layer hardware are up, whether certain protocols (such as ARP and ICMP) are enabled, and the current trust/security level of the interface.
- Router protocol statistics, including the number of packets received and sent by the following protocols (plus errors experienced): IP, TCP, UDP, RIP, EIGRP, OSPF, IS-IS, BGP, ARP, etc.
- Logging of all transactions carried out by the router protocols: ICMP, RIP, EIGRP, OSPF, IS-IS, RIP, TCP, UDP, etc.

The RIP Process itself can be divided into subcomponents with well-defined interfaces between them as illustrated in Figure 5.6. These subcomponents are described in greater detail in the subsections below.

5.6.2.1 The Route Store Process

The Route Store subcomponent of the RIP Process provides the following key functions:

- *Tracks All Routes Stored in the RIP Process*: Upon startup, each RIP router initializes its Routing Table with the network address prefixes and interfaces of the networks that are directly connected to it. Then, periodically, each router sends routing updates advertising its entire Routing Table over all of its interfaces to other RIP routers. Whenever such routing updates are received by a RIP router, it installs all of the new routes into its Routing Table, generates a Forwarding Table and starts forwarding packets.

  The process of advertising routing updates ensures all routers in the network eventually learn about the network destinations, and the routes over which they can be reached. A router that does not continue to receive regular/periodic routing updates for a remote network would eventually time out that route and stop using it for forwarding packets.

  Each RIP router transmits periodic routing updates containing its entire Routing Table to its neighbor routers every 30 seconds. When a router receives such a periodic routing update, it updates its own Routing Table with the information received, and then in turn, advertises its updated Routing Table to its neighbors. This process is repeated at each router until all routers achieve convergence and have a consistent view of the entire network topology.

- *Implements the Soft State Mechanism for Timing out Old Routes*: RIP routers use various protocol timers to control the exchange of routing updates and maintain routes in the Routing Tables. These protocol timers are described in Chapter 2 and include a routing update timer, route invalid timer, and a route flush timer. The invalid or timeout timer is used to mark/tag invalid routes in the Routing Table.

  A route that is not refreshed for a given period of time (i.e., the route-invalid timer setting) is most likely invalid because of some change in the network. For this reason, a RIP router maintains an invalid timer for each route in the Routing Table. When a route's invalid timer expires, the router declares the route as invalid but the route is still kept in the Routing Table, and only removed entirely when its route-flush timer expires.

  Each router sends routing updates every 30 seconds (default routing update timer value), a process referred to as flooding. If a router does not receive a routing update for a particular route after 180 seconds (default route invalid timer), it marks that route as unusable. If the router still does not receive a routing update for that route after 240 seconds, the router removes the route from its Routing Table.

- *Calculates the Updates Needed to Keep RIP Peers and Routing Table Manager Process in Sync with All Changes in the Network Topology*: Using the

Bellman–Ford algorithm, each RIP router builds a unique Routing Table that holds the best routes from itself to all other routers in the network. RIP routers prefer shorter routes (smaller hop counts) to longer routes when deciding which route among multiple routes to the same destination to install in the Routing Table.
- *Provides the RIP Route MIB for Querying the Routes in the RIP Routing Database*: A router running RIPv2 typically supports RFC 1724 (which defines *RIP Version 2 MIB Extensions*) **[RFC1724]**. RFC 1724 is an IETF standard that defines Management Information Base (MIB) objects that allow a network administrator to use SNMP to monitor RIPv2 performance. The RFC 1724 RIPv2 MIB extensions allow network administrators to use SNMP to monitor RIPv2 by allowing the addition of new table objects and global counters that are not defined in the older RFC 1389 RIPv2 MIB.

The new table objects and global counters have been added to help network administrators quickly debug failing RIPv2 neighbor routers or changing routes. The RIPv2 MIB stores global counters which are useful and can facilitate the detection of the harmful effects of incompatibilities in RIP implementations; two "interfaces" tables, which stores interface-specific configuration information and statistics; and an optional "peer" table which holds information that is useful for debugging/troubleshooting RIP neighbor relationships.

5.6.2.2 The Interface Manager

The Interface Manager subcomponent of the RIP Process is responsible for the following key tasks:

- *Monitors the State of the Router Interfaces and Updates Routes Accordingly*: RIP uses a number of techniques to monitor and control routing updates to and from other routers with the goal of preventing routing loops, and speeding up network convergence. The different techniques available are count to infinity, split horizon, poison reverse, and the use of various timers that include holddown timers as discussed in Chapter 2.

  The use of triggered updates in RIP increases its efficiency, particularly, on low speed point-to-point, serial interfaces. With triggered updates, a RIP router sends routing information on the low speed point-to-point link only when there has been an update to its Routing Table. For this to work, the router will need to suppress the transmission of periodic updates are over the point-to-point interface in order to reduce RIP routing traffic is on the interface.

  Triggered updates are transmitted on the point-to-point serial interface only if one of the following occurs:
  o A router receives a RIP Request message from a specific neighbor router asking for a routing update, which causes the full Routing Table to be sent. In the absence of RIP Request messages, a RIP router will broadcast or multicast routing updates every 30 seconds to all RIP neighbors.
  o A router receives routing information from an interface and updates its Routing Table, which causes only the latest changes to be sent.

Routing Information Protocol (RIP)

- o An interface on a router goes down or up, which causes a partial Routing Table update to be sent.
- o A router is powered on for the first time, which causes the full initial Routing Table to be sent.
- *Adds Connected Routes to the RIP Routing Database*: When a router receives a routing update that contains a new route to a network destination (due to a network topology change), the router updates the entry for that destination in its Routing Table to the new route. The router increments the metric value (hop count) indicated in the routing update by 1, and enters the new route in its Routing Table. The router sets the next hop IP address to the destination to be the IP address of the sending router's interface, and the local outgoing interface to be the interface through which the update was received. The router then immediately starts sending routing updates to neighbor routers to inform them about the new Routing Table update.
- *Provides the Interface Configuration MIB*: The Interface configuration table in the RFC 1724 RIPv2 MIB defines objects that can be used (by a network administrator) to track router interface configuration (on a per interface basis [see next bullet]).
- *Records Per-Interface and Per-Peer Statistics and Presents These in the Interface Statistics and Peer MIBs*: The following are required on a router to enable RFC 1724 RIPv2 MIB Extensions monitoring to be carried out with SNMP:
  - o The router must be configured with RIPv2.
  - o RFC 1724 RIPv2 MIB must be installed on the SNMP Network Management Station (NMS).
  - o The following MIBs must be installed on the SNMP NMS to allow RFC 1724 RIPv2 MIB to import data types and object Identifiers (OIDs) from them:
    - RFC1213-MIB
    - SNMPv2-CONF
    - SNMPv2-TC
    - SNMPv2-SMI

The following managed objects are defined in RFC 1724 RIPv2 MIB [**RFC1724**]:

- o **Global Counters**: These counters are used to keep track of neighbor changes (i.e., the number of responses a router sends to RIP queries received from neighbors), or route changes (i.e., the number of route changes made by RIPv2 to the IP Routing Table).
- o **Interface Status Table**: This table defines MIB objects that can be used to keep track of router interface-specific statistics (information on a per interface basis).
- o **Interface Configuration Table**: This defines MIB objects that can be used to keep track of configuration statistics for the router interfaces (on a per interface basis).
- o **Peer Table**: This table defines MIB objects that can be used to monitor RIPv2 neighbor relationships. This provides information about active RIPv2

neighbor relationships that can be useful in debugging/troubleshooting. An active RIPv2 peer refers to a RIPv2 router from which a valid RIPv2 routing update has been received in an interval of 180 seconds (i.e., the default route invalid timer setting).

5.6.2.3 The Sockets Manager

The Sockets Manager subcomponent of the RIP Process provides the following functions:

- *Sends and Receives RIP Messages (Requests and Responses)*: RIP routers use RIP Request and RIP Response messages to exchange routing information in a network. A RIP router sends a RIP Request to ask another RIP router to send back part or all of its Routing Table. The RIP Response can be a reply to a specific RIP request, or an unsolicited regular or periodic (broadcast or multicast) RIP routing advertisement. RIP Response messages contain the Routing Table entries of the sending router. Multiple RIP Response messages can be used to carry the contents of large Routing Tables.
- *Adds Routes Learnt from RIP into the Routing Table*: RIP routers transmit routing updates at regular intervals (periodic updates), and when the network topology changes (triggered updates). When a router receives a routing update that advertises a new route to a network destination, it updates the corresponding entry for that destination in its Routing Table to reflect the new route. In addition, each entry in a RIPv2 Routing Table includes a Route Tag field which stores additional information about the installed route. The Route Tag stores information that can be used to distinguish between routes learned by RIPv2 (internal routes) and routes learned from other routing protocols (external routes). Route tagging is discussed in greater detail in Chapter 7.
- *Carries out RIPv2 Packet Verification and Authentication*: Security is one of the primary concerns of network designers today. Network security not only deals with securing the data exchanged between end-users, but includes securing the routing information that is exchanged between routers. This is to ensure that the routing information that routers enter into their Routing Tables is valid, and not tampered or sourced by an entity that is trying to disrupt the operation of the network. An attacker might try to inject malicious routing information by introducing invalid routing advertisements in order to seriously degrade network performance, or fool routers into sending data to wrong network destinations. For these reasons, modern routing protocols support authentication mechanisms to prevent invalid routing updates from ending up in the Routing Tables.

  RIPv2 supports authentication while RIPv1 does not. RIPv2 support mechanisms that can be used to authenticate RIP messages **[RFC4822]**. Unlike RIPv1, RIPv2 authentication mechanisms enhance routing security by preventing the insertion of fraudulent routing information into the Routing Tables. With RIPv2 authentication, only routing updates that pass authentication testing can be inserted into the Routing Tables. A RIPv2 router applies authentication on a per RIP interface basis. Since RIPv2 authentication can be configured on a per interface basis, authentication can be configured on certain interfaces, while other interfaces will be operating without any authentication. One of the following authentication methods can be used:

o **Simple Authentication**: In this method, the sending router includes a plaintext password (or key-string) in the routing update message before it is transmitted. The receiving router receives the routing update message and uses the same password (authentication key-string) to verify the message. In plaintext authentication, a simple password must be shared by the neighbor routers (sending and receiving routers) exchanging routing information messages. If a router does not share the same password with a neighbor, it will reject all routing updates from that neighbor. It is not recommended to use plaintext authentication when strong security is required, because routers using this method will include the shared secret password (or authentication key) unencrypted in every RIPv2 message they send.

With plaintext authentication, both the sender and receiver must share an authenticating key which is specified at each device during system configuration. The network administrator can specify multiple authentication keys with each key identified by a key number or identifier (Key ID). In general, the following authentication sequence take place when a routing update is transmitted:

1. The sending router transmits a routing update message with a specific unencrypted password or authentication key and its corresponding Key ID to a receiving neighbor router. The neighbor router receives the routing update message, retrieves the authentication key in it, and checks it against the local authentication key that was configured at system configuration time.
2. If the received authentication key and the local key match, the receiving router will accept the routing update message. But, if the two authentication keys do not match, the router will reject the routing update message.

To use RIPv2 authentication, both the sending and receiving routers must be configured to use the same method of authentication (i.e., plaintext or cryptographic), and the same authentication keys. The network administrator can also configure a key chain at both the sending and receiving router interfaces that require authentication. The authentication key chain stores a number of different authentication keys. The authentication key chain allows authentication keys to be changed periodically without interrupting the exchange of routing information between routers. Each authentication key in the key chain has an index number and a corresponding key string. Each key also has a lifetime specified for it to indicate how long that key is valid. The lifetime can be specified as "accept-lifetime" and "send-lifetime" values. The accept-lifetime specifies to a router, the start and end times of the interval during which the key is considered valid (during the authenticated routing exchange with another router). The send-lifetime specifies the start and end times of the interval during which it is valid to send the key.

o **Cryptographic Authentication**: In addition to plaintext authentication, RIPv2 supports a number of cryptographic authentication algorithms including the Message Digest Algorithm Version 5 (MD5) authentication [RFC1321]. With MD5 authentication, a RIPv2 router sends a routing message with a "message digest" included in it, instead of the authentication key itself. The router creates the message digest (also called a "hash") using the authentication key and a message, but does not send the authentication key itself, thereby preventing the key from being read while it is in transit to the receiver.

A router using plaintext authentication includes the authenticating key itself in the routing message sent. But in MD5 authentication, the router sends the message digest – the authentication key is never sent. The sender creates a message digest that is included in the transmitted routing update. The receiving router uses a secret authentication key (password) to verify the routing update"s message digest. In MD5 authentication, both the sender and receiver use a shared secret authentication key that is not transmitted, which prevents an attacker from eavesdropping on the transmission medium and reading the authentication key during transmission.

The following steps can be used to configure RIPv2 plaintext or MD5 authentication in Cisco routers [CISCID13719]:

1. Define a key chain with a name. The key chain stores the set of keys that the RIPv2 router will use on the interface requiring authentication. If the interface has no key chain configured for it, then no authentication will be performed.
2. Define a key, or multiple keys for the named key chain.
3. Specify the key string (or password) for each key in the key chain. Each key string is the authentication string that the RIPv2 router must use to send and receive routing update messages.
4. The interface requiring authentication must be enabled, and the key chain to be used must be specified. RIPv2 authentication can be configured on a per interface basis. The network administrator must ensure that the router interfaces on both sides of the link are correctly configured for the authentication method used, ensuring that the key number and key string match on both interfaces.
5. Specify the type of authentication to be used on the interface (plaintext or MD5).
6. Key management (which is optional) can be configured as an added feature. Key management provides a method for controlling how authentication keys (within the key chain) are used. This allows the RIPv2 router to shift from one authentication key to another.

- *Calls Customizable Functions for Applying Routing Policy on Incoming or Outgoing Routes*: Routing policy determines how a router handles the routes it receives from and sends to neighboring routers. The router decides which routes to include in its Routing Table based on access filters (or *access-lists*) that are defined for it. Interface-specific access-list (e.g., distribute-list) and globally defined distribute-list (for inbound and outbound routes) can be defined for a given RIP process.

Routing Information Protocol (RIP)

- *Configures UDP Sockets for Use by the RIP Process*: RIP routers send and receive routing using UDP as the Transport Layer protocol, and with both source and destination UDP port numbers set to 520. RIPv1 uses broadcasts while RIPv2 uses multicasts to exchange routing information, all encapsulated in UDP segments.

5.6.2.4 The Redistribution Manager

The Redistribution Manager subcomponent provides the following functions:

- *Communicates with the Routing Table Manager Process across the Routing Protocol Input (RPI) Interface, and Sends and Receives All RPI Interface Messages*: This module communicates with the Routing Table Manager (RTM) to present routes for installation in the IP Routing Table, or export routes to other routing protocols.
- *Adds Redistributed Routes into the RIP Routing Database*: There are situations where routes may need to be transferred between two routing domains running different routing protocols (e.g., RIP and EIGRP). This requires routes generated by one routing protocol to be redistributed into the second routing protocol environment. Redistribution refers to the process of using a routing protocol to advertise routes that are learned or provided by another routing protocol or source, including static routes, or directly connected routes. Route redistribution provides the network administrator the ability to run different routing protocols in routing domains in a network where each is particularly suitable and effective. Most modern routers support route redistribution between the common routing protocols used in networks (RIP, EIGRP, OSPF, and IS-IS) in addition to static routes.

5.6.3 THE ROUTING TABLE MANAGER PROCESS

Figure 5.7 shows the RTM and how it fits in the general IP routing architecture. This architecture also applies to the other unicast routing protocol discussed in Chapter 6, and in the companion Volume 2 of this book. The RTM Process performs the following functions:

- **Generates and Maintains a Multiprotocol Sourced Master or Integrated Routing Table**: The RTM integrates routing information from multiple unicast routing protocols (RIP, OSPF, IS-IS, BGP) to create a combined or master Routing Table. When a routing protocol determines the best route to a particular destination is, it sends the route information to the RTM. The RTM may receive multiple best routes to the same destination from different routing protocols. If multiple routes exist to the same destination, only one route is selected. However, given that the metrics used by different routing protocols are not necessarily equivalent or comparable, the RTM uses the Administrative Distance (or route preference) value of the routing information sources to select a single best route for the Routing Table. The route from the routing source with the lowest Administrative Distance is selected.

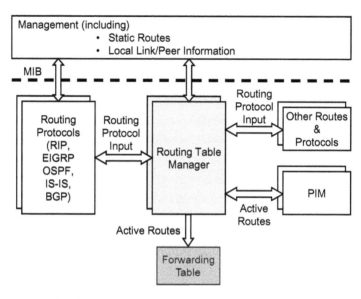

FIGURE 5.7   Routing Table Manager

The RTM also generates the unicast routing routes required for Reverse Path Forwarding (RPF) checks in the Multicast Routing Information Base (MRIB), a multicast topology table, used by multicast protocols such as Protocol Independent Multicast (PIM). The RTM provides a mechanism to combine information from different routing sources into a unified Routing Table from which a separate Forwarding Table for high-speed network address prefix lookups can be generated.

- **Populates and Maintains the Forwarding Table with Active Routes**: The RTM Process stores the best routes provided by multiple routing protocols in the master Routing Table and selects a single active route for each destination to be installed in the Forwarding Table. Multiple active routes to the same destination may be selected if equal-cost multipath (ECMP) routing is configured. The RTM enters the best routes in the Forwarding Table (also known as the Forwarding Information Base (FIB), and uses this table for all forwarding decisions in the router.

  When policy-based routing is configured, the RTM may select an active route based on policy information configured through the RTM Process Management function. The RTM transfers active routes through the Active Routes Interface to the Forwarding Table. An active route is one that has not been marked as unreachable but is still retained in the Routing Table (route flush timer has not expired). Note that even if the invalid timer of a route expires, the route may still be used to forward packets.

- **Exports or Imports Routes between Routing Protocols**: The RTM can be configured to export routes that have been learned by RIP to other routing protocols. The routes exported can also include static routes. Static routes can be configured through the management interface shown in Figure 5.7. For example, the RTM can be configured to export routing information from RIP into EIGRP, OSPF, IS-IS, or BGP. OSPF in turn, can redistribute the routing information using AS-external LSAs. IS-IS can do so in an IS-IS routing domain as external routes. The network administrator can configure via the RTM Management function, which routes to export to other routing protocols. The RTM can also be configured to import routes from other routing protocols (e.g., EIGRP, OSPF, IS-IS, or BGP).

Multiple instances of RTM can be used to provide the following functions:

- **Create Virtual Router Forwarding (VRF) Tables to Support Virtual Private Network (VPN) services**: The RTM Process derives a single active route for each network destination (based on the Administrative Distance or Route Preference value on the routing information source), and copies this information to the Forwarding Tables. A VPN refers to the extension of a private network across a public or shared network (e.g., service provider network) to enable end-users to communicate as if they are directly connected to the private network. VPNs typically support data security and management mechanisms, and use virtual connections routed (tunneled) through service provider networks. Layer 3 VPNs can be created using MPLS tunnels over a service provider network **[RFC4364] [RFC4577]**.

    A service provider can carry traffic from many VPNs. To separate the routes of each VPN from routes of other VPNs, or those in the service provider network, the Provider Edge (PE) router of the service provider constructs a separate Routing Table (called a Virtual Routing and Forwarding [VRF] table) for each VPN. The VRF table is also referred to as a VRF table. A PE router of the service provider network connects directly to a Customer Edge (CE) device belonging to a particular customer. For each VPN that is connected to the CE router, the PE router will create a VRF table for that VPN.

    Users of a particular VPN can only communicate over the routes maintained in the VRF table created for that VPN. The PE router will populate a VRF table with routes received from the CE sites that are directly connected to it and associated with that VRF routing instance (or VPN). The PE router will also enter in the VRF table, routes received from other PE routers and belong to the same VPN. For example, when two PE routers are connected to a CE router that is supporting three different VPNs, each PE router will construct three VRF tables, one for each VPN. Each PE router also creates a global Routing Table that maintains routes to other routers within the service provider network and to outside routers.
- **Simultaneous Support of IPv4 and IPv6**: IPv4 and IPv6 can each use an RTM instance to support its routing operations. Similar to the IPv4 RTM functions discussed above, the IPv6 RTM can support similar functions for IPv6 routing protocols such as RIPng, OSPFv3, EIGRP for IPv6, IS-IS for IPv6 and MP-BGP4 (Multiprotocol BGP-4).
- **Scalability and Fault-Tolerance**: The use of multiple RTM instances allow the distribution of the routing functions in a router across multiple control

planes. This can be used to create routing systems with high scalability and fault-tolerance. For instance, in a system with two distributed RTM instances, one RTM can be configured to operate as the primary (active) RTM, and the other RTM as the secondary (standby) unit. Different redundancy architectures can be conceived depending on the router processor architectures used and the level of complexity acceptable.

5.7 FILTERING ROUTING UPDATES IN RIP

This section describes some methods of filtering routes in distance-vector routing protocols such as RIP and EIGRP [CISCID9105]. Chapter 7 discusses in greater detail, the different route filtering methods used in IP routers. Distance-vector routing protocols and link-state routing protocols use different methods to advertise routing information and determine best routes, so the effects of route filtering will have different effects on these classes of routing protocols. This is because route filtering controls the routes that a router will enter into (import), or advertise (export) out of its route table. A distance-vector protocol like RIP advertises routes by sending its entire Routing Table to neighbor routers periodically (periodic updates), or sends a partial Routing Table when a network topology change occurs (triggered updates). As a result, applying route filtering to a router running RIP will influence greatly which routes get imported or exported by the router.

A link-state routing protocol like OSPF and IS-IS, on the other hand, determines best routes by sending links state advertisements (LSAs), and applying the Dykstra's algorithm on the routing information maintained in the link-state database (LSDB), which holds a complete topology map of the entire network. OSPF and IS-IS routers do not advertise routes by sending a complete Routing Table to neighbor routers.

Link-state routing protocols exchange routing information with the goal of establishing identical LSDBs in all routers in the network. All LSDBs must be synchronized to allow all routers in the network to have a common view of the overall network. As a result, route filtering is not applied on the LSAs the routers flood. It will have no effect on the LSAs sent or received by link-state routing protocol routers, or on their LSDB. Route filtering can only be applied to the information the link-state routers maintain in their Routing Tables.

As will be discussed in Chapter 7, route filters can also be applied to prevent routing updates (by the various routing protocols RIP, EIGRP, OSPF, IS-IS, and BGP) from passing through a router interface, control the routes that can be advertised in routing updates, and control how a router processes routing update messages.

5.7.1 CONFIGURATION OF PASSIVE INTERFACE TO PREVENT OR RESTRICT ROUTING UPDATES

A passive interface can be configured on a RIP router to prevent the router from sending routing updates through the specified interface [CISCID9105]. Stopping the router from sending routing updates through the specified interface prevents other routers on that network from learning about routes advertised by the sending dynamic routing protocol. The passive interface on a RIP router, prevents the router from

Routing Information Protocol (RIP)

advertising routing updates out of a particular interface (outgoing advertisements) to the corresponding neighbor router, but the router can still continue to receive and process routing updates from that neighbor (incoming advertisements).

A passive interface on an EIGRP router works differently from that on a RIP router [CISCID13675]. In a RIP router, configuring a passive interface suppresses the advertisement of routing updates out of the specified interface while still permitting routing updates to be sent and received normally over other router interfaces. In an EIGRP router, configuring a passive interface only suppresses the exchange of EIGRP Hello messages with a neighbor router over the specified interface. This prevents a router from being able to establish a neighbor relationship over the specified interface, and suppresses not only the advertisement of routing updates over the interface, but also blocks incoming routing updates from the neighbor. In essence, the passive interface prevents the EIGRP router from forming a neighbor adjacency over the interface, or exchanging routing updates.

5.7.2 Filtering Routes in Incoming and Outgoing Routing Updates

Access-lists can be defined to control the routes a router advertises (in outgoing routing updates), and the routes it receives (in incoming routing updates) and installs in its Routing Table. Cisco IOS routers like most other routers, are designed use the "`distribute-list in`" and "`distribute-list out`" commands, among other tools, to control which routes a router can accept from incoming routing updates, and which routes it can advertise in outgoing routing updates, respectively [CISCID9105].

Chapter 7 discusses in greater detail the different tools available for network path control and route filtering (e.g., route maps, prefix lists) [CCIESOLK03] [CISCASA8.5] [CISCID8606] [CISCID13669] [CISCID49111].

5.7.2.1 Distribute-List In

An access-list can be configured on an interface to filter out certain routes contained in incoming routing updates. The access-list is applied to the routes carried in the routing updates and not to the routing information source or destination. A router receiving routing updates will decide which routes to add to its Routing Table based on the configured access-lists on its interfaces. Any incoming routing update on an interface is checked against the configured access-list, and only routes that match any network address prefix entry in the list is added to the Routing Table.

In Cisco routers, a network administrator can define one inbound interface-specific access-list (distribute-list) per interface, and one globally defined distribute-list [CISCID9105]. A Cisco router uses the following algorithm to process routing updates when multiple distribute-lists are configured:

1. Receive an inbound routing update and extract the next route (i.e., network address prefix) contained in it.
2. Note the interface on which the routing update was received.
3. Does that interface have a distribute list configured for it?
    - o Yes: Is there a deny action in that list for the extracted route?
        - i. Yes: Do not enter the route into the Routing Table and return to Step 1

ii. No: Enter the route into the Routing Table and go to Step 4.
o No: Go to Step 4.
4. Is a global distribute list configured for the router?
o Yes: Is there a deny action in that list for the extracted route?
i. Yes: Do not enter the route into the Routing Table and return to Step 1.
ii. No: Enter the route into the Routing Table and return to Step 1.
o No: Enter the route into the Routing Table and return to Step 1.

5.7.2.2 Distribute-List Out

An access-list can be defined on an interface to determine which routes from the local Routing Table will be carried in outgoing routing updates **[CISCID9105]**. Cisco routers use following algorithm to export routes when multiple distribute-lists are configured:

1. Select the next route to be added to an outbound routing update.
2. Check which router interface on which the routing update should be sent out.
3. Does that interface have a distribute list configured for it?
   o Yes: Is there a deny action in that list for the route?
      i. Yes: Do not export this route and return to Step 1.
      ii. No: Export this route and go to Step 4.
   o No: Go to Step 4.
4. Check the routing protocol process or Autonomous System Number (ASN) from which the route was derived.
5. Does the routing process or ASN have a distribute list defined for it?
   o Yes: Is there a deny action in that list for the route?
      i. Yes: Do not export this route and return to Step 1.
      ii. No: Export this route and go to Step 6.
   o No: Go to Step 6.
6. Is a global distribute list defined?
   o Yes: Is there a deny action in that list for the route?
      i. Yes: Do not export this route and return to Step 1.
      ii. No: Export this route and go to Step 1.
   o No: Export this route and go to Step 1.

The router will check the distribute lists as one of many other checks that it has to perform for a distance-vector routing protocol route before it adds it in its Routing Table or in an outbound routing update. The router also checks to see if split horizon, poison reverse, routing policies, and other criteria should be applied to the route before redistributing it.

5.8 SUMMARY OF RIPV2 FEATURES

The following describe briefly how RIPv2 works:

- When a RIP router first starts up, it initializes its Routing Table with a list of its directly connected networks.

Routing Information Protocol (RIP)

- Each router then periodically sends routing updates over all of its interfaces that are RIP-enabled advertising the complete contents of its Routing Table.
  - Whenever a RIP router receives such a routing update, it installs all of the appropriate routes announced into its Routing Table and starts using it to forward packets. The ultimate goal of this process is to ensure that all routers eventually become aware of every network connected to every router in the network.
  - RIP is a "soft state" routing protocol in that, if a RIP router does not continue to receive periodic routing updates for a remote route (already installed in the Routing Table), the router eventually times out that unrefreshed route and stops forwarding packets over it.
- Every route has a "distance" metric associated with it, which indicates how far (in hop counts) it is to the destination.
  - Each time a RIP router receives a routing advertisement, it increments the hop count.
  - RIP routers give preference to shorter routes (smaller hop counts) over longer routes (larger hop counts) when deciding which of two routes (to the same destination) to install in the Routing Table.
  - The maximum hop count (or network diameter) permitted by RIPv1/v2 is 15, which means that a route with hop count greater than 15 is considered unreachable. This limits RIP to networks which have not more than 15 hops to a given destination.
  - RIP uses the hop count as the routing metric and the Bellman–Ford algorithm to determine the best route to a destination.

RIP also includes a number of mechanisms that help to improve network convergence time and stability, and eliminate routing loops:

- When a router detects a network topology change (e.g., a failed link) which gets reflected in its Routing Table, the router sends a triggered update immediately to its neighbors. The triggered update helps to speeds up network convergence and stability, and prevents routing loops from occurring.
- When a router determines a route to be unreachable, it does not remove that route from its Routing Table right away. Instead the router will continue to advertise the route in routing updates with a hop count of 16 (meaning unreachable) to other routers. This is to ensure that the neighbor routers are quickly notified of the unreachable route, and not have to wait for a soft state timeout (using the invalid timer).
- When router A learns that a route from another router B has failed or unavailable, it will advertise the same route back on the same interface to the source router B with a routing update having hop count of 16 (unreachable). This technique known as "split horizon with poison reverse" ensures that both the source and receiving routers know that the route is unavailable for reaching the destination.
- A RIP router sends a Request message to another router to request all or part of its Routing Table to be sent back to it. Thus, a newly started RIP router can send a RIP Request message to allow it to quickly query all of its neighbor routers for their Routing Tables.

REVIEW QUESTIONS

1. What Transport Layer Protocol does RIPv2 use?
2. Explain the maximum hop count in RIPv2 and how it is used.
3. What are the main message types used by RIPv2?
4. What are periodic updates in RIP and how are they sent?
5. What are triggered updates in RIP and how are they sent?
6. What is the infinity metric in RIP?
7. What is the significance of a Request message that has a single entry in it with a routing metric (hop count) of 16, and an AFI field that consists of all zeros?
8. Which field in the RIPv2 message allows it to support Variable-Length Subnet Masks (VLSMs)?
9. What is the maximum number of route entries in a RIPv2 update?
10. Explain the main difference between Plaintext Authentication and Cryptographic Authentication in RIPv2.

REFERENCES

[CCIESOLK03]. Karl Solie and Leah Lynch, CCIE Practical Studies: Configuring Route-Maps and Policy-based Routing, Sample Chapter, Cisco Press, Nov 26, 2003.
[CISCASA8.5]. Cisco ASA Services Module CLI Configuration Guide, 8.5, Chapter: Defining Route Maps, November 17, 2013.
[CISCID8606]. Cisco Systems, "Redistributing Routing Protocols", Document ID: 8606, March 22, 2012.
[CISCID9105]. Cisco Systems, "Filtering Routing Updates on Distance Vector IP Routing Protocols", Document ID: 9105, August 10, 2005.
[CISCID13669]. Cisco Systems, "Introduction to EIGRP", Document ID: 13669, August 10, 2005.
[CISCID13675]. Cisco Systems, "How Does the Passive Interface Feature Work in EIGRP?", Document ID: 13675, March 28, 2005.
[CISCID13719]. Cisco Systems, "Sample Configuration for Authentication in RIPv2", Document ID: 13719, August 10, 2005.
[CISCID16406]. Cisco Systems, "Enhanced Interior Gateway Routing Protocol", Document ID: 16406, September 5, 2017.
[CISCID49111]. Cisco Systems, "Route-Maps for IP Routing Protocol Redistribution Configuration", Document ID: 49111, August 10, 2005.
[RFC1058]. C. Hedrick, "Routing Information Protocol", IETF RFC 1058, June 1988.
[RFC1321]. R. Rivest, "The MD5 Message-Digest Algorithm", IETF RFC 1321, April 1992.
[RFC1388]. G. Malkin, "RIP Version 2 Carrying Additional Information", IETF RFC 1388, January 1993.
[RFC1582]. G. Meyer, "Extensions to RIP to Support Demand Circuits", IETF RFC 1582, February 1994.
[RFC1723]. G. Malkin, "RIP Version 2 Carrying Additional Information", IETF RFC 1723, November 1994.
[RFC1724]. G. Malkin and F. Baker, "RIP Version 2 MIB Extension", IETF RFC 1724, November 1994.

[RFC2080].   G. Malkin and R. Minnear, "RIPng for IPv6", IETF RFC 2080, January 1997.
[RFC2091].   G. Meyer and S. Sherry, "Triggered Extensions to RIP to Support Demand Circuits", IETF RFC 2091, January 1997.
[RFC2453].   G. Malkin, "RIP Version 2", IETF RFC 2453, November 1998.
[RFC4364].   E. Rosen and Y. Rekhter, "BGP/MPLS IP Virtual Private Networks (VPNs)", IETF RFC 4364, February 2006.
[RFC4577].   E. Rosen, P. Psenak, and P. Pillay-Esnault, "OSPF as the Provider/Customer Edge Protocol for BGP/MPLS IP Virtual Private Networks (VPNs)", IETF RFC 4577, June 2006.
[RFC4822].   R. Atkinson and M. Fanto, "RIPv2 Cryptographic Authentication", IETF RFC 4822, February 2007.
[RFC7868].   D. Savage, J. Ng, S. Moore, D. Slice, P. Paluch, and R. White, "Cisco's Enhanced Interior Gateway Routing Protocol (EIGRP)", IETF RFC 7868, May 2016.

6 Enhanced Interior Gateway Routing Protocol (EIGRP)

6.1 INTRODUCTION

EIGRP was developed by Cisco Systems as an improvement to IGRP (a Cisco distance-vector routing protocol now considered obsolete), and as a more scalable IGP for large networks. Routers using EIGRP do not send routing updates periodically as with RIP and IGRP. EIGRP provides a number of enhancements that allows rapid network convergence and operating efficiency. Routers running traditional distance-vector routing protocols like RIP send full routing updates periodically to their neighbors which can result in unnecessary network bandwidth consumption.

EIGRP routers transmits packets (when sending routing updates and queries) as multicasts and unicasts wherever necessary, thereby, resulting in more efficient use of network bandwidth (compared to IGRP). This chapter discusses the main EIGRP concepts, message formats, and other important issues such as neighbor discovery and maintenance, network topology discovery, best path computations, route summarization, and authentication.

6.2 EIGRP OVERVIEW

EIGRP is a classless routing protocol that supports Variable-Length Subnet Masking (VLSM) and Classless Inter-Domain Routing (CIDR) similar to RIPv2, allowing efficient use and scalable allocation of IP addresses than IGRP. EIGRP supports authentication with simple passwords and cryptographic authentication (using MD5 and SHA-2) as in RIPv2 **[RFC7868]**. EIGRP supports both manual and automatic summarization of networks, and load-balancing over equal or unequal cost paths. EIGRP messages are sent directly over IP similar to OSPF, and use protocol number 88 (0x58 in hexadecimal). EIGRP is predominantly used in Cisco routers with very limited support in other vendor routers.

EIGRP is considered an advanced distance-vector routing protocol, or a hybrid routing protocol that has a number of characteristics similar to those typically associated with link-state routing protocols (such as dynamic neighbor discovery). A hybrid routing protocol like EIGRP takes some key features and advantages of link-state routing protocols, and integrates them into a distance-vector routing protocol. Put in another way, it tries to pick the best features of the two categories of routing protocols to create a new protocol.

Compared to other distance-vector routing protocols such as RIP, EIGRP is designed to provide rapid network convergence and guaranteed loop-free topology at all times. EIGRP has traditional distance-vector routing protocol features (e.g., easy configuration, Split Horizon, Poison Reverse, auto-summarization of routes), and link-state routing protocol features (e.g., dynamic neighbor discovery). Another distance-vector routing protocol feature that EIGRP uses is, if a neighbor router advertises a (route to a) destination, it must also be using that route to forward packets to that destination.

Routers using EIGRP advertise their Routing Tables to their neighbors similar to routers using distance-vector routing protocols; however, they use HELLO messages and form neighbor relationships just as link-state routing protocol routers. EIGRP routers do not send their full Routing Tables in periodic routing updates as distance-vector routing protocol routers do. Instead, they send partial updates only when network topology or routing metric changes occur.

Particularly, EIGRP routers send only triggered partial routing updates to neighbor routers instead of full periodic routing updates as in RIP. A router that detects a change in a path, or the metric associated with a route, will advertise only those updates. The partial updates sent contain information about only the changes rather than the entire Routing Table. EIGRP automatically bounds the propagation of these partial routing updates so that only the EIGRP routers in the routing domain that require this information are updated. This results in EIGRP consuming significantly less bandwidth than RIP and IGRP. Unlike OSPF, which advertises a routing update/change to all routers within an area, EIGRP sends change updates to only routers that are affected by the change.

EIGRP routers do not send Link-State Advertisements (LSAs) as OSPF routers do, but instead, send updates like traditional distance-vector routing protocol routers do (via distance vectors). The routers send updates containing routing information to remote networks plus the cost of reaching those networks, from the perspective of the advertising routers. Additionally, as discussed above, the EIGRP router will propagate routing updates only to neighbor routers that require them, instead of all routers within the Autonomous System as with link-state routing protocols.

EIGRP routers synchronize their Routing Tables with neighbors at startup, after which they advertise specific routing updates when a network topology or metric change occurs. EIGRP uses the Diffusing Update Algorithm (DUAL) to determine loop-free best paths to remote network destinations, in addition to a number of other functions.

To allow rapid network convergence, routers using DUAL not only compute the best loop-free routes to network destinations, but also computes, in advanced, backup routes to be used when needed. A router using DUAL will store all these backup routes to allow the network to react quickly when network topology changes occur. If no backup route exists to a network destination, the router will send queries to its neighbors until a new route to the destination is found.

DUAL performs route computations and provides a loop free topology each time a network change occurs (i.e., path or route metric changes), and allows only the routers involved in the change to synchronize their Routing Tables. The routers that are not affected by topology change do not participate in the route recomputation.

Enhanced Interior Gateway Routing Protocol (EIGRP)

The default Cisco Administrative Distances for EIGRP are as follows:

- EIGRP summary route has a default Administrative Distance of 5
- EIGRP internal route has a default Administrative Distance of 90
- EIGRP external route (from redistribution) has a default Administrative Distance of 170 (i.e., EIGRP routing information advertised by an EIGRP router outside the Autonomous System)

Unlike OSPF which requires different configurations to work over different Layer 2 protocols such as Ethernet and ATM (e.g., point-to-point links, dedicated links, and non-broadcast multiaccess [NBMA] networks), EIGRP does not require such special configurations. EIGRP works effectively over all these network types and also in both LAN and WAN environments **[CISCTEAPA06]**. In multiaccess networks, such as those based on Ethernet, and have built-in broadcast/multicast capabilities, EIGRP routers form and maintain neighbor relationships using a reliable multicasting mechanism. EIGRP also handles, effectively, differences in media speeds and types, when neighbor adjacencies are formed across WAN links.

6.3 EIGRP CONCEPTS

This section discusses the main components of EIGRP, including the processes and databases used by EIGRP (see Figure 6.1). This section introduces some of the fundamental concepts underlining the design and operations of EIGRP. These concepts are required to understand how EIGRP works. The rest of the EIGRP section focuses on the details surrounding these fundamental concepts.

6.3.1 RELIABLE TRANSPORT PROTOCOL

EIGRP uses the Reliable Transport Protocol (RTP) to transmit and receive the most important messages and to ensure that these EIGRP messages are successfully delivered to all neighbors and in the right order **[MARTABE02] [RFC7868]**. Only certain EIGRP packets carrying routing information are sent via reliable transmission, and these require explicit acknowledgment (ACK) by the recipient(s). EIGRP routers use RTP to ensure reliable delivery of QUERY, REPLY, and UPDATE packets, to neighbor routers in order to maintain loop-free routing of traffic. RTP transports these EIGRP routing packets and includes a number of variables such as Sequence Numbers to ensure ordered delivery of packets. EIGRP routers acknowledge routing packets using unicast HELLO messages in to ensure reliable delivery, but these messages do not carry data.

EIGRP supports both multicast and unicast transmission of EIGRP packets. When multicast transmission is used, packets are sent to the multicast address 224.0.0.10. RTP ensures guaranteed delivery of EIGRP packets using a Cisco-proprietary algorithm known as Reliable Multicast which uses the reserved multicast address 224.0.0.10. Each neighbor router that receives a Reliable Multicast packet will respond with an acknowledgment packet in a unicast transmission to the sender.

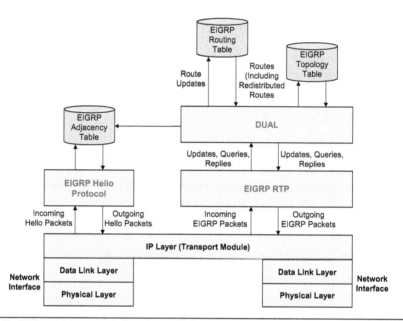

- The **DUAL module** is a protocol-independent module that implements the diffusing algorithm. It is responsible for maintaining the EIGRP Topology Table and computing the shortest-paths to all know network destinations. The DUAL module sends and receives EIGRP Query, Update, and Reply messages passing through the EIGRP RTP module. module computes best paths using the information in the Topology Table, and installs (via an associated routing table interface) the best paths in the IP Routing Table. The routing table interface (not shown) provides routing table management services for installing and deleting routing information, in addition to importing routes from the routing table (during routere distribution).
- The **EIGRP Hello protocol module** is a protocol-independent module responsible for discovering new neigbors and determining those that are unreachable. EIGRP routers send Hello messages periodically containing a number of parameters.
- The **EIGRP RTP module** is a protocol-independent module that provides reliable EIGRP message delivery. It work sover IP, and provides unicast and multicast transmissions.
- The **Transport module** provides transport services used by RTP to send and receive EIGRP messages. It is a Network Layer client of EIGRP, and provides EIGRP message encapsulation, transmission, and reception functions.
- The **Adjacency or Neighbor Table** holds information about the EIGRP neighbors of a router. It is maintained by the Hello protocol module, and the DUAL module uses information from this database.
- The EIGRP Topology Table maintains network prefixes associated with paths through directly connected neighbors. The contents of this table are updated by the DUAL module.

FIGURE 6.1 EIGRP Modules

Using RTP with a number of mechanisms, EIGRP is able to guarantee the delivery of EIGRP packets, and to ensure that they are delivered in the correct sequence.

The main features of RTP are summarized as follows:

- **Sequence Number and Acknowledge Number fields**: An EIGRP router will add a sending Sequence Number and an Acknowledgment Number to every "reliable" packet it sends, and requires an explicit acknowledgment for that packet (i.e., the Sequence Number last seen by the receiver). A receiving router does not necessarily need to send acknowledgments via separate ACK packets, since the ACK field available in every RTP unicast packet is sufficient to acknowledge the receipt of a reliable EIGRP packet (see "EIGRP Message Types" section below).

- **Piggyback Acknowledgements**: An EIGRP router will not send an ACK packet if it has a unicast packet that is ready for transmission to the sender. Since the ACK field in any RTP unicast packet is sufficient to acknowledge a neighbor's EIGRP packet, the router can suppress an ACK packet to save network bandwidth. This feature is particularly beneficial for point-to-point links and NBMA networks, because EIGRP packet transmission on these networks are sent as unicasts, and each carries an acknowledgment (also referred to as a piggyback ACK). The router does not need to send a separate ACK packet.
- **No Congestion Control or Windowing Mechanism**: RTP sends EIGRP packets carrying Sequence Number and Acknowledge Number fields, but without using any elaborate congestion control or windowing mechanism. This is because RTP sends only one packet at a time. The router uses Sequence Numbers to match acknowledgments to determine which packets have been successfully received.
  - o To ensure ordered delivery of packets, the sender will include two Sequence Numbers in each packet. The sender will assign a Sequence Number to each packet, and this Sequence Number will be incremented by one each time a new packet is sent. The sender will also write in each packet the Sequence Number (i.e., Acknowledgment Number) of the last packet correctly received from the destination router.
- **Some EIGRP Packets Do Not Use RTP**: Some EIGRP packets do not require reliable transmissions and delivery (RTP is not needed). The sender does not require an acknowledgment from the intended recipient for such packets. In these cases, a Sequence Number of 0 will be included in this EIGRP packet. EIGRP HELLO packets do not require acknowledgement when received, and are always sent as unreliable multicast transmissions. Both HELLO and ACK packets are not sent via reliable transmission (i.e., RTP) and always carry a Sequence Number of 0.

6.3.2 Main EIGRP Databases

An EIGRP router maintains three main databases (see Figure 6.1): Neighbor Table, Topology Table, and Routing Table **[CISCID13669] [CISCID16406] [RFC7868]**.

6.3.2.1 Neighbor Table

Neighbor Discovery and Recovery is the process by which EIGRP routers dynamically learn about other routers they are directly connected to. An EIGRP router must also be able to know when any neighbor becomes inoperative or unreachable. The routers achieve this by exchanging, periodically, low overhead small EIGRP HELLO packets. As long as a router receives HELLO packets from a neighbor, it can assume that it is operational and functioning. Once the router has made that determination, it can exchange routing information with the neighbor router.

Each EIGRP router maintains state information about its neighbors in a Neighbor Table. An EIGRP router will use HELLO packets to discover neighbors and form adjacencies with these new neighbors, and then include their information in the

Neighbor Table. The information includes among other parameters, when a neighbor was newly discovered, the IP address and interface through which it is can be reached, and the Hold Time as advertised in its HELLO packets.

The Hold Time (see appropriate discussion below) indicates to the router the amount of time beyond which it should treats the neighbor as invalid, that is, unreachable. If the router does not receive a HELLO packet from the neighbor within the Hold Time and the Hold Time expires, it should consider the neighbor as unreachable. When the Hold Time expires, the neighbor is declared invalid and DUAL is informed of the network change.

The Neighbor Table also contains information required by RTP for its operations, such as allowing the router to match ACKs with packets, record the Sequence Number of the last EIGRP packet received from the neighbor, and detect packets that are out-of-order [CISCTEAPA06]. The router also maintains a transmission list on a per neighbor basis that is used to queue packets for possible retransmission when a timeout occurs, or when packets are not acknowledged. For this, the router maintains a number of round-trip timers in the Neighbor Table to estimate optimal retransmission intervals for such packets.

6.3.2.2 Topology Table

This is a database used by an EIGRP router to store routing information it has learned from neighbor routers in the network (to all known destinations). When an EIGRP router first discovers a new neighbor, it will send a routing update informing the neighbor about the routes it has learned, and will in turn, receive routing updates from that new neighbor. The router will then use these new routing updates to populate its Topology Table.

The Topology Table of a router maintains all known destinations advertised by neighbor routers. Essentially, each EIGRP router stores the Routing Tables of its neighbors in its EIGRP Topology Table [CISCTEAPA06]. This follows a key feature of all distance-vector routing protocols which says that if a neighbor router is advertising a route to a network destination, it must also have that route in its Routing Table (or equivalently, be using that route to forward packets to the destination).

At a high level, an EIGRP router will be associated with each entry in the EIGRP Topology Table the following:

- The destination IP address and a list of neighbor routers that have advertised the destination.
- The router will also record for each neighbor, the advertised routing metric or cost to the destination. The neighbor's advertised routing metric, also referred to as the neighbor's Reported Distance or Advertised Distance, is the cost value that the neighbor advertises to its own neighbors and also stores in its Routing Table (see "Reported (or Advertised) Distance" section below).
- The information maintained in each entry includes at a minimum, the destination network prefix and prefix length (or network mask), Reported Distance for each EIGRP neighbor advertising reachability to the destination, and the local router's Feasible Distance (which is determined from the local router's history of Computed Distances to the destination). The Feasible

Distance of a destination is explained in the "Feasible Distance" section below.
- o The Computed Distance **[RFC7868]** to the destination is the total distance (metric) along a path from the local router to the destination through a *particular* neighbor router. This is the sum of the Reported Distance advertised by the neighbor and the link cost to that neighbor. The local EIGRP router will compute a different Computed Distance for each available neighbor.
- o The lowest Computed Distance is the local router's own Reported or Advertised Distance (metric) that it stores in its Routing Table, and in turn, advertises to its own neighbor routers.
- The state of the route to the destination (i.e., whether ACTIVE or PASSIVE).

An EIGRP router will update its Topology Table whenever the state of an interface or directly connected route changes, or when a neighbor router reports a route metric or topology change.

The term "Topology Table" can be misleading, as this database does not actually hold or represent the complete network topology, but rather the Routing Tables as advertised by the directly connected neighbor routers. The different EIGRP distance metrics are described in greater detail in appropriate sections below.

6.3.2.3 Routing Table

EIGRP routers use DUAL for all route computations, and for tracking all routes advertised by neighbors in the network. DUAL uses a distance metric or cost to compute loop free paths to network destinations, and then selects the best routes to be inserted in the Routing Table based on the concept of Successor and Feasible Successors (described in appropriate sections below). A router calculates the lowest-cost route to a destination when multiple routes exist to that destination. For each route, the router adds the cost between itself and the next-hop router, to the cost between the next-hop (neighbor) router and the destination (the latter cost is also referred to as the neighbor's Reported or Advertised Distance). The cost that results from adding these two costs can be used to determine the Feasible Distance of the local router.

A Successor is an EIGRP neighbor router (or the best next hop router) to which the local router forwards packet to be sent to a network destination, and has the least-cost route to that destination that is guaranteed to be not part of a routing loop in the network. When a router determines that there are no Feasible Successors (or backup routes, or equivalently, next best alternative loop-free routes) to a network destination, but there are neighbors advertising routes to that destination, the router will perform a route recomputation for the destination. The result of the route recomputation is the determination of a new Successor for the destination, which will be inserted in the Routing Table.

All least cost routes to a network destination form a local set of potentially useable routes. The following summarizes how these routes are selected and used:

- The local EIGRP router will select from this set, the neighbor routers that have a Reported Distance that is less than the router's own local Feasible Distance to be Feasible Successors (backup routes).

o The router will move a destination entry from the Topology Table to the Routing Table when there is a Feasible Successor (*and the route with the overall least cost to that destination will become the Successor*).
- Each Feasible Successor (or neighbor) has associated with it an advertised routing metric (the Reported Distance) that is stored in the Topology Table. When a topology change occurs in the network, or a neighbor router changes the routing metric it has been advertising (the Reported Distance), the local router may have to re-evaluate the set of Feasible Successors to use.
- When a topology change occurs, the router via DUAL will check if there are any Feasible Successors available for any affected destination. If there are Feasible Successors, the router will promote one (from the Topology Table) to be a new Successor for the destination (and enter it into the Routing Table) in order to avoid any unnecessary route recomputations, plus also speed up network reconvergence.
- Successors (best next-hops) are stored in the Routing Table while Feasible Successors (backup next-hops) are stored in the Topology Table (to be used immediately if a Successor fails), unless the Feasible Successors are installed in the Routing Table when the router is performing unequal cost load balancing (over multiple routes to a network destination).

The Neighbor Table, Topology Table and Routing Table are discussed in greater detail toward the end of this chapter.

6.3.2.4 Other EIGRP Concepts

This section presents a few other concepts pertaining to EIGRP [**RFC7868**]:

- **Base Topology**: The Base Topology refers to a routing domain that represents the actual physical (non-logical) view of the network topology (consisting of various routing devices and network segments). The EIGRP routers form neighbor relationships based on the Base Topology. EIGRP routers exchange reachability information about the network destinations (network prefixes) within the Base Topology with a Topology Identifier (TID) value of 0.
- **Subtopology**: A Subtopology in a given Base Topology represents an independent collection of links and routers on which EIGRP routers perform independent route computations. By creating subtopologies, a network engineer can implement network topology subgroups each carrying specific network traffic (class-specific network topologies). The network prefixes belonging to each specific subtopology within the Base Topology is given a unique TID.

6.3.3 NEIGHBOR FORMATION

EIGRP routers do not send periodic updates as in RIP. This means EIGRP routers have to support other mechanisms and processes to discover and track the directly connected networks on neighbor routers, and on other EIGRP routers in the network. EIGRP routers must form a neighbor or adjacency relationship before they can exchange routing information. EIGRP routers can be configured to discover

Enhanced Interior Gateway Routing Protocol (EIGRP)

neighbor routers dynamically (default Cisco configuration) or statically (manual configuration).

EIGRP routers discover neighbors dynamically by sending EIGRP HELLO packets to the multicast address 224.0.0.10. Routers that have successfully established a neighbor relationship, then exchange routing update packets that contain their full Routing Tables. These full routing update packets contain information about all routes that the sending router has discovered so far. The update packets are sent reliably (via RTP) and must be explicitly acknowledged by the receiving router. After the neighbors have completed this initial exchange of their Routing Tables (via the full updates), each EIGRP routers will advertise only incremental updates to inform neighbors when network topology or link metric changes occur. The routers will not advertise again their full Routing Tables (to the neighbors).

EIGRP neighbor relationships can be formed statically via manual configuration on neighbor routers. When an EIGRP router forms a static relationship with a neighbor, it uses the unicast address of the receiving interface of the neighbor to send packets to that neighbor, and not the EIGRP reserved multicast address 224.0.0.10. Static neighbor relationships are mostly configured on neighbors that are connected by media that do not support broadcast or multicast capabilities, such as ATM networks and other Non-broadcast Multiple Access (NBMA) networks. It is important to note that two EIGRP routers cannot establish a neighbor adjacency if one router is configured to use multicast (dynamic) transmission while the another uses unicast (static) transmission.

It also takes more than simply enabling EIGRP on any two routers for them to establish a neighbor adjacency. Two routers may not establish an EIGRP neighbor adjacency due any one of the following reasons:

- Mismatched/incompatible EIGRP composite metric K values
- Mismatched EIGRP Autonomous System Numbers (ASNs)
- If authentication is configured, mismatched/incompatible EIGRP authentication parameters
- When the neighbor routers are not using the same network prefix on their interconnecting interfaces (i.e., interfaces not on the same IP subnet)
- When the neighbor routers are using secondary IP addresses on their interconnecting interfaces for EIGRP neighbor relationships formation

EIGRP routers cannot establish neighbor relationships using secondary IP addresses, as only primary IP addresses can be used as the source IP addresses of EIGRP packets. Two EIGRP routers can form a neighbor relationship when their primary IP addresses belong to the same subnet (i.e., they share the same network address prefix); they reside in the same Autonomous System, and have compatible EIGRP composite routing metric K values.

6.4 EIGRP MESSAGE TYPES

In this section, we describe the five different packet types used by EIGRP. The following packet types are described in [RFC7868]: HELLO packets (including ACK packets); UPDATE packets; QUERY packets (including SIA-QUERY packets); REPLY packets (including SIA-REPLY packets); REQUEST Packets. These different packet types are used for EIGRP session management and for exchanging DUAL messages. EIGRP packets are carried (encapsulated) directly in a Layer 3 protocol packets such as IPv4 or IPv6, with no transport protocol such as TCP or UDP used (Figure 6.2).

Each EIGRP packet transmitted is addressed with either a multicast or a unicast Layer 3 destination address. When a Layer 3 multicast address is used, this address must be mapped to a corresponding Layer 2 multicast address when available. Where applicable, the source address of the IP packet carrying the EIGRP message is set to the IP address of the sending interface. The maximum size of an EIGRP packet is set equal to the IP (or Layer 3) Maximum Transmission Unit (MTU) of the interface on which the packet is to be transmitted (e.g., usually 1500 bytes for an Ethernet interface).

EIGRP packets are sent to the following Layer 3 multicast addresses: 224.0.0.10, the IPv4 All EIGRP Routers address; FF02:0:0:0:0:0:0:A, the IPv6 All EIGRP Routers address. If Layer 2 multicast is supported, these Layer 3 multicast addresses have to be mapped to the corresponding Layer 2 multicast addresses.

Both IPv4 and IPv6 have the same basic EIGRP packet format (Figure 6.3), although there are some protocol-specific differences when it comes to the contents of the various fields in the packet. The EIGRP packet consists of a main header containing a number of fields, followed by a variable number of Type/Length/Value (TLV) fields (Figure 6.4). Similar to IS-IS, each EIGRP packet can contain a variable number of TLV fields. When a router exchanges protocol messages with other routers, a TLV provides an encoding format for embedding information elements in those messages without having to send those information elements in their own special-purpose messages. The TLVs can be written in an EIGRP packet in any order, and they appear in the packet with no interdependencies among them.

There are three generic fields in each TLV-formatted information element in a message. The Type field identifies the kind of information carried in the information element, the Length field describes the length of the entire TLV information element, and the Value field contains the actual information (i.e., the content). The Value field is a multibyte field that contains the actual data or payload for the TLV. The Value field may be further structured into smaller elements depending on the type of TLV information element being exchanged. EIGRP routers use TLVs to exchange EIGRP control and management information such as EIGRP composite metric K values, Hold Time values, authentication type and data, sequence type, and software version type.

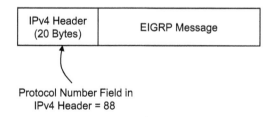

FIGURE 6.2 EIGRP Message in an IPv4 Packet

Enhanced Interior Gateway Routing Protocol (EIGRP)

```
0           7           15          23          31
| Header Version | Opcode |        Checksum        |
|                    Flags                         |
|                Sequence Number                   |
|             Acknowledgment Number                |
|  Virtual Router ID  |  Autonomous System Number  |
|                    TLVs                          |
```

Field	Meaning
Header Version (8 Bits)	Specifies the protocol version of the EIGRP packet header format. Current version is 2 (RFC 7868).
Opcode (8 Bits)	Indicates the type of EIGRP message (UPDATE = 1; REQUEST = 2; QUERY = 3; REPLY = 4; HELLO = 5; SIA-Query = 10; SIA-Reply = 11)
Checksum (16 Bits)	Contains a checksum for the entire EIGRP packet. The checksum is the standard ones' complement of the ones' complement sum. To compute the checksum, the value of the checksum field is set to zero. A received EIGRP packet is discarded if the packet checksum fails.
Flags (32 Bits)	The flag bits define special handling conditions for the EIGRP packet. Four flag bits are currently defined in RFC 7868.
Sequence Number (32 Bits)	Specifies a sequence number that is unique with respect to the sending router. This is used by RTP for orderly delivery of EIGRP packets. A value of 0 means that an acknowledgment is not required.
Acknowledgment Number (32 Bits)	Specifies the last sequence number seen from the neighbor router to which this packet is being sent. If the value is 0, then the packet carries no acknowledgment. A HELLO packet with a non-zero ACK field should be treated as an ACK packet rather than a HELLO packet. A non-zero value (meaning an ACK packet) can only be sent as unicast-addressed packets.
Virtual Router ID (16 Bits)	Specifies a number that identifies the virtual router with which this packet is associated. Packets received with an unknown, or unsupported value are discarded.
Autonomous System Number (32 Bits)	Specifies a number that identifies the sending EIGRP system. This identifies the routing domain to which the sending EIGRP router belongs. An EIGRP router is only allowed to process packets from routers belonging to the same routing domain. This field is also indirectly used as an authentication value. That is, a router that receives and accepts a packet from a neighbor must have the same AS number or the packet is ignored. The range of valid AS numbers is 1 through 65,535.

FIGURE 6.3   EIGRP Packet Header

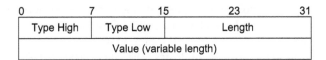

FIGURE 6.4   EIGRP TLV Encoding Format

6.4.1 HELLO Packets

EIGRP routers use HELLO packets to discover neighbor routers and to maintain adjacencies. Specifically, routers use HELLO packets to identify neighbor routers, and once they are identified, the transmission of HELLO packets serves as a keepalive mechanism between the neighbor routers. IPv4 EIGRP routers transmit HELLO packets addressed to either the IPv4 unicast address of a neighbor router, or a specific IPv4 multicast group address.

The following summarize some important properties of EIGRP HELLO packets:

- **Standard HELLO Packets**: HELLO packets carrying a Sequence Number of 0 do not require acknowledgment upon receipt. These are standard EIGRP

HELLO packets. Thus, HELLO packets are not transmitted via RTP, and carry a Sequence Number field value that is always set to 0. An EIGRP router that sends a HELLO packet does not require an ACK that it has been received. Because HELLO packets do not require explicit acknowledgment, they are referred to simply as "unreliable" EIGRP transmissions/packets.
- **Sending ACK Packets**: HELLO packets with a non-zero Acknowledgment Number is treated as an ACK packet rather as a standard HELLO packet. The non-zero value carried in the Acknowledgment Number field is the Sequence Number of a reliable EIIGRP packet last seen by the sender. An Acknowledgment Number of 0 indicates the packet does not carry any acknowledgment information.
    o   An EIGRP router sends an ACK packet to a neighbor to acknowledge the receipt of an EIGRP reliable transmission such as an UPDATE packet. An ACK packet is simply a HELLO packet that carries no data, but always carries a non-zero Acknowledgment Number.
    o   An EIGRP router sends the ACK packet to the unicast address of the neighbor router that sent the UPDATE packet and not to the EIGRP multicast group address.
- **Addressing HELLO Packets**: HELLO packets are multicast to the IPv4 All EIGRP router address 224.0.0.10. IPv6 EIGRP routers transmit HELLO packets with the source address of the transporting IPv6 packet being the IPv6 link-local address of the router's transmitting interface. When multicasting HELLO packets, the IPv6 EIGRP multicast address (FF02:0:0:0:0:0:A) is used as the destination address of the IPv6 packet. Unicast packets are transmitted to a specific neighbor router with the destination address of the IPv6 packet being the IPv6 link-local address of the neighbor router.
    o   Unlike IPv4 EIGRP neighbors, two IPv6 EIGRP neighbors are not required to share a common network prefix on the interfaces connecting them. IPv6 EIGRP routers will still check that HELLO packets received on their interfaces carry valid IPv6 link-local source addresses.
- **HELLO Packet Transmission Rates**: The rate at which EIGRP routers transmit HELLO packets to neighbors is dictated by the Hello Time or Interval. EIGRP routers multicast HELLO packets with a default setting of every 5 seconds for most network types, and 60 seconds for WAN links with bandwidth of 1544 Mb/s speeds or less.

An EIGRP router sends HELLO packets to discover neighbors and form adjacencies with them so that they can exchange routing updates. A router exchanges routing information with only adjacent routers (i.e., directly reachable neighbors). Each EIGRP router in the adjacency will then construct a Neighbor Table from the information carried in the HELLO packets it receives from the adjacent routers.

Additional HELLO packet processing at an EIGRP router will include common EIGRP checks such as, if the neighbors have matching ASNs and matching EIGRP composite metric K values (see discussion below).

Enhanced Interior Gateway Routing Protocol (EIGRP)

6.4.2 UPDATE Packets

EIGRP routers use UPDATE packets to convey routing and reachability information about network destinations to neighbor routers. An EIGRP router will send UPDATE packets in unicast transmissions to newly discovered neighbors, but will multicasts UPDATE packets to the multicast address 224.0.0.10 when a network topology or link metric change occurs. When an EIGRP router discovers a new neighbor, it will send UPDATE packets in a unicast transmission carrying a full Routing Table to that new neighbor so that it can construct its EIGRP Topology Table.

UPDATE packets are always reliably transmitted via RTP, and require explicit acknowledgement from neighbors to ensure reliable transmission of routing updates/information. An UPDATE packet when transmitted, includes a Sequence Number that allows the receiver to acknowledge receipt of the packet by responding with an ACK packet specifying the packet's Sequence Number. If an UPDATE or ACK packet is lost and does not get to its intended recipient, the UPDATE packet will be retransmitted.

6.4.3 QUERY Packets

EIGRP routers send QUERY packets to neighbors to request routing information to network destinations. An EIGRP router will send QUERY packets to neighbors when a route (to a destination) becomes unavailable, and the router requires immediate knowledge of alternate routes to allow fast network convergence. If the router sends QUERY packets and does not receive a response from any particular neighbors, it will resend the QUERY packets as unicast transmissions directly to the non-responsive neighbor(s).

The following summarize some important properties of EIGRP QUERY packets:

- **Sending a QUERY Packet when a Route Is Placed in the ACTIVE State**: An EIGRP router sends a QUERY packet to advertise that a destination is in the ACTIVE state and the sender is requesting alternate routing information to that destination from its neighbors. The sender encodes an infinite metric in the UPDATE message by setting the Delay field of the routing metric to its maximum value (which is infinity).
  o EIGRP routers send QUERY packets to their neighbors to find Feasible Successors to a network destination. This happens when DUAL is recomputing a route to a network destination to which the router does not have a Feasible Successor – destination is in the ACTIVE state (see below).
- **Sending QUERY Packets via RTP**: EIGRP routers use the RTP (since TCP is not used) which means that QUERY packets are also guaranteed be reliably delivered.
- **Addressing QUERY Packets**: QUERY packets are always sent to the multicast address 224.0.0.10. A router will always multicast a QUERY packet unless it is sent in response to a QUERY it has received. The router, in this case, will send a QUERY packet in a unicast transmission back to the Successor (neighbor) that sent the QUERY.

When there is a network topology change that causes a router to mark multiple destinations to be in the ACTIVE state, the router will construct one or more QUERY packets to be sent requesting routing information to all these destinations (in ACTIVE state). The sending router will record the state of each destination individually, so that a responding neighbor does not need to respond with a single QUERY or REPLY packet that contains all these destinations in that single packet.

The SIA-QUERY message is a subtype of the QUERY message that a router sends when a REPLY message has not been received within one-half of the Stuck-In-Active (SIA) interval (which has the Cisco default setting of 90 seconds). SIA and related matters are discussed in appropriate sections below.

6.4.4 REPLY Packets

EIGRP routers send REPLY packets in response to QUERY or SIA-QUERY packets they have received. A neighbor can send a REPLY packet to provide a Feasible Successor to the router that sent the QUERY packet.

The following summarize some important properties of EIGRP REPLY packets:

- **Sending a REPLY Packet in Response to a QUERY Packet**: A neighbor may send a REPLY packet in response to the originator of a QUERY indicating that it does not need to place a destination into the ACTIVE state because there are Feasible Successors available for that destination.
  - o The EIGRP neighbor router will respond with a REPLY packet that includes a TLV that carries information for each destination plus the associated vector metric that is maintained in the router's Topology Table. The responding neighbor router does not have to wait until it has processed the entire contents of the received QUERY packet before sending a REPLY packet. The router can send an acknowledgment immediately, and then process the QUERY packet. The acknowledgment is sent either as a self-contained ACK packet, or by piggybacking the acknowledgment information in another packet that the router has prepared and ready to be transmitted to the sender of the QUERY.
  - o When the router receives a QUERY message for a destination that does not exist in its local Topology Table, it will send a REPLY message with an infinite metric. The router will also add an entry in its Topology Table reflecting the metric it has sent in the QUERY if the existing metric is not already an infinite value. If any router receives a REPLY packet for a destination that is not in the ACTIVE state or in its Topology Table, the router will still acknowledge the REPLY packet, but will also discard it.
- **Sending REPLY Packets via RTP**: Both QUERY and REPLY packets are transmitted reliably via RTP.
- **Addressing REPLY Packets**: The neighbor will send REPLY packets to the unicast address of the sender/originator of the QUERY packet.

The SIA-REPLY message is a subtype of the REPLY message that a router sends in response to a SIA-QUERY message. It indicates to the recipient that the sender is still attempting to resolve a loop-free route for the destination that is still in ACTIVE state. A SIA-REPLY packet includes a TLV that carries information for each destination and its associated vector metric maintained in the sender's Topology Table. The SIA-REPLY indicates that the sending neighbor is still treating the given destination as being in the ACTIVE state, and still actively engaged in finding a loop-free route. Use of the SIA-REPLY message is discussed in a section below.

6.4.5 REQUEST PACKETS

EIGRP REQUEST packets are used to seek specific routing information from one or more neighbor routers. This EIGRP packet type is typically used in route server applications. REQUEST packets can be sent as either multicast or unicast packets, and are always unreliable transmissions with no guarantees of reaching the recipient.

6.4.6 EIGRP TLVs

As shown in Figure 6.3, the payload of an EIGRP message consists of one or more TLVs. Figure 6.4 shows the general structure of a TLV. This section describes some of the main TLVs defined in **[RFC7868]**:

- **Parameter Type TLV (0x0001)**: This TLV is carried in EIGRP HELLO messages to indicate the K parameters and Hold Time of the sending router (Figure 6.5). The K values are used for EIGRP composite metric computations. This TLV is also carried in the initial EIGRP UPDATE packet when a neighbor is discovered. Any two EIGRP neighbors must agree on K values for them to form an adjacency. This is done in order to avoid the creation of routing loops in the EIGRP domain.
- **Authentication Type TLV (0x0002)**: This TLV may be included in any EIGRP packet to indicate to the receiving neighbor, the authentication type and data the sender supports (Figure 6.6). A neighbor router that receives this TLV containing a mismatch in authentication information must discard the EIGRP packet.

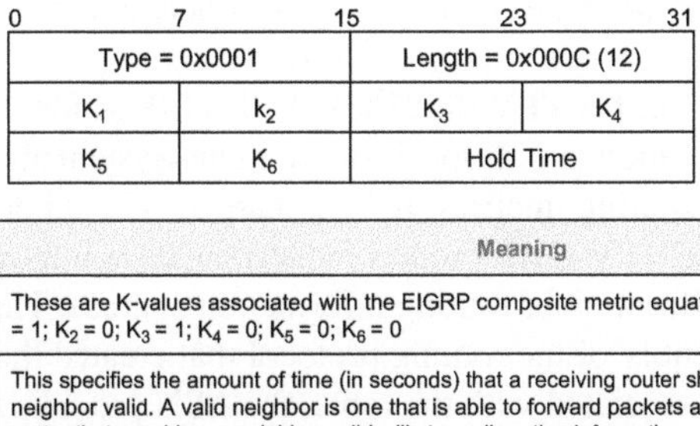

0	7	15	23	31
Type = 0x0001		Length = 0x000C (12)		
K_1	K_2	K_3	K_4	
K_5	K_6	Hold Time		

Field	Meaning
K-values (6 Bytes)	These are K-values associated with the EIGRP composite metric equation. The default values are: K_1 = 1; K_2 = 0; K_3 = 1; K_4 = 0; K_5 = 0; K_6 = 0
Hold Time (2 Bytes)	This specifies the amount of time (in seconds) that a receiving router should consider the sending neighbor valid. A valid neighbor is one that is able to forward packets and participates in EIGRP. A router that considers a neighbor valid will store all routing information advertised by the neighbor.

FIGURE 6.5 EIGRP Parameter Type TLV

0	7	15	23	31
Type = 0x0002			Length	
Authentication Type	Authentication Length	Authentication Data (Variable)		

Field	Meaning
Authentication Type (1 Byte)	This specifies the type of authentication used
Authentication Length (1 Byte)	This specifies the length, measured in bytes, of the Authentication Data.
Authentication Data (variable)	Depends on the type of authentication used. Multiple authentication types can be present in a single Authentication Type TLV. • **MD5 Authentication Type (0x02)**: MD5 Authentication will use Authentication Type code 0x02, and the Authentication Data will be the MD5 Hash value. • **SHA2 Authentication Type (0x03)**: SHA2-256 Authentication will use Authentication Type code 0x03, and the Authentication Data will be the 256-bit SHA2 Hash value.

FIGURE 6.6 EIGRP Authentication Type TLV

- **Sequence Type TLV (0x0003)**: An EIGRP router sends this TLV to inform receivers to not accept EIGRP packets with the CR-Flag set. A router uses this TLV to order/sequence multicast and unicast addressed packets.
- **Software Version Type TLV (0x0004)**: An EIGRP router uses this field to indicate to other routers the EIGRP TLV format versions it is using.
- **Multicast Sequence Type TLV (0x0005)**: An EIGRP uses this TLV to announce/report the Sequence Number of the next multicast packet that has the CR bit set.
- **Peer Termination Type TLV (0x0007)**: An EIGRP router includes this TLV in HELLO packets to notify a list of neighbors that it has reset the adjacency. A router sends this TLV anytime it needs to reset an adjacency, or signal that an adjacency is going down.

Reference **[RFC7868]** specifies the following IPv4-specific TLVs:

- **IPv4 Internal Routes TLV (0x0102)**: EIGRP routers send this TLV to advertise routes that are internal to the EIGRP routing domain (Figure 6.7). A router sends this TLV to announce the IPv4 destination addresses that it has learned and the routing metrics for those IPv4 networks. An EIGRP router sends this TLV to advertise the routes configured on its network interfaces as well as the networks that are learned via other EIGRP routers.
- **IPv4 External Routes TLV (0x0103)**: An EIGRP sends this TLV to describe and inject information into its local Autonomous System about IPv4 destination addresses and routing metrics it has learned via other routing protocols (Figure 6.8). This TLV carries routes redistributed (or imported) into an EIGRP routing domain from other routing information sources. The advertising router provides the identity of the routing protocol that created the route, the external routing metric associated with the route, the ASN providing the route, an indicator if the route should be marked as part of the EIGRP Autonomous System, and an administrative tag (also called a route tag) specified by the network operator to be used for route filtering at EIGRP Autonomous System boundaries.

Enhanced Interior Gateway Routing Protocol (EIGRP)

```
0                7              15              23              31
|   Type (0x0102)                |           Length              |
|              Next-Hop Forwarding Address                       |
|                    Scaled Delay                                |
|                  Scaled Bandwidth                              |
|              MTU              |           Hop Count            |
|  Reliability  |    Load       |  Internal Tag |    Flags       |
| Prefix Length |          Destination (variable)                |
```

Field	Meaning
Next-Hop Forwarding Address (4 Bytes)	If the value is zero (0), the IPv4 address from the received IPv4 header is used as the next hop for the route. Otherwise, the specified IPv4 address will be use
Scaled Delay (4 Bytes)	This is an administrative parameter assigned statically on a per-interface-type basis and specifies the total path delay. A delay of 0xFFFFFFFF represents an unreachable network.
Scaled Bandwidth (4 Bytes)	The effective bandwidth of the slowest link measured in units of 2,560,000,000/kbps.
MTU (3 Bytes)	This specifies the minimum MTU value for the path to the destination
Hop Count (1 Byte)	This specifies the distance to the destination in number of routers traversed (hops). The initial value for locally originated routes is 0, and each router increments this value by 1. A route with a hop counter greater than the maximum allowed is considered unreachable.
Reliability (1 Byte)	This specifies the current error rate for the path, measured as an error percentage. This is expressed as a fraction of 255. A value of 255 indicates 100% reliability.
Load (1 Byte)	This specifies the load utilization of the path to the destination, measured as a percentage. This is expressed as a fraction of 255. A value of 255 indicates 100% load.
Internal Tag (1 Byte)	This is a tag assigned by the network administrator that is untouched by EIGRP. This allows a network administrator to filter routes in other EIGRP border routers based on this value.
Flags (1 Byte)	• **Source Withdraw (Bit 0)**: Indicates if the router that is the original source of the destination is withdrawing the route from the network or if the destination is lost due as a result of a network failure. • **Candidate Default (CD) (Bit 1)**: Set to indicate the destination should be regarded as a candidate for the default route. An EIGRP default route is selected from all the advertised candidate default routes with the smallest metric. • **ACTIVE (Bit 2)**: Indicates if the route is in the ACTIVE State.
Prefix Length (1 Byte)	This specifies the length of the IP prefix being announced (i.e., number of 1s in the route mask).
Destination (variable)	This specifies the IP prefix being announced (i.e., IP subnet address, IP network address, or IP supernet address)

FIGURE 6.7 EIGRP IPv4 Internal Routes TLV

- **IPv4 Community Type (0x0104)**: A router sends this TLV to provide community tags for specific IPv4 destinations in a network. This TLV contains an IPv4 Destination field describing the IPv4 address associated with the community information, a 2-byte Community Length field specifying an unsigned number that indicates the length of the Community List, and a variable-length Community List that contains one or more 8-bytes EIGRP communities.

6.4.7 EIGRP FLAGS FIELD

Reference [RFC7868] defined the following four flag bits that describe special handling conditions of an EIGRP packet:

- **INIT-Flag (0x01)**: This bit is set in the initial EIGRP UPDATE message a router sends to a newly discovered neighbor. The sending router sets this bit to instruct the EIGRP neighbor to advertise the full set of routes in its Routing Table.

0	7	15	23	31
Type (0x0103)			Length	
Next-Hop Forwarding Address				
Router Identifier				
External Autonomous System (AS) Number				
Administrative Tag				
External Protocol Metric				
Reserved		External Protocol	Flags	
Delay				
Bandwidth				
MTU			Hop Count	
Reliability	Load	Internal Tag	Flags	
Prefix Length	Destination (variable)			

Field	Meaning
Router Identifier (4 Bytes)	This specifies the Router ID of the router that injected the external route into the EIGRP domain. This field is checked by EIGRP when an external route TLV is received.
External AS Number (4 Bytes)	This indicates the external AS in which the sending router is a member. If the source protocol is EIGRP, this field will be the [VRID, AS] pair. If the external protocol does not have an AS, other information can be used (for example, Cisco uses process-id for OSPF).
Administrative Tag (4 Bytes)	This is a tag assigned by the network administrator that is untouched by EIGRP. This allows a network administrator to filter routes in other EIGRP border routers based on this value. This tag may be used for route filtering or other policy management.
External Protocol Metric (4 Bytes)	This contains the composite metric that resides in the routing table as learned by the foreign protocol. If the External Protocol is another EIGRP routing process, the value can optionally be the composite metric or 0.
External Protocol (1 Byte)	This specifies the ID of the routing information source that originated the redistributed route. The following codes have been defined: 1 = IGRP; 2 = EIGRP; 3 = Static; 4 = RIP; 5 = Hello; 6 = OSPF; 7 = IS-IS; 8 = BGP; 9 = IDRP; 10 = Connected

FIGURE 6.8 EIGRP IPv4 External Routes TLV

- **CR-Flag (0x02)**: An EIGRP router sets this bit to indicate to any neighbor that receives the packet to only accept it if the neighbor is in Conditionally Received mode. An EIGRP router is considered to have entered the Conditionally Received mode when it receives and processes an EIGRP HELLO packet containing the SEQUENCE TLV.
- **RS-Flag (0x04)**: An EIGRP router sets the Restart flag (RS-Flag) in the EIGRP HELLO and UPDATE packets it sends during the restart period. The receiving router checks the RS-Flag to detect if a sending router is restarting. From the perspective of the (sending) restarting router, a neighbor router upon detecting that the RS-Flag has been set, will maintain the adjacency, and will set the RS-Flag in the EIGRP UPDATE packet it sends, to indicate it is performing a soft restart. During a soft restart, the router uses stored prefix information to reconfigure and activate its Routing Table without tearing down the existing adjacency.
- **EOT-Flag (0x08)**: An EIGRP router sets the End-of-Table flag to mark the end of the startup process with a neighbor. If a router receives an UPDATE message with this flag set, it indicates the sending neighbor has completed sending all UPDATE packets. Upon receiving the flag bit, the receiving router will

remove any stale routes learned from the neighbor (setting the End-of-Table flag) prior to performing the restart. A stale route is any route that existed in the Routing Table of the receiving router before the restart, and has not been refreshed by the neighbor via any UPDATE packets.

6.5 EIGRP METRICS

EIGRP uses a classic composite metric that is based on bandwidth (BW), traffic load, delay, and reliability on the path to a network destination **[RFC7868]**. EIGRP also supports wide metrics as defined in **[RFC7868]**, but these are not discussed further in this chapter. EIGRP does not use the Maximum Transmission Unit (MTU) for calculating the classic composite metric. The EIGRP composite metric is calculated using the following formula:

$$\text{EIGRP}_{\text{Metric}} = \left[\left(K_1 \cdot \text{BW}_E + \frac{K_2 \cdot \text{BW}_E}{256 - \text{Load}_E} + K_3 \cdot \text{Delay}_E\right) \cdot \frac{K_5}{K_4 + \text{Reliability}_E}\right] \cdot 256$$

where BW_E is the minimum link bandwidth on the path to a destination, and Delay_E is the total path delays (i.e., total of all outbound interface delays). The K values allow network operators to modify and tune the EIGRP composite metric for EIGRP different deployments, and for achieving different EIGRP network behaviors. Cisco recommends that network operators choose the K values after careful planning because improperly selected K values can prevent EIGRP routers from forming neighbor relationship, and cause the network to not converge.

- K_1 **Parameter**: The K_1 parameter is used to scale the bandwidth BW_E available along a path when computing the EIGRP composite metric. One of the following two variations of "bandwidth" can be used:
  o *Minimum Theoretical Bandwidth*: This is the minimum theoretical interface bandwidth along the path up to the destination. When used, path selection is based on the lowest reported router interface bandwidth along the path. By default, EIGRP uses "bandwidth" based the minimum theoretical bandwidth.
  o *Network Throughput*: This is, simply, the reported (actual) load on the router interface. When used, paths selection is based on the lowest "available" bandwidth along all interfaces on the path, adjusted for network congestion effects.
- K_2 **Parameter**: The K_2 parameter is used to scale the ratio of the "bandwidth" and the path load utilization (i.e., the worst or heaviest load on a link/interface up to the destination).
- K_3 **Parameter**: The K_3 parameter is used to scale the total one-way cumulative delay along the path up to the destination.
- K_4 **and** K_5 **Parameters**: These parameters are used to scale a quantity that represents path reliability, which can be based on link quality and packet loss.

K_5 has special (conditional) handling or interpretation such that if $K_5 = 0$, then the reliability term (i.e., $K_5/(K_4 + \text{Reliability}_E)$) is defined to be 1. Thus, if $K_5 = 0$, the composite metric reduces to:

$$\text{EIGRP}_{\text{Metric}} = \left(K_1 \cdot \text{BW}_E + \frac{K_2 \cdot \text{BW}_E}{256 - \text{Load}_E} + K_3 \cdot \text{Delay}_E \right) \cdot 256$$

The EIGRP router does not dynamically measures reliability (Reliability$_E$) and load (Load$_E$), but only does so at the time a path change happens **[RFC7868]**. Reliability$_E$ represents the current error rate on the path, and is expressed as a value from 1 to 255. With this, 100% reliability is expressed as 255/255, while 90% reliability is expressed as 229/255. Load$_E$ is the load utilization on the path, and is also expressed as a value from 1 to 255. A Load$_E$ of 255/255 represents a completely utilized link (100% utilization), while a value of 127/255 represents a 50% utilized link.

Typically, Cisco EIGRP routers use only the minimum link bandwidth and total path delay to compute composite metrics. The bandwidth and delay values used by an EIGRP router are those configured on the router's interfaces connected to its next hop on the path leading to the network destination. Although other metrics can be configured, this is not recommended by Cisco as this can lead to routing loops in the network. To compute the EIGRP composite metric, the bandwidth and delay metrics are first scaled. The bandwidth is scaled using the following formula:

$$\text{BW}_E = \frac{10^7}{\text{BW}_{\min}} \cdot 256,$$

where $\text{BW}_{\min}$ is the smallest bandwidth (in kilobits per second) among all outgoing interfaces on the route to the network destination. The delay is scaled using the following formula:

$\text{Delay}_E = \text{Delay}_{\text{tot}} \cdot 256$, where $\text{Delay}_{\text{tot}}$ is the sum of all interface delays (in tens of microseconds) on the route to the network destination. Delay_E can take a maximum value of 0xFFFFFFFF (in hexadecimal) which also indicates that the destination is unreachable.

The bandwidth and delay for a Fast Ethernet interface are 100 Mb/s (100,000 kb/s) and 0.1 milliseconds, respectively, and are scaled as follows:

The EIGRP bandwidth metric (BW_E) is computed as:

$$\text{BW}_{Fast\_Eth} = \frac{10^7}{100000} \cdot 256 = 25600$$

The EIGRP delay metric (Delay_E) is computed as:

$$\text{Delay}_{Fast\_Eth} = 256 \times 10 \times 10 \text{ microseconds}$$
$$= 2560 \text{ (in tens of microseconds)}$$

The default Cisco setting of the K values are, $K_1 = K_3 = 1$ and $K_2 = K_4 = K_5 = 0$. This gives a default composite metric, adjusted for scaling factors, as follows:

$$\begin{aligned} \text{EIGRP}_{\text{Metric}} &= \left(\text{BW}_E + \text{Delay}_E\right) \cdot 256 \\ &= \left(\frac{10^7}{\text{BW}_{\min}} + \text{Delay}_{\text{tot}}\right) \cdot 256 \end{aligned}$$

The default K values were carefully selected by Cisco to provide optimal EIGRP performance in most network deployment scenarios.

RFC 7868 **[RFC7868]** defines additional route metrics called Wide Metrics that enable EIGRP routers to perform the route selection in networks that support high bandwidth interfaces. To use these metrics, both the EIGRP message format and composite metric formula have been modified. These changes allow EIGRP routers to select routes based on delay values (measured in picoseconds) in the network. Readers can refer to **[RFC7868]** for details on the use of Wide Metrics (which are based on five vector metrics: load, latency, minimum throughput, reliability, and MTU).

6.6 FEASIBLE AND REPORTED DISTANCE

EIGRP uses the concept of Feasible Distance and Reported Distance to determine best routes to network destinations.

6.6.1 Feasible Distance

The Feasible Distance is the least-known total cost from a particular EIGRP router to a network destination since the last time the destination entry in the EIGRP Topology Table transitioned from ACTIVE state (i.e., no usable route available) to PASSIVE state (i.e., usable route available) **[RFC7868]**. The Feasible Distance is not necessarily the cost of the current best path to the destination, but instead, is a record of the smallest known path cost since the last time the destination entered the PASSIVE state (see Figure 6.9).

Each time DUAL runs, it selects a best path (through a next-hop referred to as the Successor), and one or more second best paths (through next-hops referred to as the Feasible Successors), if available, to reach each network destination. The Feasible Distance is the lowest calculated total metric (among all paths) from the local router to a particular network destination since the last time the destination was in the PASSIVE state. Each EIGRP router computes only one Feasible Distance for each network destination. The concept of Successor and Feasible Successor are discussed below.

6.6.2 Reported (or Advertised) Distance

The Reported Distance (sometimes referred to as the Advertised Distance) is the cost from a particular router to a network destination which the router advertises in all

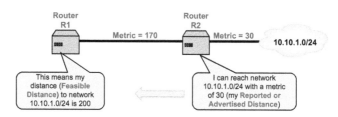

FIGURE 6.9 Understanding EIGRP Feasible and Reported Distance

EIGRP messages carrying routing information **[RFC7868]** (see Figure 6.9). The Reported Distance a router advertises is not equivalent to the router's current cost or distance to the network destination, and may be different when the router goes through path recomputation for that destination. Each EIGRP router computes only one (local) Reported Distance for each network destination.

The Reported Distance is used only to compute the Feasible Distance to a given network destination. There could be cases where the Reported Distance associated with a route to a network destination as reported by a neighbor could be promising, but the overall routing metric through that neighbor ends up not being the best overall route. Hence, the route through the reporting neighbor would not be selected as it does not constitute the current Feasible Distance. Theoretically, the Feasible Distance of a router to a given destination is equal to the lowest Reported Distance computed by the router itself (among all available paths).

6.7 SUCCESSOR AND FEASIBLE SUCCESSOR

EIGRP uses the concept of Successor and Feasible Successor to maintain best routes to network destinations.

6.7.1 SUCCESSOR

A Successor is an EIGRP neighbor that satisfies the Feasibility Condition for a specific network destination and, at the same time, provides the least-cost route to that destination **[RFC7868]**. A Successor is the best next-hop router from the local router to the destination, has the least-cost route to the destination, and is guaranteed not to be part of a routing loop in the network (see Figures 6.10 and 6.11).

An EIGRP router chooses the Successor based on the least-cost route available to reach that destination, in addition to the Feasibility Condition that requires that a neighbor's Reported Distance to the destination be less than the router's own Feasible Distance to that destination (see Feasibility Condition below) **[RFC7868]**. The Successor for a given destination entry in the Topology Table is simply the best next

Enhanced Interior Gateway Routing Protocol (EIGRP)

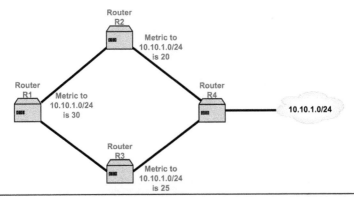

- **Successor**: This is the route with the best metric to reach a particular network. The router will install this route in its Routing Table.
- **Feasible Successor**: This is an alternative routes to a particular network that can be used immediately if the currently best route (the Successor) fails, without causing a routing loop. The router will store these routes in its EIGRP Topology Table.
Not all alternative routes to a particular network can become Feasible Successors. In order for a route to become a feasible successor, the following condition must be satisfied (the Feasibility Condition): The neighbor's Advertised Distance for the route must be less than the local router's Feasible Distance (through the Successor).

- In the network, Router R1 has two routes to reach the network 10.10.1.0/24.
- The route through R2 has the best metric (30) and it is stored in the R1's Routing Table.
- The other route, through R3, is a Feasible Successor route, because the Feasibility Condition has been met. That is, R3's Advertised Distance of 25 is less than R1's Feasible Distance of 30. So, R1 stores that route in its EIGRP Topology Table. This route can be immediately used if the primary route fails.

FIGURE 6.10 Example 1: Understanding EIGRP Successor and Feasible Successor

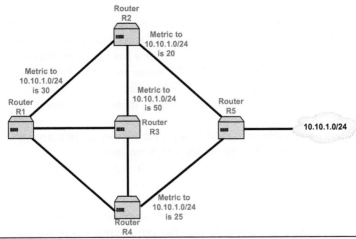

Router R5 has advertised the network **10.10.1.0/24**. Router R1 has three paths to reach 10.10.1.0/24:
- **Route R2 to R5**: Let us assume that this is the best route (the Successor route). Router R1 will install this route in its Routing Table, with the metric of 30.
- **Route R3 to R2 to R5**: For a route to become a Feasible Successor, the neighbor's Advertised Distance for the route must be less than the Feasible Distance through the Successor . This is not satisfied here since R3 has advertised the metric of 50 to reach **10.10.1.0/24**, which is greater that the Feasible Distance of R1 (30).
- **Route R4 to R5**: This route will become a Feasible Successor route, since R4's Advertised Distance is less than the Feasible Distance through the Successor (that is, **25 < 30**). Router R1 will place this route in its EIGRP Topology Table to be used immediately if the best route fails. Best routes (the Successors) from the Topology Table are stored in the Routing Table Feasible Successors are only stored in the Topology Table and can be used immediately if the primary route fails.

FIGURE 6.11 Example 2: Understanding EIGRP Successor and Feasible Successor

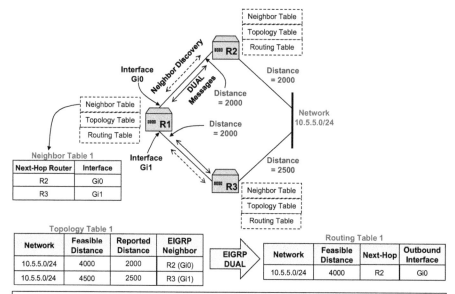

FIGURE 6.12 EIGRP Neighbor Table, Topology Table, and Routing Table

hop router to which data traffic is forwarded on its way to the destination. Successors are installed in the Routing Table to be used for forwarding packets, and multiple successors can exist for a given destination if they have the same least cost.

Like all IP routing protocols, an EIGRP router only installs the next-hop router (Successor) information in its Routing Table (Figure 6.12). The router does not install information about the subsequent routers on the path (after the next-hop) to the network destination in the Routing Table. Each router relies on the next-hop router (which also maintains its own Routing Table) to make the correct routing decision when forwarding packets to the network destination.

Each router computes the best path to reach a given network destination and installs the IP address and outbound interface of the corresponding best next-hop router along the path in its Routing Table. Each router trusts the best next-hop router (i.e., the Successor) to forward packets toward the intended destination. This hop-by-hop forwarding of packets along the path through a network is performed one router to the next. The EIGRP Routing Table, essentially, contains a subset of the information in the EIGRP Topology Table. An EIGRP router uses the Topology Table to maintain more detailed information about the route(s) to each network destination, backup routes (if any exists), and information used by EIGRP DUAL.

6.7.2 Feasible Successor

A Feasible Successor is an EIGRP neighbor router that satisfies the Feasibility Condition for a specific network destination and, hence, guarantees a loop-free route

Enhanced Interior Gateway Routing Protocol (EIGRP)

through that neighbor to the destination, but may not be the least-cost route available **[RFC7868]**. A Feasible Successor can be described as a next hop router that leads to the destination, but may not lie on the least-cost route to that destination.

The following are the main characteristics of a Feasible Successor:

- A Feasible Successor is a neighbor router to the destination that is guaranteed not to be a part of a routing loop in the network.
  o A Feasible Successor is a neighbor router that is closer to the destination, but does not lie on the least-cost route to that destination and, thus, is maintained only in the Topology Table and not used to forward packets.
- For a next-hop router to qualify as a Feasible Successor, it must have a Reported Distance that is less than the Feasible Distance of the local router (via the current Successor).

The requirement that the Reported Distance of the backup route be less than the Feasible Distance of the Successor ensures that the Feasible Successor lies on a loop-free path to the destination. In this case, the Feasible Successor cannot use a route through the local router to the destination thereby causing a routing loop. When the Reported Distance of the next-best route is greater than or equal to the Feasible Distance through the Successor, a Feasible Successor for the local router cannot be chosen. In this case, the router must use EIGRP QUERY and REPLY messages to find any alternative paths to the network destination.

The following are other properties of Feasible Successor:

- **Selection of Feasible Successors**: An EIGRP router selects Feasible successors at the same time it selects Successors but they are installed only in the Topology Table. Because a Feasible Successor does not lie on the least-cost path to the destination, it is not installed in the Routing Table, and is not used as the primary path for forwarding packets to the destination. The primary path to the destination is through a Successor. The router may install multiple Feasible Successors in the Topology Table for a destination.
- **Feasible Successors for Unequal Cost Load Balancing**: A router could still use the Feasible Successor for forwarding packets in the event unequal cost load balancing is used. When using unequal-cost load-balancing, the Feasible Successors serve as next hop routers and are installed in the EIGRP router's Routing Table to be used for traffic forwarding to the network destination.
- **Feasible Successors as Always Available Backup Routes**: In the event, the current Successor becomes unreachable (because of a change in the network topology or a route metric change), a Feasible Successor, if one exists, can become (promoted to) the next Successor (and installed in the Routing Table), thereby avoiding route recomputations. If no Feasible Successor exists in the Topology Table, the destination is placed in ACTIVE state (see below), and the router must perform route recomputation to determine a new Successor.
- **Speeding up Network Convergence**: Route recomputation may not be processor-intensive, but it can affect network convergence time, so it is beneficial for the EIGRP routers to avoid route recomputations whenever possible. The availability of a Feasible Successor allows a router to recover immediately

upon network topology or route metric changes, thereby reducing the number of route recomputations and speeding up network convergence. Unlike the Successor, the EIGRP router always maintains the Feasible Successors to a network destination in its Topology Table.

The Feasible Successor provides, essentially, a backup route to a network destination (ready to be used at any time) in the event the existing Successor become unavailable. A Feasible Successor provides an alternative loop-free backup route to the same destination as the Successor but at a higher routing cost. A router can forward packets to a destination marked in the PASSIVE state (see discussion below) at any time through any of its Successors or Feasible Successors without having to notify them; they will accept and forward packets any time. The router trusts the Successor or Feasible Successors to forward traffic toward the destination.

6.8 ROUTE STATES: ACTIVE AND PASSIVE STATES

Each network destination entry in the EIGRP Topology Table a can be marked as in one of two states, PASSIVE state or ACTIVE state. These states describe whether a route to a network destination (in the Topology Table) is guaranteed to be both the lowest cost route available and is loop free (the PASSIVE state), or whether no such guarantee can be given (the ACTIVE state).

The router reevaluates the state of a destination entry and possibly, changes it every time it detects a network topology change. A network topology change refers to any event that causes the EIGRP router to add or change the Reported Distance to a particular destination, or remove a route to that destination from the router's Topology Table. For example, a topology change occurs when an EIGRP router detects a change in the link cost to a network destination, or receives an EIGRP message from a neighbor advertising a new Reported Distance to that destination.

6.8.1 PASSIVE State

A destination entry in the Topology Table is in the PASSIVE state when there is at least one EIGRP neighbor that can provide the current least-cost or best route, and also satisfies the Feasibility Condition check that guarantees a loop-free network topology (discussed below) [**RFC7868**]. A route entry for the destination is marked in PASSIVE state, when it is considered usable for packet forwarding, and the EIGRP router is not performing a route recomputation. With the destination entry in the PASSIVE state (i.e., the stable operational state), the router does not coordinate with its neighbors to perform any route recomputation because no such action is needed:

- The next hop EIGRP router for the destination marked in the PASSIVE state (usable route) is still available for use in packet forwarding.
- For a destination to be in PASSIVE state, there must be at least one Successor. The Feasible Successor that provides the least-cost route is also called a Successor. Put in another way, for a destination to remain in the PASSIVE state, there must be at least one neighbor that is a Feasible Successor that provides the least-total-cost route.

6.8.2 ACTIVE State

A destination in the Topology Table is considered to be in the ACTIVE state if EIGRP neighbors provide lowest-cost routes that do not pass the Feasibility Condition check, therefore, meaning the routes cannot guarantee a loop free topology **[RFC7868]**. Specifically, in the ACTIVE state, any of these routes to the destination is considered unusable for packet forwarding, and the EIGRP router must coordinate with its neighbors to come out with a new loop-free least-cost route.

With the destination in the ACTIVE state, the router is actively coordinating with its neighbors to recompute a new least-cost loop-free route for the destination. If the least cost route is provided by any router that cannot be a Feasible Successor (and therefore not a Successor), the destination entry in the Topology Table remains in ACTIVE state, thereby requiring a new least-cost loop-free route to be recomputed. During the ACTIVE state, the router uses EIGRP QUERY and REPLY messages to actively coordinate with its neighbors to compute a new least-cost loop-free route.

The properties of a destination in the ACTIVE State can be summarized as follows:

- **Destination Never Enters the ACTIVE State when Feasible Successors Are Available**: The destination is placed in the ACTIVE state when the router is performing a route recomputation, however, if there are always Feasible Successors available, the router will never place the destination in the ACTIVE state, and will avoid a route recomputation. When no Feasible Successor is available, the destination will go into ACTIVE state and a route recomputation will commence. When the router detects that a link to a neighbor router that is the only available Feasible Successor goes down, it will enter all destinations reachable through that neighbor into the ACTIVE state, and will commence route recomputations for these destinations.
- **Commencing Route Recomputation**: Recall that the router will perform route recomputation when the Successor (i.e., the current best route) to a destination becomes unavailable and there are no Feasible Successors available for the destination. The router commences a route recomputation for a destination by sending QUERY packets to all of its neighbors (except the Successor) inquiring if there is an alternative route leading to the given destination. A neighbor router can either reply indicating it has an alternative route for the affected destination, or return a QUERY packet indicating that it is also performing a route recomputation for the destination.
    o If the neighbor router has an alternative route to the destination, it will respond to the QUERY packet with a REPLY packet containing the alternate route and will not propagate the QUERY packet further. If the neighbor does not have an alternative route, it will send QUERY packets to each of its own neighbors seeking an alternative route to the destination.
    o These QUERY packets are propagated through the routing domain, creating an expanding tree of QUERY packets. Whenever a router replies to a QUERY, it stops the propagation of QUERY packets through that branch of the QUERY tree. However, QUERY packets can still propagate through

other parts of the routing domain as other EIGRP routers attempt to find alternative routes to the destination, which might not be available or exist.
o While the destination is in the ACTIVE state, the router cannot change the current next-hop router or Successor it is using to the destination. Once the router receives all replies for a given QUERY from the neighbors, it can select a new Successor and transition the destination to PASSIVE state.
- **Failing the Feasibility Condition Check**: A situation where the least-cost route provided by a neighbor may fail the Feasibility Condition check is during a network topology change. For instance, during a topology change when the current least-cost route in the EIGRP Topology Table fails, the next least-cost route available may traverse a neighbor that is not a Feasible Successor.
- **Stuck-In-Active (SIA) Condition**: The condition of SIA refers to when a destination in the EIGRP router's Topology Table has stayed in the ACTIVE state for more than a predefined time period (the default Cisco setting for SIA is 90 seconds). When EIGRP routers are searching for an alternative route to a destination, the reliable multicast protocol used by EIGRP, requires that a REPLY be received for each QUERY packet sent in the network. This means when a destination is placed in the ACTIVE state and a router sends QUERY packets, the only way the destination can be removed from the ACTIVE state and placed in the PASSIVE state is by receiving a REPLY for every QUERY packet sent, a process which can take some time. Therefore, if the router does not receive REPLY packets for all the outstanding QUERY packets it has sent within 90 seconds, the destination is placed SIA state

6.8.3 COMMENTS ON FEASIBLE SUCCESSORS WHEN A ROUTE IS IN THE PASSIVE OR ACTIVE STATE

For both the PASSIVE and ACTIVE states discussed above, it is not really important or critical if the Feasible Successors available do not provide the lowest total-cost route to a destination. What matters is if these EIGRP neighbors are guaranteed to provide a loop-free route to the destination, even if the route is potentially not the shortest one available.

During the time an EIGRP router places a destination in the Topology Table in the ACTIVE state, it must not change the Successor for that destination (i.e., change the current Successor-Directed Acyclic Graph [SDAG]), nor change its own Feasible Distance or Reported Distance until the destination returns to the PASSIVE state **[RFC7868]**. Any further information the EIGRP router receives about this destination during the time it is in the ACTIVE state is reflected only in Computed Distances. The router postpones any updates to the Successor, Feasible Distance, and Reported Distance until the destination returns to PASSIVE state.

A SDAG is a graph constructed from the contents of the Routing Tables of individual routers in the network for a particular destination, such that nodes of this graph represent the routers in the network topology, and a directed edge connects router/node A to router/node B only if router/node B is a Successor to router/node A. In the absence of network topology changes and after network convergence, the SDAG becomes a tree **[RFC7868]**.

6.9 FEASIBILITY CONDITION

The Feasibility Condition forms an integral part of DUAL. EIGRP routers use the Feasibility Condition (which is a sufficient but not a necessary condition) to check whether a neighbor EIGRP router provides a loop-free route to a network destination. Every route to a network destination that satisfies the Feasibility Condition is guaranteed to not cause a routing loop. However, not all routes that are loop-free satisfy the Feasibility Condition thereby, making the Feasibility Condition only a sufficient but not a necessary condition [RFC7868].

All route selection in DUAL must undergo the Feasibility Condition check. After a network topology or link metric change is detected, the result of the Feasibility Condition check determines the state (PASSIVE or ACTIVE) of a destination entry in the EIGRP Topology Table. A destination remains in the PASSIVE state if, after the network change, the neighbor router providing the least cost route satisfies the Feasibility Condition. The destination entry is marked in the ACTIVE state if after the network change, no neighbor provides a least cost route that satisfies the Feasibility Condition.

DUAL uses the Feasibility Condition as part of the diffusing computation process to determine how soon (where and when) to terminate the computations. Routers that are unaffected by a network topology or metric change are not required to participate in a (DUAL) diffusing computation and may not even be aware that a network change has occurred. A router may not perform a DUAL computation in the following two cases [RFC7868]:

1. If a router is informed about a network change, it may place a destination entry (in the Topology Table) in the PASSIVE state if it is aware of other routes that are downstream toward the destination (i.e., routes satisfying the Feasibility Condition). By meeting the Feasibility Condition, a route is guaranteed to be loop free and is a path downstream between the router and the network destination.
2. If a router is informed about a network change that affects a network destination for which it does not currently have any reachability routing information, the router is not required to place the destination entry into the ACTIVE state, nor is the router required to participate in the DUAL process.

The Feasibility Condition is satisfied when the Reported Distance of a neighbor router (i.e., its advertised cost) to a network destination is less than the Feasible Distance of the local router to that destination [RFC7868]. This can be explained simply as:

- The Feasibility Condition is satisfied when the neighbor reports that it is closer to the destination than the router itself has determined it is, since the last time that destination in the Topology Table was placed in the PASSIVE state.
- An EIGRP router does this by using the Source Node Condition which states that a neighbor router satisfies the Feasibility Condition if the neighbor's Reported Distance to a network destination is less than the router's own Feasible Distance to that destination.

The Feasible Distance is not necessarily the current minimum distance (i.e., the least-cost route) to a network destination, but instead, is a record of the least-cost distance the router has known (historical minimum) since the last diffusing computation for that destination was completed. This means, the Feasible Distance noted by a router can be either equal or less than the current best distance computed.

- A neighbor router that advertises a route to a particular destination with a specific cost (i.e., Reported Distance) that does not meet the Feasibility Condition may be upstream to the current router. Thus, this neighbor cannot be guaranteed to be the next hop that provides a route that is loop-free to that destination. The router, therefore, does not record such routes advertised by upstream neighbors in its Routing Table but instead will save them in its EIGRP Topology Table.

6.10 EIGRP DIFFUSING UPDATE ALGORITHM (DUAL)

DUAL is the routing algorithm used by EIGRP for computing the routes that get installed in the IP Routing Table (see Figure 6.1). DUAL works by diffusing or spreading out the computational tasks across the routers in the network, while implementing mechanisms that provide loop-free routing. DUAL allows routers to construct least-cost routes to all reachable network destinations, and guarantees that each route computed does not cause routing loops. DUAL operates to provide loop-free routing during all times of network operation, including periods of network topology or link metric changes, and periods of network convergence.

In DUAL, only routers that are affected by a network change compute the new best routes required. The routers do this by coordinating with each other using diffusing routing computations and the Feasibility Condition to check if prospective routes do not create routing loops. Only the routers affected by a network change need to advertise and act on the new routing information advertised. Routers that are not affected by a network change do not participate in any related route recomputations. DUAL was designed to be used with routing protocols that use distance vectors or link-states routing information. The benefit of DUAL is that, it provides fast loop-free routing even in the presence of multiple network topology changes with low messaging overhead.

6.10.1 High-Level Description of DUAL

A diffusing computation is a distributed computation method in which a single node starts the computation, and then delegates subtasks of the computation to its neighbors, who in turn, may delegate sub-subtasks to their neighbors, and so on. This method includes a signaling scheme that allows the starting node to detect when the computation (including all subtasks) has properly finished while avoiding false endings. DUAL uses diffusing computation to perform the task of best path computations as well as coordinated updates of Routing Tables in the network.

Using diffusing computation, DUAL allows the routers affected by the network change to coordinate among themselves to compute new least-cost routes to network destinations. The diffusing computation expands by allowing routers to query additional routers for their current Reported Distances to the destinations affected by the network change. The computation shrinks as routers receive replies to their queries. Routers not affected by the network change simply send replies to queries immediately, thereby, ending the expansion of the diffusing computation through them.

These inherent properties of DUAL produce a diffusing computation process that can adjust the scope its computation, and terminate at the right point and time. DUAL is able to control the termination point of diffusion of route computation by managing how reachability information is distributed through the network.

By using methods such as route summarization (or aggregation) and filtering, DUAL is able to hide the reachability information of networks, and as a result control the scope of the diffusing process. This provides a mechanism for creating effective failure regions/domains within a single Autonomous System, and a mechanism through which the route processing and convergence characteristics of a network can be managed.

6.10.2 MESSAGE TYPES USED BY DUAL

Three basic message types are used by DUAL in its operations. These are the QUERY, UPDATE, and REPLY messages:

- **UPDATE Message**: A router sends this message to neighbors to indicate that a metric has changed or a network destination has been added to its Topology Table.
- **QUERY Message**: A router sends this message when a route fails the Feasibility Condition, which may happen, for example, when a destination becomes unreachable, or the path metric increases to a value that is greater than the router's current Feasible Distance.
- **REPLY Message**: A router sends this message in response to a QUERY or SIA-QUERY message that it has received.

The SIA-QUERY and SIA-REPLY are two additional subtypes (as explained above) that have been defined for the QUERY and REPLY messages, respectively, for use with DUAL and EIGRP routers.

When a destination is in the PASSIVE state, a neighbor router may propagate a received QUERY message if it is not a Feasible Successor (i.e., it has no alternate route to that destination). However, if the neighbor router finds that it is a Feasible Successor (i.e., it has an alternate route), it will not propagate the QUERY, and will send a REPLY message for that destination with a metric set equal to the current metric in its Routing Table. When the local router receives a QUERY message from a neighbor that is not a Successor for a destination that it has placed in the ACTIVE state, it will send a REPLY message to that neighbor, and will not propagate further that QUERY message. The local router will send a REPLY message for that destination containing a metric equal to its current Routing Table metric.

6.10.3 Some Behaviors of DUAL

EIGRP uses DUAL (described in **[GARCIALA89] [GARCIALA93]**) to compute best paths to network destinations that are guaranteed to prevent routing loops from occurring in the network. DUAL has the ability to select best paths that result in a loop-free network topology without any need for periodic updates or Holddown timer mechanisms for maintaining routes in the routing table (as in RIP). Such route maintenance timer mechanisms can result in slower network convergence as discussed in Chapter 2.

The following summarize some important behaviors of DUAL:

- Recall that an EIGRP router places a destination entry (in the Topology Table) in PASSIVE state when it is not performing any route recomputations for that destination. If the Successor for the destination becomes unreachable (or fails) and the destination entry has Feasible Successors, then the router does not need to perform any route recomputations and the destination entry does not go into ACTIVE state.
- DUAL places the destination entry in ACTIVE state if the Successor fails and there are no Feasible Successors available. The EIGRP router will send EIGRP QUERY messages to its neighbor routers looking for a Feasible Successor to the destination. A neighbor router can respond with a REPLY message indicating it has a Feasible Successor, or a QUERY message indicating that it does not have a Feasible Successor and will participate in the route recomputation.
- The router does not return the destination entry to the PASSIVE state until it has received a REPLY messages from all neighbors. However, if the router does not receive REPLY messages from all neighbors before the "active-time" timer expires (default timer setting is 90 seconds), it will declare the destination entry (in the Topology Table) as SIA.

The DUAL finite state machine described in **[RFC7868]** captures very well the decision processes involved in route computations. The finite state machine allows a router to tracks all the routes advertised by neighbor routers as well as the distance information (metrics) used by DUAL to compute loop-free routes to network destinations. The DUAL finite state machine operates on a per destination basis in the Topology Table, and also handles each destination independently. It should be noted that when a single link fails, multiple destinations in the Topology Table may go into ACTIVE state. When a network change occurs, each router affected will compute a separate Successor-Directed Acyclic Graph (SDAG) for each destination, allowing a loop-free route can be computed for each reachable destination.

DUAL selects routes to be installed in the EIGRP Routing Table based on the Feasible Successors available to each network destination. When a router finds no Feasible Successors but there are neighbor routers still advertising reachability information to the destination, the router will perform route recomputation to determine a new Successor.

Enhanced Interior Gateway Routing Protocol (EIGRP)   179

The time it takes for routers in the network to compute new routes impacts the network convergence time. Thus, to avoid unnecessary route recomputations, when a network change occurs, DUAL will try to find if there are any Feasible Successors available. If one Feasible Successor is available, the router will use it in order to avoid performing unnecessary route recomputation.

6.10.4 STUCK-IN-ACTIVE (SIA) AND THE USE OF SIA-QUERY AND SIA-REPLY MESSAGES

When a particular destination in the Topology Table of an EIGRP router transitions to the ACTIVE state, the router will send a QUERY packet to its neighbor, and the ACTIVE timer will be started to limit the amount of time the destination will stay in the ACTIVE state. A destination is declared as SIA when the router does not receive a REPLY within a predetermined time period.

When a router places a destination in the SIA state, it resets the neighbor relationships for all the neighbor routers that did not respond with REPLY packets. The router then recomputes routes to all destinations discovered through these neighbors, and readvertises all these routes. The most common reasons a router enters destinations in SIA are as follows [CISCTEAPA06]:

- The router is not able to respond to a QUERY packet, which can happen as a result of high CPU utilization or memory shortage (where the router cannot allocate enough router memory to process a QUERY packet or construct a REPLY packet).
- The link between a router and its neighbor is faulty resulting in some EIGRP packets getting lost and not reaching the intended recipient. A neighbor router may receive enough EIGRP packets to maintain the neighbor relationship, but does not receive all QUERY or REPLY packets sent.
- A link failure between a router and its neighbor which causes traffic to flow in only one direction (results in a unidirectional link).

6.10.4.1 Stuck-In-Active (SIA)

The SIA decision process and corresponding time interval is divided into two equal intervals and this takes place immediately after a router sends a QUERY to its neighbors. The process of determining if a destination is SIA starts immediately after a router sends a QUERY to its neighbors. As soon as one-half of the SIA time interval is up (default for Cisco routers is 90 seconds), the router will send an SIA-QUERY to its neighbors which must respond with either a REPLY or SIA-REPLY message. Any neighbor that fails to respond with either a REPLY or SIA-REPLY within the second one-half of the preset SIA interval will result in the sending router placing the particular destination in SIA state.

Cisco limits the number of SIA-REPLY messages expected by the originating router to a maximum of three. If the SIA timer expires after three SIA-REPLY messages have been received from the neighbors, and still the destination remains in the ACTIVE state (as indicated in the SIA-REPLY messages), the router will declare the destination as SIA.

When a destination is declared as SIA, the EIGRP router will take one of following two actions **[RFC7868]**:

a. The router will delete the route through that neighbor to that destination, treating the neighbor as if it had responded with a REPLY message that says the destination is unreachable.
b. The router will delete all routes advertised by that neighbor and reset the adjacency with that neighbor, treating the neighbor as if it had responded to indicate that all routes it had advertised are unreachable.

6.10.4.2 SIA-QUERY

When an EIGRP router is still waiting for a REPLY message from a neighbor for a QUERY message that is still outstanding, the router may not know what is preventing the REPLY from being received. The absence of a REPLY message from a neighbor could be due to a lost REPLY packet, possibly caused by network congestion, or a slow neighbor that is yet to send a REPLY (due to high CPU utilization or insufficient memory to process a received QUERY or construct the REPLY packet).

To enable a router to determine if the neighbor is still attempting to determine a loop-free route to the destination that is in the ACTIVE state, the router may send a SIA-QUERY message to that neighbor. This enables the router to ascertain if there is a communication problem with the neighbor, or if the neighbor is still in the process of finding a loop-free route to the destination through its downstream routers.

By transmitting a SIA-QUERY message, the router can extend the effective time a destination remains in the ACTIVE state. The router does this by simply resetting the ACTIVE timer that it has set for the destination, allowing the neighbor to continue to search for a loop-free path, but as long as the neighbor successfully communicates back that it is still doing so.

The router must send SIA-QUERY messages on a per-destination basis at the halfway mark of the ACTIVE timer setting. Furthermore, the router can send up to three SIA-QUERY messages for any particular destination, and each one of these must be sent at one-half of the ACTIVE timer period, and so long as each message is successfully acknowledged by the recipient of the SIA-QUERY with a SIA-REPLY. A QUERY message is also transmitted on a per-destination basis and at the halfway mark of the ACTIVE timer setting.

Upon receiving a SIA-QUERY message, the neighbor router must first respond immediately with an ACK message to the sender, and then continue to process the SIA-QUERY message. The neighbor router must respond to the originator of the SIA-QUERY message with a SIA-REPLY message indicating that it is actively processing routes for this destination by setting the ACTIVE flag in the SIA-REPLY message.

If the neighbor router receives a SIA-QUERY message for a destination for which it has not received the original QUERY message, the neighbor must treat the SIA-QUERY message as though it was a QUERY message and act as follows:

1. The neighbor should acknowledge the receipt of the SIA-QUERY message.
2. The neighbor must respond with a REPLY message if it can be a Successor.

Enhanced Interior Gateway Routing Protocol (EIGRP)   181

3. If the SIA-QUERY is from a neighbor router that cannot be a Successor (Feasibility Condition check fails), it will transition the destination to the ACTIVE state, and respond with a SIA-REPLY message to the sender with the ACTIVE bit set.

6.10.4.3 SIA-REPLY

Upon receiving a SIA-QUERY message, an EIGRP neighbor will send a SIA-REPLY message as response to the sender. The neighbor sends the SIA-REPLY message after it has processed the full SIA-QUERY message received. The SIA-REPLY message sent includes a TLV for each network destination along with its vector metric as indicated in the neighbor's Topology Table. If the destination is still in the ACTIVE state, the neighbor will send the SIA-REPLY message with the ACTIVE bit set. This indicates to the sender of the SIA-QUERY message that the neighbor processed the message and is still in the process of finding a loop-free path to the destination (most likely waiting for replies to QUERY messages it has sent to its own downstream neighbor routers).

The SIA-REPLY message informs the sender of the SIA-QUERY that the neighbor is still searching for a loop-free path or has completed that search. In the former, the SIA-REPLY is an explicit notification that the neighbor is still actively performing route recomputations. This allows the sender of the SIA-QUERY message to determine if it should continue to keep the destination in the ACTIVE state, or it should reset its adjacency relationship with the neighbor, and delete all routes through this neighbor from its Topology Table.

6.11 EIGRP NEIGHBOR DISCOVERY AND MAINTENANCE

A neighbor (or peer) of a particular router is another router with which the router can establish an adjacency relationship through an EIGRP session (also referred to as an adjacency). Two routers can only become neighbors if they have direct Layer 2 connectivity (no intervening Layer 3 devices), and have been configured with compatible EIGRP configuration parameters. Two routers with interfaces connected to a common VLAN or subnet are deemed as physically adjacent or are neighbors. Two routers that are not directly connected but are multiple hops apart cannot establish an adjacency relationship because they have one or more Layer 3 devices between them.

Each router in a network has to discover all the other routers that are its neighbors, and then establish an adjacency relationship with each one of them. During neighbor discovery/recovery, each router dynamically learns of the other routers that it has direct Layer 2 connectivity to. These are routers that are attached to the networks directly connected via Layer 2 to its interfaces.

Each router must also be able to know when any neighbor becomes inoperative or unreachable. EIGRP routers achieve this through periodic transmission of small, low overhead HELLO packets to potential neighbors. As long as a router receives HELLO packets regularly from a neighbor, it can determine if the neighbor is up and operational. Only after two neighbor routers establish an adjacency can they start exchanging routing information.

Unlike RIP, Routers running EIGRP discover and maintain information about their immediate neighbors. The routers multicast HELLO messages to the multicast address 224.0.0.10 (the All EIGRP routers address) every 5 seconds for most network types. Each EIGRP router then constructs a Routing Table with the learned neighbor information. EIGRP routers send routing updates only when necessary (i.e., when topology changes occur) and advertises them only to neighbor routers. As a result, no periodic update timers are used. Similar to OSPF and IS-IS, EIGRP uses HELLO messages to discover and maintain adjacency relationships with neighbor routers.

6.11.1 Neighbor Hold Time

After neighbor discovery, each router stores state information about all of its discovered neighbors. A router stores the IP address of each discovered neighbors, the interface on which it is connected, and the Hold Time. The Hold Time refers to the length of time during which a router would treat the neighbor (i.e., sender of the Hold Time) as valid, that is, as still operational and reachable. A valid neighbor refers to a neighbor that is able participate in EIGRP DUAL processing plus forward and receive packets. When a router marks a neighbor as a valid neighbor, it will keep all routing information advertised by that neighbor.

When a router transmits a HELLO packet, it also includes its Hold Time in the EIGRP Parameter Type TLV (Type 0x0001) that it advertises to neighbors (see TLV format in Figure 6.5). The Hold time indicates to the receiving routers the maximum time they should wait to receive subsequent HELLO packets. If a router receives any packet (including HELLO packets) within the Hold Time interval, then it will reset the Hold Time period. However, if the Hold Time expires without any packet being received, then the EIGRP route computation engine (DUAL) will be notified of the network topology change.

The default Hold Time a Cisco EIGRP router uses to determine if a neighbor is reachable and operational is 15 seconds, which is three times the default HELLO Time (of 5 seconds). If a router does not receive a HELLO from a neighbor within the Hold Time (15 seconds), it declares the neighbor as invalid. The router then removes the neighbor from its Routing Table, and informs DUAL about this network change.

On multipoint WAN interfaces (e.g., ATM), with speeds not more than 1544 Mb/s, EIGRP routers multicast HELLO messages every 60 seconds. The corresponding default neighbor Hold Time on these types of interfaces is 180 seconds. Specifically, the default HELLO Time/Hold Time timer settings are 60/180 seconds for low-speed links and 5/15 seconds for high-speed links. Two EIGRP routers can still establish a neighbor relationship even if their HELLO and Hold Times do not match. Each router will still include its Hold Time in its HELLO packets so that the neighbors can stay alive even if their HELLO and Hold Times do not match.

6.11.2 Use of HELLO Packets

When an EIGRP router is properly configured and starts up, it will start multicasting HELLO packets on its EIGRP enabled interfaces to the multicast address 224.0.0.10.

Enhanced Interior Gateway Routing Protocol (EIGRP)

The router will send the HELLO packets containing, in addition, the EIGRP metric K values configured on it. Two routers can only become neighbors if they are configured with the same EIGRP metric K values. This ensures that the EIGRP metric used throughout the network of EIGRP routers is consistent.

6.11.3 Use of UPDATE Packets

A router discovers a newly active neighbor when it receives a HELLO packet from that neighbor. The router will then transmit a unicast UPDATE packet to the new neighbor (i.e., a NULL UPDATE packet) containing no routing information. This initial NULL UPDATE packet contains no network topology information and must have its INIT-Flag set (which instructs the neighbor to advertise the routes it has learned). Both the NULL UPDATE and its ACK packet are sent in unicast transmissions. The router cannot send further UPDATE packets until the initial NULL UPDATE packet is acknowledged by the neighbor.

It is noted in [RFC7868] that during the initial formation of adjacency relationship, the neighbor routers use a three-way handshake process to verify if both routers have multicast and unicast connectivity working properly. This is to prevent the routers from sending packets to each other before verifying if multicast and unicast packet delivery is successfully configured at both ends.

However, during normal neighbor/adjacency formation, routers multicast HELLO packets to discover their neighbors, after which the routers will enter information about the new neighbors into their Neighbor Tables. Each router will then exchange unicast UPDATE packets with each neighbor to learn about the routing information held by the neighbor to allow it complete the neighbor relationship formation.

6.11.4 Use of QUERY Packets

To prevent a router from sending packets to a newly discovered neighbor that has failed to established correct bidirectional multicast and unicast connectivity, or a neighbor that has restarted while establishing the adjacency relationship, the router must place/mark the new neighbor in a "pending" state. During this period of adjacency formation, the new neighbor is placed in the pending state, and it is not eligible to participate in the route computation process (i.e., network convergence process).

Thus, when a router receives any QUERY packet from another router, this reception will not cause the receiver in turn to send a QUERY packet to the new (and pending) neighbor. The router would perform route computation process without involving the new neighbor in the process. When the router (i.e., the one in the process of establishing an adjacency relationship with the new neighbor) receives a QUERY packet from a different router (i.e., a fully established neighbor), it will perform the normal Feasibility Condition check to determine whether it needs to respond by sending a REPLY message indicating a valid path to the network destination, or place the destination in the ACTIVE state.

If the router determines that it must place the destination in ACTIVE state, it will send a QUERY packet to each fully established neighbor that is participating in the route computation process, and will expect REPLY packets from each. The router will not send a QUERY packet directly to any pending neighbor, and therefore, will not expect a REPLY packet from any. However, if a pending neighbor resides on an interface of the router that contains a mix of pending neighbors and fully established neighbors (e.g., on a multiaccess broadcast network segment), the router might receive a QUERY packet, but it is not expected to respond by sending a REPLY packet.

6.12 BUILDING THE EIGRP TOPOLOGY TABLE

Each entry in an EIGRP router's Topology Table has associated with it a destination IP address (prefix), list of neighbor routers that have advertised routing information to this destination, and the advertised routing metric or cost associated with each path/route to that destination (i.e., the neighbor's Reported or Advertised Distance).

- Recall that the Computed Distance is the total metric (cost) a router computes for a path from itself to a destination network through a particular neighbor using the link cost between itself and the neighbor, and the neighbor's Reported Distance.
- Each EIGRP router computes and maintains one Computed Distance for each destination/advertising neighbor pair.
- Thus, the Reported/Advertised Distance of the local router to a given destination is the lowest Computed Distance to that destination.
- The Feasible Successor for a particular destination will advertise its lowest Computed Distance to that destination as its local Reported Distance to own neighbors.
  - o In other words, when a neighbor advertises its lowest Computed Distance to a destination, this becomes its own Reported Distance, and is the metric that the neighbor uses to represent or describe how far it is to that destination.

It should be noted that when a router advertises a route to a destination, then the router must be using that route to forward packets to that destination (a feature that distance-vector routing protocols use).

6.12.1 ROUTE MANAGEMENT

A route in the EIGRP Topology Table is considered to be either an internal or external route. Internal routes have precedence over external routes independent of their associated routing metric or cost. If a router receives an internal route for a given network destination, it will run the diffusing computation considering only the internal routes it has learned. The router will select the Successor from external routes, if available, only when there are no internal routes for that network destination.

6.12.1.1 Internal Routes

Internal routes refer to routes to network destinations that have been learned from an EIGRP Autonomous System and advertised within the same Autonomous System.

Enhanced Interior Gateway Routing Protocol (EIGRP)

Simply, these are routes that are originated and propagated in the same Autonomous System. For example, a directly attached network on the interface of an EIGRP router is considered an internal route within the router's Autonomous System, and this information is propagated throughout that Autonomous System. An EIGRP router tags internal routes in its Topology Table with the following information:

- Router Identifier (ID) of the router within the EIGRP Autonomous System that originated the route.
- An administrative tag (also referred to as a route tag that is configurable) that can be used to define and apply network policies (e.g., for customized routing and path selection, traffic engineering). By using administrating tagging of EIGRP routes, a network administrator has a flexible means of tagging routes and implementing network policy controls for customizing routing in a network (see Chapter 7).

6.12.1.2 External Routes

External routes are routes to network destinations that have been learned by a different routing protocol (e.g., RIP, OSPF), discovered in another EIGRP Autonomous System, or originally configured manually as static routes. An external route can be a route to a destination outside a particular EIGRP Autonomous System that has been redistributed into that EIGRP Autonomous System. An EIGRP router marks each external route in the Topology Table with the routing source (dynamic routing protocol or static routing) that originated the route. An EIGRP router tags external routes with the following information **[CISCID13669]**:

- Router ID of the EIGRP router in an external Autonomous System that originated the external route. The Router ID is a 32-bit number that uniquely identifies the source of the routing information being redistributed.
- External ASN which is a 32-bit number identifying the external Autonomous System in which the routing information source is a member.
- An administrative tag (that is configurable) that is used by a network administrator to tag a route when redistributing it. Routers can set administrative tags value to routes which can then be used to apply specific routing polices within a network (see Chapter 7). The tag allows a network administrator to implement policies to filter routes in EIGRP border routers. Administrative tagging is especially useful in border routers of an Autonomous System where EIGRP would interoperate with an interdomain routing protocol implementing global-wide network policies and routing. A border router may be running multiple routing protocols (e.g., EIGRP as well as OSPF and RIP). In transit Autonomous Systems, EIGRP routers would typically interact with other interdomain routing protocols that could be implementing network policies that have global scope. Thus, administrative tagging can provide a mechanism for very scalable policy-based routing. This kind of tagging is particularly useful when the network manager wants to implement network policies across different Autonomous Systems.

- Protocol ID of the external routing protocol (e.g., RIP, OSPF) that learned the external route being redistributed (e.g., routes learned by an OSPF router can be redistributed into an EIGRP Autonomous System as discussed in Chapter 7).
- The routing metric (or default metric) assigned to the route being redistributed from the external routing protocol into EIGRP. This is a 32-bit composite metric value in entered in the local routing table for the external routing protocol (see Chapter 7).
- Flag bit indicating if route can be used for default routing.

EIGRP running in a border router may advertise OSPF-learned routes within the EIGRP Autonomous System. In this case, EIGRP would advertise these routes and tag them as OSPF-learned routes with each route associated with a routing metric that is set to the default metric assigned to OSPF.

6.12.2 Use of Split Horizon and Poison Reverse

According to the definition of the split horizon rule as discussed in Chapter 2, a router should never readvertise a route out of the router interface on which it was learned. So, EIGRP implements this rule as follows: if an EGRP router has learned a Successor to a network destination on a particular interface, it should never readvertise that (Successor) route on that same interface.

According to the poison reverse rule, if a router learns a route on one of its interfaces (to a network destination), it should advertise that route as unreachable through that same router interface [**RFC7868**]. The EIGRP router uses the interface on which the route is learned to reach the advertised destination. So, the router applies this rule only to the interface it is using to reach that particular destination. If there are routes that the EIGRP router has learned on that interface that it is not using to reach the advertised destination, it may advertise these routes out that interface (see Chapter 2).

There are some circumstances where an EIGRP router will suppress or poison the transmission of QUERY and UPDATE messages on its interface to prevent routing loops from occurring as network changes are propagated in the network:

- With EIGRP, a router implementing split horizon will suppress a QUERY or UPDATE about a route on the router interface on which it was learned, while it will advertise a destination as unreachable on that interface when implementing poison reverse.

Split horizon blocks a router from advertising routing information about a network destination out of any of its interfaces that the router itself uses to route packets to that destination. This optimizes communications among routers in the network, particularly when links or routers fail. Also, when a router makes changes in its Topology Table that results in a change in the interface (or route) through which the router uses to reach a destination, the router will turn off split horizon and perform a poison reverse by advertising the old route with infinity metric out of all of its interfaces indicating that the route is unreachable. This is to ensure that other routers in the network will not attempt to use the old route that is now invalid.

Enhanced Interior Gateway Routing Protocol (EIGRP)

The EIGRP router will do so for a destination under any of the following conditions:

- **When Two EIGRP Routers Are in the Startup or Restart Mode**: When two EIGRP routers first establish a neighbor or adjacency relationship, they start by exchanging their Topology Tables (the startup mode). Two EIGRP (neighbor) routers are in the startup mode when they are exchanging their EIGRP Topology Tables for the first time. During the startup mode, when a router receives routing information for each destination, it will advertise the same routing information for that destination back to its new neighbor but with the routing metric set to the maximum value which is infinity (a process referred to as Route Poisoning). Neighbor startup also occurs when a link cost changes between the two routers.
- **When an EIGRP Router Is Advertising a Topology Table Change**: If an EIGRP router uses a specific neighbor router as the Successor (route) to a given network destination, it will transmit an UPDATE packet for that destination carrying a routing metric of infinity when that Successor is unavailable or equivalently, the destination is placed in the ACTIVE state.
- **When an EIGRP Router Is Sending a QUERY or UPDATE Message**: In most circumstances, an EIGRP router follows the split-horizon rule as explained above by never readvertising a route that it has learned out of the router interface on which it was learned. When a router receives a routing metric change from the Successor (to a given network destination) via a QUERY or UPDATE packet that causes the destination to be placed in the ACTIVE state, the router will send QUERY packets to its neighbors on all of its interfaces except the interface leading to the Successor. The router does not send a QUERY packet out of the inbound interface on which the routing information was received (that causes that destination to be marked as ACTIVE). Simply, the router applies the split horizon rule when it receives a QUERY or UPDATE packet from the Successor it is using to reach the destination reported in the QUERY packet.

6.13 INITIAL NEIGHBOR AND ROUTE DISCOVERY

When EIGRP routers discover new neighbors, they also go through the process of learning routes that those neighbors learned on their own. In this section, we describe the initial neighbor and route discovery process that provides the information EIGRP routers use to build their EIGRP Neighbor, Topology, and Routing Tables. The following steps describes the process that EIGRP routers go through when they perform their initial route discovery **[CISCTEAPA06]**. Let us consider a new EIGRP router (Router-Nw) and an existing neighbor EIGRP router (Router-Ex) on a shared network segment (e.g., a link, Layer 2 multiaccess network such as an Ethernet network):

1. EIGRP Router-Nw comes up on the shared network segment and transmits a HELLO packet out on all of its interfaces.
2. Each EIGRP router (including Router-Ex) that receives the HELLO packet on its receive interface, will reply with EIGRP UPDATE packets that contain all

the routes it has in its local Routing Table, except the routes the router has learned through the receive interface (i.e., by applying the split horizon rule).
   a. Router-Ex will send an EIGRP UPDATE packet to Router-Nw, even though the two have not established a neighbor relationship (which happens when Router-Ex sends a HELLO packet to Router-Nw).
   b. The EIGRP UPDATE packet sent from Router-Ex has its initial bit (INIT-Flag [0x01]) in the Flags field (Figure 6.2) set, indicating that this is the beginning of the EIGRP startup process with a neighbor process **[RFC7868]**.
   c. The EIGRP UPDATE packet conveys to Router-Nw information about the routes that the neighbor Router-Ex has discovered/learned, in addition to the routing metric that the Router-Ex is advertising for each route/destination.
3. After Router-Nw and Router-Ex have exchanged HELLO packets, and have established a neighbor adjacency, Router-Nw will reply to Router-Ex with an ACK packet, indicating that it has received the EIGRP UPDATE packet containing the routing information.
4. Router-Nw installs the routing information in the EIGRP UPDATE packet in its Topology Table. The Topology Table of Router-Nw contains the routes advertised to all network destinations by its neighbor/adjacent routers. The Topology Table is organized to list each network destination along with all the neighbors through which the destination can be reached and their associated routing metrics.
5. Router-Nw will then send an EIGRP UPDATE packet to Router-Ex after updating its Topology Table.
6. Upon receiving the EIGRP UPDATE packet from Router-Nw, Router-Ex will send an ACK packet to Router-Nw.

After Router-Nw and Router-Ex have successfully exchanged the EIGRP UPDATE packets and updated their Topology Tables, they are now ready to transition to the next process which is to make route computations, and then select the Successor (best or primary route(s)) and Feasible Successor (backup route(s)) for their Topology Table. Only the Successor routes will then be installed in the Routing Table. An EIGRP router selects Feasible Successors at the same time the Successors are selected. Multiple Feasible Successors for a given destination (if any exist) can be maintained in the Topology Table.

The EIGRP routers use DUAL to keep track of all routes advertised by neighbors and to compute the Successor to each network destination. DUAL computes the Successor using an EIGRP composite metric, and also ensures that each selected Successor provides a route that is loop-free. DUAL also computes Feasible Successors (or backup routes), if available, to each destination that provide loop-free paths. This allows an EIGRP router to immediately uses Feasible Successor if the Successor fails, without having to initiate any Holddown timer as in RIP (see Chapter 2). The Feasible Successor (if one is available) provides a loop-free path and results in fast network convergence.

Enhanced Interior Gateway Routing Protocol (EIGRP)

The EIGRP router selects the Successor to each network destination and installs these in the local Routing Table, along with the destination network address, the routing metric to the destination, the IP address of the Successor, and the outbound interface leading to the Successor. If the EIGRP Topology Table contains more than one route that have the same least-cost (i.e., multiple Successors) to a given network destination, then all of these Successors (up to four by default [CISCTEAPA06]) will be installed in the Routing Table.

6.14 EIGRP LOAD BALANCING

Load balancing in a router refers to the distribution of network traffic across a group or set of routes leading to a common network destination. In equal-cost load balancing, a router distributes traffic across multiple routes that have the same routing metric (cost) to a destination. With unequal-cost load balancing, the router distributes traffic across multiple routes that have different routing metrics to a destination.

Load balancing aims to optimize path bandwidth utilization, increases the effective network bandwidth to a destination, avoids the overload of any single path, maximize the throughput of end-user applications, and minimize end-user application response time. Using multiple routes with load balancing instead of a single route from a router, provides route redundancy, and increases the reliability and availability of routes through a network. Multiple alternative paths in a network provide redundancy, avoiding a single point of failure.

An EIGRP router can perform equal-cost load balancing when it has multiple Successors to a destination in its Routing Table (Figure 6.13), and unequal-cost load

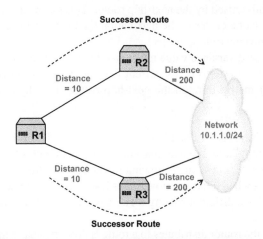

- Routers R2 and R3 are both connected to network 10.1.1.0/24 and advertise routes to reach that network to R1.
- Router R1 receives two routing updates for network 10.1.1.0/24 with the same Reported Distance.
- Router R1 installs both routes in its Routing Table and load balances across these Successor routes.

FIGURE 6.13 EIGRP Equal-Cost Load Balancing

balancing when it has a Successor and one or more Feasible Successors in its Routing Table. Load balancing can be performed on a per-packet, per-destination, per-source, or per-flow basis. A router may maintain, for instance, up to 16 best routes in its Routing Table but use a maximum of 4 routes for load balancing.

An EIGRP router uses a *variance* value for performing load balancing. The variance is a multiplier that controls the degree to which the router performs unequal-cost load balancing [CISCTEAPA06]. The variance value or multiplier is a value in the range of 1 to 128, and defines the range of routing metric (or cost) values that are acceptable/permissible for load balancing, and also provides a mechanism for controlling the degree of load balancing.

The default variance value of 1 indicates equal-cost load balancing. When a variance is configured, it is used to compute the range of the routing metric values (i.e., the Feasible Distances) the router uses to get to a given network destination [CISCTEAPA06]. The router uses this range of routing metric values to determine the feasibility of using a potential route in the load balancing procedure (i.e., a Feasible Route with Variance).

An EIGRP router considers a route to a particular network destination to be Feasible, if the next-hop router on the path is closer to that destination than the router itself, and if the routing metric for the entire alternate path is within the range of the routing metric values computed using the configured variance. The EIGRP router installs only Feasible paths in its Routing Table and uses only these paths (that are Feasible) for load balancing. The two Feasibility Conditions that govern load balancing are as follows [CISCTEAPA06]:

1. The best routing metric computed locally by the router (i.e., its current Feasible Distance) must be greater than the best routing metric (i.e., the Reported Distance) advertised by the next-hop router. This means the next-hop router on the path must be closer to the destination than the local router itself, a condition that prevents routing loops.
2. The configured variance value multiplied by the best routing metric computed locally by the router (i.e., its current Feasible Distance) must be greater than the routing metric through the next-hop router (i.e., the alternative Feasible Distance).

If these two conditions are satisfied, the route is declared Feasible and the router can add it to its Routing Table (see examples in Figures 6.14 and 6.15).

An EIGRP router can use the following methods to control how traffic is distributed among routes when it has multiple routes leading to a given network destination, and these routes have different routing metrics (costs) [CISCTEAPA06]:

1. The traffic the router distributes to a route is proportional to the inverse ratio of the routing cost associated with that route.
2. The router installs all routes that are Feasible, and have routing metrics within the range of routing metric values computed using the configured variance in its Routing Table, but uses only the routes with the minimum cost in load balancing.

Enhanced Interior Gateway Routing Protocol (EIGRP)

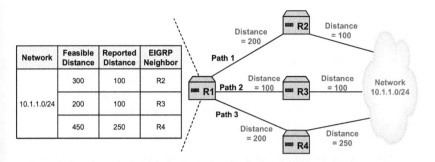

FIGURE 6.14 Example 1 of EIGRP Unequal-Cost Load Balancing with a Variance

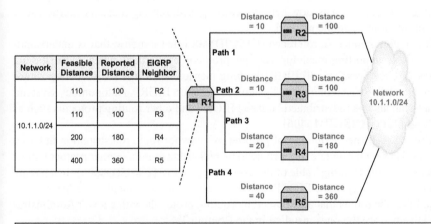

FIGURE 6.15 Example 2 of EIGRP Unequal-Cost Load Balancing with a Variance

6.15 EIGRP ROUTE REDISTRIBUTION

Route redistribution is the process of using a specific routing protocol to readvertise routes that are learned by another routing protocol, or are configured as static routes (see Chapter 7). Running multiple routing protocols in a network is often dictated by factors such as using equipment from different vendors not necessarily running the same routing protocols, integrating different networks as a result of company mergers, integrating different networks managed by different network administrators, and so on. A single router might also run multiple routing protocol processes on it, thereby, requiring route redistribution between them.

Many of these multiple protocol environments call for the use of route redistribution to allow greater internetworking. However, differences in the characteristics of the various routing protocols and methods used, such as, classful and classless routing capabilities, Administrative Distances, and routing metrics can affect route redistribution.

- Different routing protocols use different routing metrics and so, when routes are redistributed from one routing protocol into another, the metrics used by each protocol can affect how the redistribution works.
- Different routing protocol have different Administrative Distances which can be used to determine which route among routes from different routing protocols to the same network destination should be the preferred route. But the different Administrative Distances can cause problems when route redistribution is being performed.

Route redistribution, therefore, has to be used with care since it can potentially cause routing problems such as slow network convergence, routing loops, or inefficient or suboptimal routing in the network.

It is recommended in reference **[CISCID8606]** that a metric that is understandable to the importing/receiving routing protocol be defined when redistributing routes. EIGRP preserves all EIGRP routing metrics when it is redistributing routes from one EIGRP Autonomous System to another EIGRP Autonomous System. Routes can also be redistributed between EIGRP and other routing protocols such as RIP and OSPF **[CISCID16406]**.

Cisco has defined a number of proprietary rules for route redistribution in Cisco routers. These rules dictate that a route to be redistributed into another protocol, must be present in the Routing Table of the exporting protocol **[CISCID8606]**. Just having a route present only in the EIGRP Topology Table is not sufficient for redistribution. Recall that the routes that are sourced by routing protocols with a lower Administrative Distance are the ones installed in the Routing Table.

When an EIGRP router is redistributing routes from one EIGRP Autonomous System to another, it will reply to QUERY messages from the originating Autonomous System using the normal QUERY processing rules, and will also transmit a new QUERY message into the other Autonomous System **[CISCID8606]**. The original QUERY message does not propagate through to the second Autonomous System, and is bounded by the router at the Autonomous System border. Instead, the EIGRP router at the border leaks the original QUERY message into the second Autonomous System through the transmission of a new QUERY. A benefit of this technique is that, it can prevent SIA problems in a network because the number of EIGRP routers a QUERY must propagate through before a reply is received, is bounded or limited **[CISCID8606]**.

External administrative tags can also be used for breaking potential routing loops that may occur when redistributing routes between EIGRP and other routing protocols

Enhanced Interior Gateway Routing Protocol (EIGRP)

[CISCID8606]. A route can be tagged when it is redistributed (or imported) into EIGRP, allowing it to be blocked when redistributing (or exporting) routes from EIGRP into the originating external routing protocol. A route map can be used to set and/or filter the administrative tag on a route. Route maps and Access Control Lists (ACLs) have many features in common. A route map can be used to define which of the routes from a particular routing protocol (source protocol) are allowed to be redistributed into another routing protocol (target protocol) **[CCIESOLK03] [CISCASA8.5] [CISCID49111]**.

6.16 EIGRP ROUTE SUMMARIZATION

Route summarization, also called route aggregation, is a method used to represent multiple networks (with contiguous IP addresses) as a single summary address that can be advertised to other routers (see Chapter 7). This is often use in networks with multiple subnets with contiguous addresses, and where there is the need to reduce the number of routes that a router must maintain in its Routing Table, and/or advertise in routing updates. EIGRP supports two forms of summarization; automatic (or auto-) summarization and manual summarization **[CISCID16406]**.

6.16.1 AUTO-SUMMARIZATION

With auto-summarization, EIGRP will summarize routes to their classful address boundaries (Class A, B, or C) automatically when sending routing updates across network boundaries. That is, each time EIGRP uses auto-summarization and sends routing updates cross the border between two different major networks.

For example, let us assume an EIGRP router is connected to the networks 172.18.1.0/24 and 10.3.1.0/24. When the router sends updates from the network 10.3.1.0/24 into the network 172.18.1.0, it will send them with the address auto-summarized to 10.0.0.0/8, which is the classful address boundary or mask. The default network mask for the Class A address 10.x.x.x is /8. Updates from 172.18.1.0/24 will summarize the address as 172.18.0.0/16 (the Class B network address with mask /16).

Auto-summarization is enabled by default, but is generally turned off as recommended by Cisco (common best practice). The auto-summarization feature may be useful in some case, but it can also cause problems when applied to networks with discontiguous addresses. For example, when two or more separate subnetworks within the same classful address block are connected together, the EIGRP auto-summary route can end up being the same. This makes it problematic to route packets in an internetwork that consists of different subnetworks within the same classful address block, because the auto-route summaries are not distinct.

Let us assume that a router RA has a directly attached subnet 10.4.11.0/24 that it advertises to a neighbor router RB. With auto-summarization, RA will summarize the subnet 10.4.11.0/24 as the classful address 10.0.0.0/8 to RB. Because RB has learned a route to the classful network 10.0.0.0/8 from RA, it will forward all packets destined for any IP address within the Class A address range 10.0.0.0 to 10.255.255.155 to RA, which can cause problems if there are other networks elsewhere with this address range. This is one among several other reasons why network administrators usually turned off the auto-summarization feature.

EIGRP was developed from IGRP which uses classful addresses, and the auto-summarization features stems from IGRP's legacy classful addressing feature. So, instead of advertising a specific route on a variable length classless address boundary, EIGRP will only advertise the route as a Class A, B, or C network. EIGRP auto-summarization creates the most optimal summary routes but these routes tend to be "over-summarized" routes. EIGRP will auto-summarize an external route only if one of its component networks is an internal route **[CISCID16406]**.

6.16.2 Manual Summarization

Using manual summarization in EIGRP, a network administrator can summarize internal and external routes on virtually any bit boundary in the IP address. For example, the networks 192.3.1.0/24, 192.3.2.0/24, and 192.3.3.0/24 can be summarized into the CIDR block 192.3.0.0/22. Manual summarization can be used to represent multiple routes as a single route, and can be performed at any router in a network. It is configurable on a per-interface basis on a router.

6.17 EIGRP AUTHENTICATION

An EIGRP router can be configured with neighbor router authentication (also referred to as route authentication) to prevent it from receiving fraudulent routing information from neighbors **[CISCTEAPA06]**. A router can be configured to participate in the exchange of routing updates with neighbors based on predefined passwords. This section describes the types of authentication methods used by EIGRP routers.

The default configuration of authentication in a Cisco EIGRP router is null authentication, where a router does not authenticate routing protocol packets exchanged with its neighbors. However, when a router interface is configured to perform neighbor authentication, it will authenticate all EIGRP routing update packets it receives from the neighbor. Any two routers involved in neighbor authentication accomplish this by using an authentication key (or password) that is known to both the sender and receiver.

EIGRP routers support two types of authentication:

- Plaintext Authentication (also known as Simple Password Authentication) – This form of authentication is also supported by RIPv2, OSPF, and IS-IS.
- MD5 Authentication – MD5 authentication is a type of cryptographic authentication, also supported by RIPv2, OSPF, IS-IS, and BGP.

Both types of authentication send authentication data in similar fashion, except that routers using MD5 authentication send a message digest instead of the shared secret authentication key (secret password) itself. The sender creates a message digest using the secret authentication key (and in some protocols, possibly, a Key ID) and the protocol data to be transmitted, but does not send the secret authentication key itself. This prevents an attacker from reading the secret key while it is in transit to the

Enhanced Interior Gateway Routing Protocol (EIGRP)

receiver. On the other hand, a router using simple password authentication sends the authenticating key (i.e., a plaintext password) itself along with the protocol data over the transmission medium to the receiver.

6.17.1 Simple Password Authentication

With simple password authentication, both router interfaces participating in the exchange of routing updates are configured with the same authentication key (a simple password that is also sent in plaintext to the receiver). Simple password authentication is vulnerable to passive attacks where an attacker with a protocol analyzer can easily read the password on the transmission medium. It is not recommended for a network that requires stronger security because it is highly vulnerable to attacks. The simple password authentication is primarily used to prevent unauthorized routers from making accidental routing changes to the network. MD5 authentication or better is recommended when stronger security is required.

6.17.2 MD5 Authentication

In MD5 authentication, a shared secret authentication key (or secret password), and a Key ID are configured on each router (interface) participating in the exchange of routing updates. The sending router applies an authentication algorithm on the routing protocol packet, the shared secret key (and possibly, the key ID) to generate a message digest (also called a hash). The router then appends the message digest to the routing protocol packet to be transmitted to the receiver.

Unlike the simple password authentication, the shared secret authentication key is not transmitted over the transmission medium to the receiver. The router sends the message digest instead of the shared authentication key, which prevents an attacker

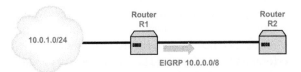

In EIGRP auto summarization, routes are summarized to classful address at network boundaries in the routing updates.

Router R1 has a locally connected network 10.0.1.0/24 that is advertised to Router R2. With auto summarization, Router R1 summarizes the network 10.0.1.0/24 before sending the route to R2. R1 sends the classful route 10.0.0.0/8 to R2 instead of the more specific 10.0.1.0/24 route.

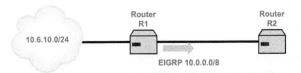

Router R1 has a directly connected network **10.6.10.0/24** that it advertises to R2. With auto-summarization, R1 will summarize the network **10.6.10.0/24** and send the classful route **10.0.0.0/8** to R2.

FIGURE 6.16 EIGRP Automatic Summarization

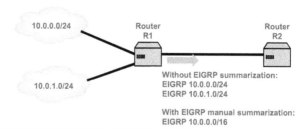

- EIGRP manual summarization can be configured on a per-interface basis. A single route can be used to represent multiple routes, which reduces the size of routing tables in a network.
- Router R1 has two directly connected networks: 10.0.0.0/24 and 10.0.1.0/24. EIGRP advertises these networks as two separate routes.
- However, R1 can be configured to advertise only one summary route for both subnets, which helps reduce R2's routing table.
- R1 can send only one route 10.0.0.0/16 to reach both networks to R2. R2 can use only one route to reach both networks on R1.
- NOTE: In this example, this summary route included the two networks on R1, but also some other addresses that are not in these networks. The range of the summarized addresses is 10.0.0.0 – 10.0.255.255, so R2 thinks that R1 has the routes for all addresses inside that range. That could cause some problems if these addresses exist somewhere else in the network.

FIGURE 6.17 EIGRP Manual Summarization

from eavesdropping on the medium, and learning the key during transmission. The message digest carried in each EIGRP packet also prevents unapproved sources from introducing unauthorized, malicious or false routing messages into the network.

With MD5 authentication, a shared secret authentication key and a Key ID is configured on both the sending and receiving router. Each authentication key is associated with its own Key ID which is stored locally at each router. The authentication algorithm and the MD5 authentication key in use at any given time, is uniquely identified by a combination of the Key ID, and the router interface over which the routing protocol packets are transmitted.

The keys used by an EIGRP router can be managed using the concept of Key Chains where each Key Chain has a number of different keys. A time interval (called a lifetime) which specifies the time interval during which a particular key will be active and valid, can be specified for each key defined within a Key Chain. This allows the routers to perform transition from one key to another (key rollover) while sending EIGRP packets. To configure EIGRP MD5 authentication, a Key ID (number), secret authentication key (or secret password), and a lifetime has to be specified.

An authentication key cannot be used outside its lifetime, that is, outside the time periods during which it is activated. To configure MD5 authentication, the Key ID and secret authentication key have to match on both the sending and receiving routers. However, the names of the Key Chain do not have to be the same on the two routers.

Best practice recommends that for a given Key Chain, the lifetimes of keys should overlap to avoid any period of time during which no valid key is available for use. If the configuration produces a time period during which no key is activated, the routers cannot perform neighbor authentication, and therefore cannot exchange routing updates. The routers also need to have a common notion of the time-of-day (via, e.g., clock synchronization using Network Time Protocol [NTP]) to allow all participating routers to rotate through their shared keys in synchronized manner. This allows all routers use the same shared secret key at the same time.

Enhanced Interior Gateway Routing Protocol (EIGRP)

During the lifetime of a given key, neighbor routers can send routing update packets with that activated key. A router will send only one authentication packet using one key at any given time, regardless of how many valid keys exist. The router will examine the Key ID (number) associated with each key in the Key Chain from lowest to highest number, and will use the first valid corresponding key it encounters [CISCTEAPA06]. The first key is determined by the Key ID, while its validity period is defined by the specified lifetime.

To configure a lifetime for a key (a period during which it is valid), an *accept-lifetime* and *send-lifetime* parameters need to be configured. The *accept-lifetime* specifies the time period during which the receiver will accept the key for use on received packets. The *send-lifetime* specifies the time period during which the sender will use the key for sending packets. Both the *accept-lifetime* and *send-lifetime* are defined using parameters that consist of a *start-time*, *end-time*, and length of time (in seconds) that a particular authentication key is valid for sending/receiving packets.

6.18 HIGH-LEVEL EIGRP ROUTER ARCHITECTURE, PROCESSES, AND DATABASES

This section focuses on the various processes and components involved in the development of the EIGRP databases. To start with, we try to depict diagrammatically, how the various processes and databases discussed above are related. To do this, we use Figure 6.18 to depict, generically, and as a high-level block diagram, the various processes and databases used in EIGRP. This diagram (like similar diagrams used

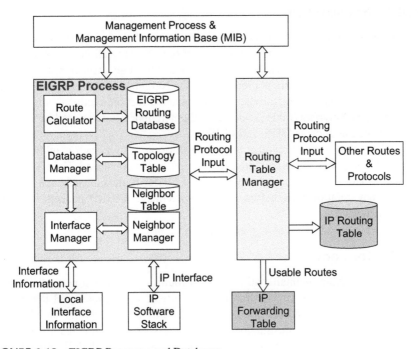

FIGURE 6.18 EIGRP Processes and Databases

else in this book) only suggest a high-level relationship between the various protocol components and not an example implementation.

The EIGRP Protocol Process component supports the core EIGRP protocol functions which **include**:

- Management of the router interfaces and HELLO protocol
- Transmission and reception of EIGRP packets which are directly encapsulated in IP
- Management of neighbors, building adjacencies and the Neighbor Table
- Maintenance of the Topology Table
- Parsing EIGRP packets and passing the new information received to DUAL
- Calculation of best EIGRP routes using DUAL and maintenance of the EIGRP Routing Database
- Distribution of best next hop routes (Successors and if required, Feasible Successors) to the Routing Table
- Implementing policies to control the redistribution of routes between EIGRP and other routing protocols
- Carries out EIGRP packet verification and authentication

As discussed earlier, an EIGRP router stores network topology and routing information in three main databases: Neighbor Table, Topology Table, and Routing Table. The design and contents of these databases are described in greater detail below. The discussion goes beyond those presented in sections above.

6.18.1 Neighbor Table

An EIGRP router uses the Neighbor Table to maintain a list of routers that are directly connected to it and have formed an adjacency (neighbor) relationship. Routers that are not directly connected to the router but rather through another router, are not considered neighbors and are not recorded in the Neighbor Table. The EIGRP router uses this table to know which routers it can have a direct bidirectional communication with. The Neighbor Table is similar to the neighbor database used in link-state routing protocols such as OSPF and IS-IS.

The Neighbor Table maintains important information for each neighbor router such as the neighbor's address, interface on which it is connected, Hold Time, and other information relevant to the neighbor. Two EIGRP routers must form a neighbor or adjacency relationships before they can exchange EIGRP packets including routing updates.

6.18.2 Topology Table

The Topology Table maintains all advertised routes to all known network destinations, plus the neighbor routers that have advertised these routes and the routing metrics to those destinations. This table maintains all Successors and Feasible Successors to all network destinations. Simply, the Topology Table maintains routes that an EIGRP router has learned from its neighbor as stored in their Routing Tables.

An EIGRP router will mark destinations in the Topology Table as being in the PASSIVE or ACTIVE state. A PASSIVE state indicates that the EIGRP router has determined a route for the particular destination and has finished route computations. ACTIVE state indicates that the EIGRP router is still trying to compute the best route for the destination. An EIGRP router never uses routes in the Topology Table for actual traffic forwarding until they are installed in the Routing Table. Also, the router will not insert routes in the Topology Table into the Routing Table if those destinations have been marked as ACTIVE, have only a Feasible Successor (with no Successor determined yet), or the routes have a higher Administrative Distance than another available equivalent route.

6.18.3 Routing Table

Each router in a network performs path computations and selects the best paths to reach each known destination and installs this in the Routing Table. An EIGRP router will first examine its Topology Table and determine the best route among all available routes to each network destination. The router does so by comparing the Feasible Distances to reach each destination, and then select the route with the lowest Feasible Distance (Successor route) and install it in its Routing Table. The Feasible Distance for the selected best route is entered in the Routing Table as the EIGRP routing metric to reach that destination.

The router installs only the IP address of the best neighbor router (also referred to as the next-hop router) along with the outbound interface (and possibly, the routing metric) to reach the network destination. The router does not install any information about the subsequent routers (after the next hop router) along the path in the Routing Table. Instead, the router relies on the next-hop router called the Successor to route traffic toward the network destination:

- An EIGRP router selects the best routes to all network destinations from the EIGRP Topology Table and places them in the Routing Table. The EIGRP router installs Successor routes in the Routing Table.
- The router inserts the relevant information required for packet forwarding from the Topology Table into the Routing Table which can then be used for traffic forwarding. The Routing Table contains a subset of the information maintained in the EIGRP Topology Table.
- The Topology Table contains more detailed routing information to each network destination, backup routes, routing metrics, plus other important information required by DUAL for route computations.

A router will select Successors and Feasible Successors from the Topology Table (when unequal-cost load balancing is used) and stored these in the Routing Table. The (combined) IP Routing Table (Figure 6.18) maintains all best routes selected from the EIGRP Topology Table and those provided by other routing protocol processes. *In Figure 6.18, the EIGRP Routing Table (or Database) and the IP routing table are simply logical components, and an EIGRP router may choose to maintain these as one integrated table and not as separate tables.*

6.18.4 Determining Successors and Feasible Successors

As discussed earlier, an EIGRP router uses DUAL to track all routes advertised by its neighbors, and enter them in their Topology Tables. The router then performs route computations on all routes available to the network destinations, and selects one lowest-cost and loop-free route to each destination, and installs these into the Routing Table. The EIGRP router will compare all Feasible Distances of all routes to reach a network destination in its Topology Table, and then install the lowest-cost route in its Routing Table (the Successor route).

EIGRP routers (via DUAL) will compute and select loop-free primary, and if available, backup routes to a network destination. When a primary route fails, the router will immediately use any available backup route which results in fast network convergence. EIGRP routers will maintains a Successor (primary route) and Feasible Successor(s) (backup route(s)) to each network destination and install these in its Topology Table (with up to six of these per network destination). The router will install the primary route (via the Successor) in its Routing Table.

Recall that a Successor is the best route with the lowest routing metric to reach a particular network destination. After an EIGRP router selects the best route to a destination, it will insert in its Routing Table, the destination network, the associated routing metric, outbound interface to reach the Successor or next-hop, and the IP address of that next-hop router. If the Topology Table contains multiple routes with equal-cost Feasible Distance to a network destination, all these Successors (maximum of 6) for each network destination will be inserted into the Routing Table.

For a neighbor to qualified as a Feasible Successor to a network destination, its Reported Distance must be less than (and not equal to) the Feasible Distance through the Successor to that destination. This ensures that the route through the Feasible Successor is loop-free, and that neighbor will never route packets to the destination through the local router, because doing so would result in a routing loop. This is because, when a neighbor's Reported Distance to a network destination is greater than or equal to the Feasible Distance of the local router to that destination, the neighbor could route packets to the destination via the local router.

It has been proven in **[GARCIALA89]** and **[GARCIALA93]** that every neighbor that satisfies the relationship, "Reported Distance is less than Feasible Distance", for a particular network destination has a loop-free route to that destination. This constitutes the Feasibility Condition required for loop-free routing using diffusing computations as in EIGRP DUAL.

- This feasibility check can be explained simply as follows: If a neighbor router advertises a distance (i.e., Reported Distance) to a destination that is strictly smaller than the Feasible Distance of the current router, then this neighbor lies on route that is not part of a routing loop to this destination.

A Feasible Successor is the best alternative loop-free route (i.e., backup route) to a network destination. This alternate route is not the least-cost route to that destination and not selected as the primary route to forward packets, and as a result is not installed in the EIGRP Routing Database. An EIGRP selects Successors and Feasible

Successors to a destination are at the same time during the DUAL computation process. The router will maintain Feasible Successor(s) in its Topology Table, as well as in the Routing Table only when it is performing unequal-cost load balancing.

Feasible Successors play an important role in EIGRP and allow an EIGRP router to recover and use alternative routes immediately upon experiencing network topology failures. This feature allows EIGRP routers to reduce the number of DUAL computations required to find new routes to network destinations and therefore improve network convergence. It is important to note that an EGRP router maintains Feasible Successor routes only in the Topology Table when unequal-cost load balancing is not used.

6.18.5 Populating and Maintaining the Neighbor Table

The Neighbor Table maintains key information about the neighbors an EIGRP router has established an adjacency relationship with. This table stores the IP address of the neighbor router, the connecting interface on which HELLO packets from the neighbor are received, as well as, other information pertinent to neighbor formation as described below, column-wise [CISCID16406] [CISCTEAPA06]:

- **H**: This is a number used internally by the Cisco IOS to lists the order in which the neighbors are learned, with the starting number being 0.
- **Address**: This is the IP address of the neighbor seen in the source IP address field of IP packets received from the neighbor. The local router uses this address when sending a unicast packet to the neighbor.
- **Interface**: This is the router interface over which the neighbor was learned (and reachable) and HELLO packets are sent.
- **Hold**: This value should not be greater than the Hold Time advertised by the neighbor and not less than the Hold Time minus the Hello Time. If the timer value reduces to 0, and the router has not received any EIGRP packets (including HELLO packets), the neighbor is declared down (invalid). This value lies between 15 and 10 seconds, for a Hold Time of 15 seconds and Hello Time of 5 seconds. When the local router receives an EIGRP message from the neighbor, it resets the Hold Timer.
- **Uptime**: This is a timer that tracks how long the neighbor relationship or adjacency has been up (in hours, minutes, and seconds). It is the elapsed time since the router established an adjacency with the neighbor.
- **SRTT**: This is the Smooth Round-Trip Time (in milliseconds), which is the time between sending an EIGRP packet and receiving an acknowledgment for it. It is the average time it takes for the router to send an EIGRP packet to the neighbor and for an acknowledgment to be received for that packet.
- **RTO**: This is the Round-Trip Timeout (in milliseconds), which is the length of time the router has to wait before retransmitting an EIGRP packet (from the retransmission queue) if an acknowledgement is not received for it. The SRTT timer is used to determine the RTO (i.e., the retransmit interval).
- **Q Cnt**: This is the output queue count representing the number of EIGRP packets (QUERY, REPLY, and UPDATE) that the EIGRP software process has

queued and waiting to be sent, or has sent but not yet acknowledged. A value equal to 0 indicates that there are no EIGRP packets in the queue. The local router uses the output queue to store EIGRP packet descriptors corresponding to the EIGRP messages that have been sent out but waiting for the receiver to acknowledge them. The router removes packet descriptors from the output queue only when acknowledged by appropriate EIGRP HELLO messages.
- **Seq Num**: This is the Sequence Number of the last EIGRP packet (sent via reliable transmission (e.g., QUERY, UPDATE, and REPLY) that was received from the neighbor. EIGRP packets, such as HELLO and ACK packets, are sent via unreliable transmission and do not carry Sequence Numbers. The Sequence Number is to ensure that reliable EIGRP packets received from the neighbor are in the right order or sequence.

As discussed earlier, EIGRP routers exchange HELLO, ACK, QUERY, REPLY, UPDATE, and REQUEST packets. The time a router has to wait for an ACK packet for a reliable EIGRP packet (e.g. QUERY, UPDATE, and REPLY) before changing from multicast to unicast transmission mode is specified by a Multicast Flow Timer. The SRTT is the average time between the moment the router transmits an EIGRP packet to a neighbor, and receives a corresponding acknowledgment. The RTO specifies the time the router has to wait before retransmitting a unicast EIGRP packet that was not acknowledged. The router will calculate both the Multicast Flow Timer and RTO for each neighbor from the SRTT. EIGRP uses methods and formulas that are proprietary for calculating the SRTT, RTO, and Multicast Flow Timer values.

6.18.5.1 Understanding the SRTT, RTO, and Q Cnt Parameters

An EIGRP routers use the H variable to record the order in which neighbors are learned. In addition to the other variables, each router will record the Hold Time advertised by the neighbor, the SRTT and the Uptime (which is the time since it added the neighbor to the Neighbor Table). A router via RTP maintains a retransmission list for each neighbor to ensure that ongoing communication between neighbor routers is maintained. The router uses this list to track all the reliable EIGRP packets that it has sent but have not been acknowledged by the receiver(s) within the RTO.

When a router sends an EIGRP QUERY, REPLY, or UPDATE packet, it will queue a copy of that packet. If the RTO expires before the router receives an ACK packet for a packet or does not receive one at all, it will send another copy of the queued packet. The router will transmit another copy of the reliable EIGRP packet every time the RTO expires, up to a maximum of 16, or until the Hold Time of the neighbor expires which will cause it to terminate the neighbor relationship **[CISCTEAPA06]**.

An EIGRP router has to wait for the Hold Time specified by the neighbor to expire (default time is 180 seconds for low-speed links and 15 seconds for high-speed links) before declaring that the EIGRP neighbor relationship or adjacency is down. When the router determines that the neighbor is down and it cannot reestablish the adjacency, it will delete all destinations that are reachable through that neighbor from its Routing Table. The router through DUAL will then attempt to find alternative routes to the affected destination as the network reconverges.

Enhanced Interior Gateway Routing Protocol (EIGRP) 203

For neighbors that are slow to respond to reliable multicast EIGRP packets, and have not unacknowledged these multicast packets when the RTO timer expires, the router will retransmit these packets as unicast packets. This allows the router to proceed with the reliable multicast transmission without delaying communications with other neighbors, to ensure faster network convergence especially in networks with variable link speeds.

The router uses the Multicast Flow Timer to determine how long it has to wait for an ACK packet to a multicast packet before switching from multicast to unicast transmission mode. The router uses the RTO to determine how long it has to wait before retransmitting a unicast packet. The router adjusts the RTO dynamically over time (with maximum value of 5,000 milliseconds or 5 seconds) based on the SRTT.

The router uses the Q Cnt variable to keep track of the number of enqueued packets. The router will also record the Sequence Number of the last QUERY, REPLY, or UPDATE packet received from the neighbor in its Neighbor Table. The router (via RTP) uses the Sequence Numbers to ensure that reliable EIGRP packets received from the neighbor are in the correct order. The router will record the last Sequence Number received from the neighbor so that it can detect out of order packets. The router will also use a transmission list maintained on a per neighbor basis to queue packets for possible retransmission when they have not been acknowledged on time or not acknowledged at all.

6.18.6 Populating and Maintaining the Routing Table

EIGRP routers populate their Routing Tables using the following procedures:

1. Each EIGRP router advertises its Routing Table (containing all routes it has learned) to all neighbors that are registered in its Neighbor Table.
2. Each EIGRP router receives and stores the Routing Tables sent by their neighbors in its Topology Table.
3. Each EIGRP router examines its local Topology Table to determine the Successor and Feasible Successor to each known network destination.
4. Each EIGRP router selects the best route to a network destination from its Topology Table and inserts this into its Routing Table

If a router does not receive an EIGRP packet before the Hold Time expires, it will delete the associated neighbor adjacency, and remove all Topology Table entries learned from that neighbor. This action is similar to what happens when a neighbor sends a routing update stating that some routes passing through it are unreachable. The router will also send out routing updates indicating that routes through that neighbor are unreachable. If the unreachable neighbor is a Successor to some network destinations, the router will remove these networks from its Routing Table, and alternative routes (Feasible Successor), if available, will be recomputed using DUAL. The network is able to quickly reconverge if an alternative route or Feasible Successor is available. The following important points are worth noting:

- As discussed above, a destination is considered to be in the PASSIVE state when the router is not performing recomputations to find a best route to that

destination. On the other hand, a destination is considered to be in an ACTIVE state when the router is performing recomputations to find a new Successor to the destination when the existing Successor is unreachable and has been declared as invalid.
- When a network change occurs (e.g., a link or a neighbor fails or is down), some destinations may become unreachable. An EIGRP router will detect these changes and will attempt to find new routes to the affected destinations. The router will remove the old routes that are no longer available from its Routing Table. Unlike routers running other distance-vector routing protocols such as RIP, the EIGRP router will not transmit all the information in its Routing Table when a network change occurs, but will only transmit the incremental changes that it has seen since the Routing Table was last updated.
- Also, the EIGRP router does not transmit its entire Routing Table periodically, but only the actual change in the Routing Table data when a change occurs. This behavior is similar to what link-state routing protocols do, thus making EIGRP more of a hybrid routing protocol shares some features distance-vector routing protocols and link-state routing protocols.

6.18.7 Handling the Loss of a Route to a Network Destination

When an EGRP router loses a route to a network destination in its Routing Table, it will first check its Topology Table to see if there is a Feasible Successor available. If one exists, that destination will not be placed in the ACTIVE state, and the best Feasible Successor available (if there are multiple ones) will be elevated to Successor and inserted into the Routing Table. The availability of a Feasible Successor means that route will be used immediately in the Routing Table without DUAL having to initiate route recomputations.

If no Feasible Successor exists in the Topology Table, the router will place the destination in the ACTIVE state. The router will determine a new Successor to the destination through the DUAL route recomputation process, and which it will begin by transmitting QUERY packets to all of its neighbors. The amount of time it takes to complete the route recomputation to discover a Successor affects the overall network convergence time. Finding a Feasible Successor in the Topology Table surely speeds up network recovergence when the router loses a route in its Routing Table.

6.18.8 Handling Queries for a Route to a Network Destination

An EIGRP router would perform the following tasks when it receives a QUERY packet from a neighbor:

1. If a router receives a QUERY packet from a neighbor and its Topology Table does not contain an entry for the queried destination, or there is no current Successor to this destination (and the destination is in ACTIVE state), it will immediately reply indicating that the destination is unreachable (i.e., it has no path leading to this destination).

2. If a router receives a QUERY packet from a neighbor and its Topology Table lists the neighbor as the Successor for the queried destination, it will attempt to find new a Successor. If the Topology table contains one or more Feasible Successors, the router will install the Feasible Successor with the lowest metric (new Successor) into its Routing Table, and then immediately reply with the information about the selected Feasible Successor to the neighbor that sent the QUERY packet.
3. If a router receives a QUERY packet from a neighbor and its Topology Table lists the neighbor as the Successor for the queried destination and contains no Feasible Successor, it will propagate the QUERY to all of its neighbors (except the neighbor that originated the QUERY) seeking an alternative Successor. The router will not send a response to the querying neighbor until it has received a reply for each QUERY packet it has propagated to its own neighbors.
4. If a router receives a QUERY packet from a neighbor that is not the Successor for the queried destination, it will immediately reply with the information of its Successor to the neighbor.

To each neighbor a router sends a QUERY packet, it will set a Reply Status Flag (r) to keep track of the outstanding replies to the QUERY packets it has sent out. The router will then complete the DUAL route computation when it has received a reply for every QUERY packet it has sent out.

When a router starts the DUAL computation, it will set an ACTIVE timer for 180 seconds. If the router does not receive a reply for each outstanding QUERY before the ACTIVE timer expires, the affected network destination will be placed in the SIA state. The router will reset its neighbor or adjacency relationship with the neighbor that failed to reply, which may then cause the router to delete the route through that neighbor to the destination, or to place all destinations known through that neighbor in the ACTIVE state. The router will follow this by re-exchanging routing information with that neighbor.

In a stable, well-designed network, when a link or a neighbor router fails, other neighbor routers should easily detect this through the expiry of the Hold Timer, instead of the ACTIVE timer. Such a failure normally should not cause network destinations to be declared in SIA state. SIA may occur due to some routers being overloaded in a network, heavily congested links, presence of low bandwidth links in mainly high bandwidth network, or looping of QUERY packets sent by routers seeking new routes to network destinations. In the worst case, SIA may lead to routers deleting adjacencies with established neighbors, and flushing of valid routes to network destinations which could end up affecting the overall stability of the network.

6.18.9 Note on Route States, Successors, and Feasible Successors

An EIGRP router marks a destination in the Topology Table as either as PASSIVE or ACTIVE. A destination is entered into the PASSIVE state when the router has finished identifying a Successor for that destination. The router changes the destination to ACTIVE state when it determines that the current Successor no longer meets the Feasibility Condition and there is no Feasible Successor (i.e., no backup route(s)) identified, or available for that destination.

The router changes the destination back from ACTIVE to PASSIVE state when it has received replies to all QUERY messages it has sent to its neighbors, indicating the availability of Feasible Successor. If the router determines that an existing Successor no longer satisfies the Feasibility Condition but there is at least one Feasible Successor available, it will elevate the Feasible Successor with the lowest total path cost (i.e., the sum of the Reported Distance as advertised by the Feasible Successor and the link cost to this neighbor) to become the new Successor, and the destination will remain in the PASSIVE state.

6.19 SUMMARY OF EIGRP FEATURES

EIGRP uses an enhanced distance-vector algorithm referred to as the Diffused Update Algorithm (DUAL) to calculate the shortest path to each network destination. EIGRP advertises network information to its neighbors using a number of enhancements that have advantages over other distance-vector algorithms:

- EIGRP uses HELLO messages and forms neighbor router relationships similar to a link-state routing protocol. The HELLO messages also serve as keepalive mechanism and are exchanged periodically to maintain adjacencies with neighbor routers.
- EIGRP does not forward routing updates to broadcast address 255.255.255.255 but, instead to the multicast address 224.0.0.10 when multicast transmission is used.
- EIGRP uses the proprietary Cisco RTP to transport routing updates to other routers.
- Like RIPv2, EIGRP supports VLSM and CIDR.
- EIGRP uses a composite routing metric (that can be derived from bandwidth, delay, path reliability, path load, and MTU size) for path calculations, instead of a single routing metric of hop count as in RIP.
- An EIGRP router advertises routing updates only when there is a change in the network (i.e., when changes in a path or metric for a route occur). The router does not transmit full Routing Table updates in a periodic fashion like RIP – but instead sends triggered partial updates where only the Routing Table changes are propagated. Routing updates are only sent to the neighbors that need them, which results in less bandwidth usage, especially in large networks with many routes.
- EIGRP uses a Feasibility Condition to select loop-free paths to network destinations.
- EIGRP uses a Topology Table to maintain all the routes received from neighbor routers. DUAL also maintains primary routes (Successors) and backup routes (Feasible Successors), when available, in the Topology Table. If a primary route (Successor) in the Routing Table fails, the router can use any available backup route (Feasible Successor). The switchover to the backup route is

immediate, and EIGRP does not coordinate with other routers to perform route recomputation.
- An EIGRP router has the option to load balance traffic to a destination network across multiple equal or unequal cost paths.

These enhancements allow EIGRP to have rapid convergence time when changes in the network topology occur. Because EIGRP has characteristics of both distance-vector routing and link-state routing protocols (e.g., forming adjacencies with neighbor routers and using multiple route metrics such as bandwidth and delay for best path calculations instead of hop count), it is sometimes referred to as a hybrid routing protocol.

REVIEW QUESTIONS

1. How are EIGRP messages sent over IP? Directly or over a Transport Layer Protocol?
2. How are EIGRP messages transmitted? Broadcast or multicast? Elaborate on your answer.
3. What is purpose of the DUAL in EIGRP?
4. What is the purpose of the RTP in EIGRP?
5. Explain the main functions of the Neighbor Table, Topology Table, and Routing Table in EIGRP.
6. Explain the difference between the Feasible Distance and Reported Distance (also known as the Advertised Distance) in EIGRP.
7. Explain the difference between a Successor and Feasible Successor in EIGRP.
8. Explain the difference between a route in ACTIVE State and one in PASSIVE State in EIGRP.
9. What is the Feasibility Condition in EIGRP?
10. What is Stuck-In-Active (SIA) in EIGRP?
11. Explain briefly how EIGRP uses Split Horizon and Poison Reverse.
12. What is the difference between equal-cost load balancing and unequal-cost load balancing in EIGRP?
13. What is the *variance* parameter in EIGRP load balancing?
14. Why is EIGRP auto-summarization not a desirable feature, and why is it best to disable it?

REFERENCES

[CCIESOLK03]. Karl Solie and Leah Lynch, CCIE Practical Studies: Configuring Route-Maps and Policy-based Routing, Sample Chapter, Cisco Press, November 26, 2003.
[CISCASA8.5]. Cisco ASA Services Module CLI Configuration Guide, 8.5, Chapter: Defining Route Maps, November 17, 2013.
[CISCID8606]. Cisco Systems, "Redistributing Routing Protocols", Document ID: 8606, March 22, 2012.
[CISCID13669]. Cisco Systems, "Introduction to EIGRP", Document ID: 13669, August 10, 2005.

[CISCID16406]. Cisco Systems, "Enhanced Interior Gateway Routing Protocol", Document ID: 16406, September 5, 2017.

[CISCID49111]. Cisco Systems, "Route-Maps for IP Routing Protocol Redistribution Configuration", Document ID: 49111, August 10, 2005.

[CISCTEAPA06]. Diane Teare and Catherine Paquet, "Configuring the Enhanced Interior Gateway Routing Protocol", Chapter in *Building Scalable Cisco Internetworks (BSCI) (Authorized Self-Study Guide)*, 3rd Edition, December 26, 2006, Cisco Press.

[GARCIALA89]. J. J. Garcia-Luna-Aceves, "A Unified Approach to Loop-Free Routing Using Distance Vectors or Link States", *SIGCOMM '89, Symposium Proceedings on Communications Architectures & Protocols*, Volume 19, 1989, pp. 212–223.

[GARCIALA93]. J. J. Garcia-Luna-Aceves, "Loop-Free Routing using Diffusing Computations", Network Information Systems Center, SRI International, appeared in IEEE/ACM Transactions on Networking, Vol. 1, No. 1, 1993.

[MARTABE02]. Abe Martey, Johnson Liu, Zaheer Aziz, and Faraz Shamim, *Troubleshooting IP Routing Protocols*, Cisco Press, May 2002, ISBN: 1587050196.

[RFC7868]. D. Savage, J. Ng, S. Moore, D. Slice, P. Paluch, and R. White, "Cisco's Enhanced Interior Gateway Routing Protocol (EIGRP)", IETF RFC 7868, May 2016.

7 Network Path Control and Factors That Affect Routing Table Properties

7.1 INTRODUCTION

Traffic within routing domains (intradomain routing) and between routing domains (interdomain routing) flow according to the routes discovered by the routing protocols used. Thus, altering the behaviors of the routing protocols, and path attributes, can translate into changes in routes and traffic flow to network destinations. This chapter discusses the various mechanisms IP routers use for controlling routing in networks. Routers can use route aggregation, route redistribution, route filtering, policy-based routing, BGP path attribute manipulation, and others, to effect changes in routing protocol behaviors and routes to network destinations.

We start by describing the main considerations for employing multiple routing protocols in a network. We also discuss the tools available for path control, and performance issues related to network path control. An IP router can use these mechanisms to address networking issues such as the following:

- Determine the specific routes from a neighbor that need to be filtered (permit/deny).
- Specify which routes a router can advertise to its neighbor routers.
- Specify which network path control tools a router can use to accomplish certain desired routing behaviors.

In this chapter, we use the Cisco Internetwork Operating System (IOS), which is a family of network operating systems used on Cisco Systems routers and network switches, as the reference platform for all our discussions on network path control, and router configuration tools and commands. Some aspects of the discussion in this chapter assume some knowledge of OSPF, IS-IS, and BGP. Interested readers can refer to Volume 2 of this two-part book for detail descriptions of these routing protocols.

7.2 RUNNING MULTIPLE ROUTING PROTOCOLS

As discussed in Chapters 1 and 2, routing protocols enable IP routers to exchange the information needed to determine how network destinations can be reached. The routers then populate their IP Routing Table with this network reachability information. When a router runs multiple routing protocols, it scans the different Routing Tables belonging to these protocols, and selects the best path(s) to each network destination based on the Administrative Distance (or the Route Preference) values of the routing protocols (see Chapter 2).

7.2.1 Running Multiple Overlapping Routing Protocols

The type of routing protocol running in a network is one of the major factors that define how best paths to network destinations are determined. As discussed in Chapters 1 and 2, different routing protocols use different routing metrics, Administrative Distances, and have different convergence times, all of which can affect how best paths are selected. Running multiple routing protocols in the same network can result in different routing behaviors and different paths being selected.

Inefficient routing may result because the different routing protocols running will most likely provide inconsistent routes to the same network destination. It is not hard to see that network planning, network management/administration, and traffic engineering will become very difficult to perform when multiple overlapping routing protocols are running in the same network.

Even careful planning is not enough to interwork these multiple protocols in such a way that consistent routing results will be obtained. Even though tools such as the use of Administrative Distances can be used to determine the routes to be installed in the Routing Table, running overlapping routing protocols in the same network only to use the Administrative Distances to select routes does not make sense and is counterproductive. For these reasons among others, it is generally recommended to run each routing protocol in its own network space or domain, and then find ways to interwork the different routing domains.

7.2.2 Running Multiple Non-Overlapping Routing Protocols

Ideally, it is generally preferable to run a single routing protocol in a network, for simplicity in network design, and ease of management and administration. Unfortunately, because of company acquisitions or mergers, administrative constraints (different networks managed by different network administrators), preference for particular routing protocols by some network administrators, internetworking with company partners, or geographical constraints, this is not always possible, making networks with multiple routing protocols more common (Figure 7.1). Although, there are still a number of issues to be addressed, in this scenario, each routing protocol runs in its own network space or routing domain. The only requirements are mechanisms for interworking the different routing domains.

Also, as networks grow over time, they become more complex, and may have to support several different routing protocols because of evolution of network protocols and advances in network device technologies. Other reasons for running multiple routing protocols include mismatch between routing devices because they run different routing protocols, the need for multivendor device interoperability, the need for application-specific routing protocols (one protocol does not meet all needs), and migration from an older IGP to a new IGP (but with the need to keep the older IGP in the interim).

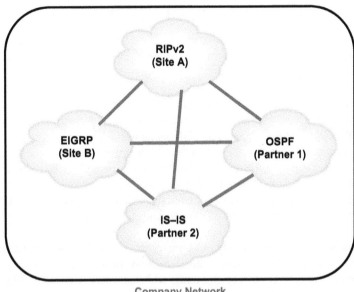

FIGURE 7.1 Running Multiple Routing Protocols

As discussed above, it is not advisable to run different routing protocols in the same network (i.e., overlapping routing protocols). Instead, it better to create distinct boundaries between networks (or routing domains) that use different routing protocols, and then use route redistribution to interwork these different routing protocols. In most networks that run more than one routing protocol, there is often the need to redistributed routing information between the protocols (see discussion below).

The different routing protocols in use today were not designed to interoperate seamlessly with one another. Each routing protocol collects routing information in a different way, and this information is very different from those collected by other routing protocols. Other than the types of information collected, each protocol reacts to network topology changes in its own way. Thus, to enable efficient internetworking of the different routing domains, proper exchange of routing information between the routing protocols is vital (i.e., proper inter-routing protocol exchange).

Route redistribution allows routes learned by one routing protocol to be injected (imported or advertised) into another routing protocol (e.g., EIGRP into OSPF or vice versa). The receiving routing protocol usually marks these redistributed routes as external (as in OSPF and EIGRP). IP routers usually regard external routes as less preferable than locally originated routes. Generally, network operators try to limit the number of routing protocols used in order to avoid excessive route redistribution. As discussed below, some of the problems that can arise when redistributing routes, include incompatible routing information, routing loops, route feedback, and inconsistent routing protocol convergence times. These issues are discussed in the "Route Redistribution" section below.

7.3 THE NEED FOR NETWORK PATH CONTROL TOOLS

To help us better understand the route filtering and path control tools discussed in this chapter, it is important to highlight the different elements involved in processing routing information in an IP router:

- Processing of routes that the IP router receives from other routers.
- Input routing policies that the IP router uses to filter received routes (and in BGP routers, to manipulate the BGP path attributes of a route).
- Best path selection algorithms that the IP router uses to determines the best path to each network destination. These best paths are installed in the IP Routing Table.
- Output routing policies that the IP router to filter routes (and in BGP routers, to manipulate BGP path attributes of outgoing routes) being sent to other IP routers.
- Processing of routes that the IP router sends to other routers.

In addition, in OSPF, route filtering of summary routes can be done at the distribution points, that is, at ABRs (Autonomous System Border Routers) and ASBRs (Autonomous System Boundary Routers), while in EIGRP, route filtering for all routes can be configured in any router. Figure 7.2 shows the flow of routing information and packets within an IP router.

7.3.1 What is a Routing Policy?

In a typical IP router, the flow and processing of information involves the following data types:

- **Routing Information**: This is the information a router learns about routes (to all known network destinations) and exchanges with its neighbors. This could be obtained via the directly connected networks on the router, static routing, and/or dynamic routing protocols.

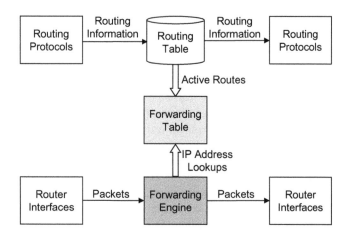

FIGURE 7.2 Flows of Routing Information and Packets

- **Transit Packets**: These are pieces of data that pass through a router as they travel from the data sources to their destinations.
- **Local Packets**: These are pieces of data that are originated by or destined for the router itself. Local packets usually contain routing information data for device configuration and network management protocols (such as Simple Network Management Protocol [SNMP], Secure Shell [SSH], and Telnet), and data for network control protocols (such as the Internet Control Message Protocol [ICMP] and Internet Group Management Protocol [IGMP]).

A routing policy is a collection of administrative and management rules that control the flow of routing information and end-user data in and out of a router. From routing information perspective, a routing policy defines conditions in an IP router for accepting, rejecting, and modifying the attributes of specific routes received in route advertisements (i.e., import policy), or conditions for extracting routes from the IP Routing Table and advertising them to neighbor routers (export policy) (see Figure 7.3). Particularly, a routing policy can define rules for controlling the exchange of routing information between routing protocols, or between routing protocols and the Routing Table in a router:

- Routing policies enable a network administrator to control (filter) which routes are imported into the Routing Table and which routes are exported from the Routing Table.
- Routing policies can be configured such that only some valid routes are received or advertised by a router. Routes discovered by another routing protocol may be imported into a routing protocol, and only routes that satisfy certain conditions will be accepted. The attributes of the routes that are accepted by a routing policy (as in BGP) may be modified to meet local policy requirements of the importing router.

The decision about which routes to accept from and/or advertise to neighbors has an impact on the traffic that crosses a network. Routing policies can be used to enforce

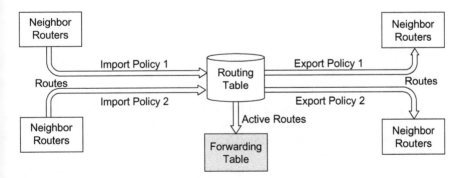

FIGURE 7.3 Importing and Exporting Routes

business agreements between two or more Internet Service Providers (ISPs) concerning the type and amount of traffic allowed to pass between them.

The following are examples where a network operator would want to define network-specific routing policies:

- **Limit the Importation of Routes into the Routing Table by a Routing Protocol**: If a router does not install certain routes in its Routing Table, these routes cannot be used to forward packets, and cannot be redistributed into other routing protocols. A routing policy can be defined specifying the criteria that a route must match, and the actions the router must perform if a match occurs. The import routing policy is applied when routes are imported into the Routing Table from the routing protocol.
- **Limit the Exportation of Routes Learned by a Routing Protocol**: A router can be configured to advertise only certain routes to its neighbors according to a defined outbound routing policy. The export routing policy is evaluated when the router exports routes from the Routing Table into a routing protocol.
- **Enable Route Redistribution between Routing Protocols**: A routing policy can be configured to import only certain routes learned by another routing protocol (a process called *route redistribution*).
- **Manipulate the Characteristics of Routes**: The network administrator may specify or modify certain route characteristics such as the Administrative Distance, routing metrics, BGP AS-Path attributes, or the BGP Communities attribute to satisfy some network routing behaviors. These route characteristics can be manipulated to control which routes are selected to network destinations.
- **Perform Per-Packet Load Balancing**: Load balancing is the process of scheduling traffic optimally over multiple paths. For example, a routing policy can be defined for per-destination load balancing. Per-destination load balancing uses multiple paths and allows a router to distribute packets based on the destination address of packets.
    - o The downside of per-destination load balancing, however, is that, packets for a given source–destination address pair are forced to take the same path and are always guaranteed to take that path even if the other paths are not busy and are available.
    - o Per-packet load balancing, on the other hand, does not require the use of a routing policy, and allows packets to be evenly distributed across multiple equal-cost paths. Per-packet load balancing allows a router to send successive packets over different paths without regard to individual user or hosts sessions. Typically, it uses the round-robin scheduling to determine which path each packet should be forwarded on. However, using per-packet load balancing to forward packets across multiple paths to a given destination, can cause out-of-sequence packets in a particular data flow. This can result in unsatisfactory performance for voice and video communications.
- **Enable Class of Service (CoS)-Based Packet Forwarding**: A routing policy can be configured at a router to go beyond the traditional "best-effort"

forwarding behavior that uses simple first-in-first-out (FIFO) queueing. A routing policy can be defined to allow the router to take into account the different forwarding requirements of different media (e.g., mapping traffic to input queues based on CoS, mapping traffic with different CoS values to different output queues, CoS-to-DSCP (DiffServ Codepoints) mapping, and IP Precedence-to-DSCP mapping).

It is important to note that the actual routing policies used in a router are influenced by the router's architecture (input queuing, output queuing, or input/output queuing capabilities), routing protocol implementation (some routing protocols are CoS-aware or support traffic engineering, while others are not), network architectures (some network support traffic engineering while others do not), the design and management practices of a particular network, contractual agreements with service providers, and so on.

7.3.2 Implementing Routing Policies

A routing policy can be implemented using a number of tools. For example, in most routers such as those using Cisco IOS, filtering tools such as route maps, distribute lists, or prefix lists can be used. These tools have associated with them configuration options and a number of policy actions that a router can take when certain conditions are met. Figure 7.4 provides a high-level description of the route filtering process in a typical IP router. Each route filtering tool has its own way of processing routes and these details will be described below for each tool.

Through route filtering, an IP router can choose which routes to receive from other IP routers and which routes to send to other IP routers. Route filtering is an integral part of a routing policy. Route filtering can be applied to routing updates traveling to and/or from a neighbor IP router. Using inbound route filtering, an IP router can screen out which routes it is willing to accept from neighbor routers. Similarly, using outbound route filtering, an IP router can specify which routes it will advertise to other routers.

IP routers also use route filtering to determine which routes can be passed from one routing protocol to the other (e.g., RIP to OSPF, EIGRP to BGP). As will be discussed below, IGP learned routes and static routes can be passed between IGPs, and between an IGP and BGP. Route filtering specifies exactly which routes can be passed from one routing protocol to the other.

As discussed above, the process of injecting routes from one routing protocol to another is referred to as *route redistribution*. As will be discussed in the "BGP Route Filtering and Path Attribute Manipulation" section below, routes that are accepted into BGP can have their BGP path attributes manipulated with the goal of influencing how routing in an internetwork of Autonomous Systems will be performed. BGP path attributes such as the AS-Path attribute, Multi-Exit Discriminator (MED) attribute, Local Preference attribute, Cisco Weight attribute and others can be manipulated to affect the BGP best path selection process.

Route filtering techniques are used to filter specific routes from being imported or exported, while on the other hand, packet access control lists (ACLs) are used to filter

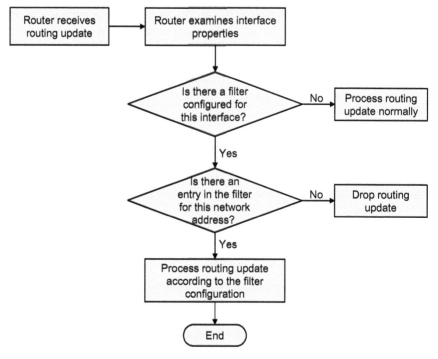

FIGURE 7.4 Route Filtering Process

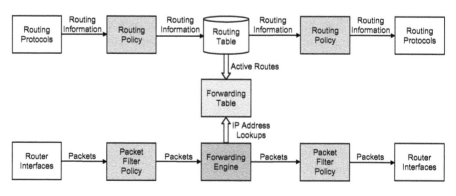

FIGURE 7.5 Policy Control Points

"data plane traffic" through the router as illustrated in Figures 7.5 and 7.6. Figure 7.6 shows the flow of routing information and user packets through an IP router. The route filters (route maps, distribute lists, or prefix lists) are implemented at the routing policy control points while the packet ACLs are implemented at the packet filter policy control points.

Network Path Control and Factors That Affect Routing Table Properties

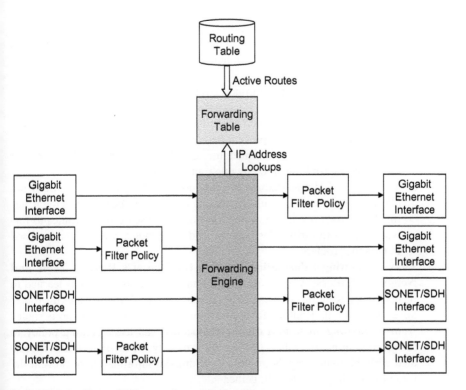

FIGURE 7.6 Firewall Filters to Control Packet Flow

As would be seen later in this chapter, route filters provide greater route filtering flexibility and can be configured for the following:

- Prevent specific routes from being received through a router interface.
- Control the advertisement of specific routes to other routers.
- Control the installation of specific routes in the local IP Routing Table of a router.

It is important to recognize that if route filters are not configured correctly or if the filters are applied to the wrong router interfaces, serious network performance issues may arise.

7.3.2.1 Routing Policy Control Points

A routing policy defines the conditions used by a router to match a route, and the actions the router performs on the route when a match occurs. Figure 7.5 shows the routing policy control points in a typical IP router. These control points are the points in the router at which routing information entering and exiting the Routing Table are filtered or their attributes manipulated. A routing policy control point can process the following types of routing information flow in the router:

- **Control of Routing Information Flow between the Routing Protocols and the Routing Table:** A router stores the routing information it receives in its

Routing Table, and subsequently advertises this to its neighbors. Inbound routing policies can be defined to control the flow of this information. The network administrator can define routing policies to control which routes get installed in the Routing Table.

  o For example, when a router imports routing information from a routing protocol into its Routing Table, it can use a routing policy to modify the route's Administrative Distance and/or routing metric, tag the route to identify it for later manipulation, or prevent the route from being installed in its Routing Table. A network administrator can also define routing policies for redistributing routes learned by one routing protocol into another.

  o The network administrator can configure a routing policy that does not allow routes associated with a particular customer to be placed in the Routing Table. As a result, the router will not use these customer routes to forward packets to various destinations, and these routes will not be advertised to neighbor routers.

- **Flow of Routing Information between the Router's Routing Table and Neighbor Routers:** A router sends information about routes it has learned to its neighbors. Routing policies can be defined to control which routes the router advertises from its Routing Table to neighbors. When a router exports routes from its Routing Table to a routing protocol, it can use a routing policy to assign routing metric values to the routes, modify the BGP attributes associated with the routes, tag the route with additional information, or prevent the route from being exported altogether.

Routing policies can be defined to change specific characteristics of a route (as in BGP attribute manipulation). The BGP attributes of routes can be modified to control which routes get selected as the active route to a destination. This can be used to effect changes to intra-autonomous system routing or inter-autonomous system routing. Routing policies can also be configured when implementing policy-based routing as discussed below.

7.3.2.2 Packet Filter Policy Control Points

Figures 7.5 and 7.6 show the packet filter policy control points at which data packets passing through the router can be controlled. These control points are used to filter data packets (including local packets) before they enter the router for destination address lookups, and/or after the exit the forwarding engine after IP Forwarding Table lookup. Packet filter policies can be defined to control the flow of data packets transiting the router as well as those originated by and/or destined for the router:

- **Control Flow of Packets Transiting the Router**: When a router receives a packet on an interface, it performs a destination address lookup in its Forwarding Table to determine which outbound interface and next hop IP address the packet should be forward to on its way to the destination. The router then forwards the packet through the appropriate interface toward its destination. Packet filter policies can be defined to control the flow of these data packet types.
- **Control Flow of Local Packets Originated by and Destined for the Router**: When a router receives a local packet (e.g., routing update, SNMP packet,

ICMP packet), it forwards it to the appropriate local process for further processing. The router can also originate packets destined for other network devices in the network. Usually, a local process running in the route processor (or routing engine) is responsible for handling these packets. Packet filter policies can be configured to control the flow of all these local packets.

Inbound packet filters can be defined to control packets that are received on a router interface. Outbound packet filters can also be configured to control packets that are sent out a router interface. Packet filter policies can also be used to protect a router from malicious traffic destined for the router itself, or passing through the router to a network destination. Packet filters can be configured to control local packets to a router to protect it from external attacks such as denial-of-service (DoS) attacks. These types of packet filters are also referred to as control plane packet filters. The packet filter policies can be defined to control the data packets the router can accept as local traffic, accept as transit traffic, and/or transmit on its interfaces to other devices.

7.3.3 Routing Policies and BGP Attribute Manipulation

A BGP route represents a unit of routing information that pairs a given network destination with a number of BGP path attributes to that destination. BGP uses a set of parameters called BGP path attributes to describe the characteristics of a route (i.e., a destination address prefix). The path attributes are carried as part of the UPDATE messages that BGP routers exchange, and describe path information associated with a network prefix (i.e., the Network Layer Reachability Information [NLRI] in the UPDATE message). The BGP best selection algorithm uses these path attributes to determine the best path to a network destination when multiple paths exist.

BGP routers use various path attributes in the BGP UPDATE messages they send to peers as a way to influence routing (such as BGP AS-Path, MED, Local Preference, and BGP Communities). We describe in the special "BGP Route Filtering and Path Attribute Manipulation" section below, how BGP routers filter routes and manipulate BGP path attribute to effect different routing behaviors. BGP route filtering can be used to filter and control network prefixes that are received from and advertised to BGP neighbors. Network prefixes that are exchanged between BGP peers can be filtered or controlled using inbound and outbound route filters that can match on BGP AS-Path attribute values, IP address prefixes (in the NLRI of BGP UPDATE messages), or any other path attributes of a BGP prefix such as BGP Communities.

7.4 WHAT IS POLICY-BASED ROUTING (PBR)?

Many organizations today demand the flexibility to implement in their networks, packet routing and forwarding decisions based on routing policies that go beyond the standard or normal IP routing and forwarding using the IP Routing Table. PBR has become a very useful tool for accomplishing this and is widely used by many organizations and service providers.

The standard or normal forwarding process in IP routers is to forward packets to their destination addresses based on the routing information maintained in the routers' IP Routing Tables. Network operators can use PBR with defined routing policies that selectively forward IP packets on network paths based on administrative policy decisions and a number of packet parameters (e.g., source IP address, destination IP address, IP protocol types, Transport Layer port numbers, quality-of-service [QoS] markings, or application types), rather than on the normal IP Routing Table maintained by the routing protocols. PBR provides routing instructions that override the normal routing behavior of the IP routers in the network.

PBR can be used as a path control tool to bypass the packet forwarding instructions provided by the IP Routing Table, and to define network paths based on user-defined policies (user configured routing information). PBR allows the IP routers to select routes based on these user-defined policies instead of using the IP Routing Table. When PBR is configured, the IP routers forward packets according to the configured rules. PBR does not change the destination of routed traffic. Instead, PBR affects the next-hop to which traffic is forwarded before it is sent to its final destination.

PBR provides an extremely powerful, flexible, yet simple tool that network operators can use to implement routing solutions in cases where contractual, geographical, legal, or political constraints demand that user traffic be routed through specific paths in a network or internetwork. PBR can be used to force traffic to specific paths and network destinations such as NAT devices, security appliances, and WAN bandwidth optimization devices. A network operator can use PBR to control path selection in a network, providing capabilities and benefits such as the following [CISCTEARDIA10]:

- **Source-Based Transit Service Provider Selection**: An organization can use PBR to route traffic based on where it originates (networks or set of users) through different ISP connections or ISPs to their destinations.
- **Path Selection Based on Ability to Guarantee QoS**: An organization can apply prioritization and/or QoS marking to differentiated traffic at the network edge so that it can be policy routed. The traffic can then be policy routed in the network core based on the QoS markings and other packet parameters such as source/destination IP addresses, TCP/UDP port numbers, application types, and so on. The edge devices set the QoS markings in the IP packet headers, and then PBR and priority queuing mechanisms is leveraged to prioritize and route traffic in the network backbone or core. PBR is used to policy route these packets in the network core to their destinations. This architecture improves network performance by eliminating the need to classify and mark traffic explicitly in the network backbone or core.
- **Path Selection Based on Cost Savings**: Using PBR, an organization can select which traffic to direct to a low-bandwidth, low-cost connection, and which to send to a higher-bandwidth, high-cost connection. Critical traffic such as video conferencing, and other real-time streaming traffic which are typically short-lived can be policy routed over the high-bandwidth, high-cost connection. Other non-critical, non-real-time traffic such as emails, and FTP sessions can be policy routed over a lower-bandwidth, low-cost connection.

- **Load Sharing across Multiple Paths**: An organization can identify traffic at the network edge devices, and then implement PBR to distribute that traffic among multiple paths based on the identified traffic characteristics.
- **Directing Traffic to Security Devices Like Firewalls**: A network operator might use PBR to direct traffic to security devices like firewalls that provide packet filtering, authentication and data encryption services. PBR may be configured on the boundary router of a network to force incoming traffic to be directed to a security device. After the device has applied appropriate security policies or encryption, the traffic is forwarded to its destination.

7.5 ROUTE SUMMARIZATION

Due to the continuous growth of corporate networks, service provider networks, and the Internet as a whole, the number of network and subnets IP addresses in Routing Tables also continue to grow rapidly. This growth has always taxed the processing and memory resources in IP routers, and the bandwidth needed to exchange routing updates and maintain the IP Routing Tables. The use of Variable-Length Subnet Masks (VLSMs), route summarization (also called route aggregation), and Classless Inter-Domain Routing (CIDR) have become very useful and powerful tools for managing the IP address growth in networks and internetworks (see discussion in Appendix B, "IPv4 Packet", of [AWEYA1BK18]).

7.5.1 Using of VLSM and CIDR

VLSM allows IP subnets to be created (subnetted) with different sizes that do not necessarily have addresses that fall on an IP classful address boundary. CIDR is based on VLSM and allows arbitrary-length IP address prefixes to be specified for networks. The introduction of CIDR allowed IP address blocks of arbitrary-length to be allocated to organizations based on their actual and short-term projected IP network addressing needs. The introduction of CIDR also allowed the aggregation of multiple contiguous IP addresses prefixes into supernets (i.e., summary or aggregate IP addresses) that can be advertised in the larger Internet using routing protocols that support VLSM and CIDR like RIPv2, EIGRP, OSPFv2, and BGPv4.

Route summarization (or aggregation) is now used (made possible by CIDR-aware routing protocols) to advertise aggregate routes in today's networks, thus reducing the number of entries in the IP Routing Table. Route summarization improves routing scalability and efficiency especially in large networks. Using CIDR, service provider networks are able to summarize IP addresses, allowing them to reduce the size of the IP Routing Tables, and the routers to handle more routes. Summarizing a set of routes to a single router advertisement reduces both the processing load on routers and the perceived complexity of the networks behind the summary route. The importance of route summarization becomes more apparent as the network size increases.

A routing protocol can summarize routes on arbitrary network address bit boundaries only if it supports (VLSMs). Some routing protocols like RIPv2 and EIGRP support route summarization at any bit boundary in a network address (in automatic summarization mode), and at major network address boundaries (in manual summarization mode). Because of some inefficiencies that can arise as a result of automatic summarization, this mode of route summarization is not recommended. Routing protocols like OSPF support only manual route summarization.

We describe in this section, some of the methods and commands (as applied to Cisco IOS) used for creating aggregate routes. The discussion covers the commonly used routing methods and protocols (static routes, RIP, EIGRP, OSPF, and IS-IS). Although, the router configuration commands used here may differ in some ways from those used in current versions of router operating systems, the goal here is to describe in a general sense, how route summarization can be carried out. In OSPF, route summarization can be configured at ABRs and ASBRs, while in EIGRP, route summarization can be configured in any router. The special section "Route Aggregation in BGP" below focuses on route summarization in BGP.

7.5.2 RIP Route Summarization

Most RIP implementations support both automatic (auto) summarization and manual summarization [CISCRIPCOMD18] [CISCRIPCONGUI]:

- **Auto Summarization**: RIP can be configured to use auto summarization to summarize routes to their IP classful network addresses automatically. The Cisco IOS `auto-summary` command (with no arguments or keywords) can be used in RIP to summarize routes automatically to their classful address boundaries. RIPv2 auto summarization feature can be disabled to avoid some of the pitfalls of this method of summarization. When disabled, routers will be able to propagate routing information and update their Routing Tables with the appropriate (optimal) summary addresses and network masks.
- **Manual Summarization**: A specific RIP router interface can be configured to advertise a summary route for a set of more-specific IP addresses on the router. For example, the network administrator may wish to summarize a number of host IP routes and advertise the resulting summary route into RIPv2. The `ip summary-address rip` command can be used in interface configuration mode to manually summarize a given set of routes learned via RIPv2 or redistributed into RIPv2.

    interface *type-number*
    ip summary-address rip *ip-address network-mask*

    o The parameter *type-number* specifies the router interface on which to advertise the summary route for the IP address set.
    o The parameter *ip-address* specifies the IP address of the routes to be summarized.
    o The parameter *network-mask* specifies the network mask that is used for summarizing the routes.

Network Path Control and Factors That Affect Routing Table Properties

The use of RIPv2 manual summarization and this command comes with some guidelines and restrictions as described in **[CISCRIPCOMD18] [CISCRIPCONGUI]**.

7.5.3 EIGRP Route Summarization

The Cisco IOS `auto-summary` command (with no arguments or keywords) can be used in EIGRP to summarize routes automatically to their IP classful address boundaries just as in RIP. Because of a number of inefficiencies that can arise as a result of using auto summarization, many EIGRP implementations allow this feature to be disabled. The "EIGRP Route Summarization" section in Chapter 6 discusses in detail route summarization in EIGRP.

The Cisco IOS `summary-address` command can be used to manually configure a summary address for EIGRP **[CISCEIGRPCOMD18]**:

> `summary-address` *ip-address mask* [*administrative-distance*]
> [`leak-map` *leak-map-tag*]]

- The parameter *ip-address* specifies the summary address for the range of IP addresses being summarized.
- The parameter *mask* specifies the IP subnet mask used for the summary route.
- The parameter *administrative-distance* (optional) specifies the Administrative Distance (with valid range 1–255) of the summary route. EIGRP summary routes are assigned a default Administrative Distance value of 5.
- The `leak-map` keyword (optional) allows a leak map to be used to advertise a component (more-specific) route that would otherwise have been suppressed by the manual summarization. Any subset of the component routes or addresses that make up the summary route can be leaked. The network administrator must define a route map and access list to source the route to be leaked.
- The parameter *leak-map-tag* (optional) specifies the name of a leak-map.

The Cisco IOS `summary-metric` command can be used to configure a fixed metric for an EIGRP summary (aggregate) address **[CISCEIGRPCOMD18]**:

> `summary-metric` *network-address subnet-mask* { *bandwidth-metric*
> *delay-metric reliability-metric effective-bandwidth-metric mtu-metric* |
> `distance` *administrative-distance*}

- The parameter *network-address* specifies the IP summary or aggregate address to apply to an interface.
- The parameter *subnet-mask* specifies the subnet mask.
- The parameter *bandwidth-metric* specifies the minimum bandwidth of the redistributed route in kilobytes per second. This can take a value from 1 to 4,294,967,295.
- The parameter *delay-metric* specifies the route delay in tens of microseconds. This can take a value of 1, or any positive number that is a multiple of 39.1 nanoseconds.
- The parameter *reliability-metric* specifies the likelihood of successful packet transmission expressed as a number from 0 to 255. The value 0 means no reliability while 255 means 100% reliability.

- The parameter *effective-bandwidth-metric* specifies the effective bandwidth of the route expressed as a number from 1 to 255 (where 255 is 100% loading).
- The parameter *mtu-bytes* specifies the MTU of the route, expressed in bytes. This can take a value from 1 to 65,535.
- The parameters *administrative-distance* specifies the Administrative Distance (with valid range 1–255) of the route.

7.5.4 OSPF Route Summarization

OSPF does not support any form of automatic summarization as in other routing protocols like RIPv2 and EIGRP. Instead, OSPF supports only manual route summarization which can be done on interarea OSPF routes at the Area Border Routers (ABRs), and OSPF external routes at the ASBRs. Chapter 1, section "OSPF Route Summarization" of Volume 2 of this two-part book discusses in detail route summarization in OSPF.

OSPF has the capability to summarize routes at IP address boundaries other than at classful IP network address boundaries. The Cisco IOS **summary-address** command can be used to manually create aggregate addresses for OSPF **[CISCOSPFCOMD19]**:

summary-address {*ip-address mask* | *prefix mask*} [**not-advertise**] [**tag** tag] [**nssa-only**]

- The parameter *ip-address* specifies the summary address for the range of IP addresses being summarized.
- The parameter *mask* specifies the IP subnet mask used for the summary route.
- The parameter *prefix* specifies the IP route prefix for the destination.
- The **not-advertise** keyword (optional) is used to suppress routes that match the specified *prefix/mask* pair.
- The parameter *tag* (optional) specifies the tag value that can be used as a "match" value for controlling redistribution via route maps.
- The **nssa-only** keyword (optional) sets the nssa-only attribute for the summary route generated (if any) for the specified *prefix*. This limits the summary to OSPF NSSAs (Not-So-Stubby-Areas).

7.5.5 IS-IS Route Summarization

Similar to OSPF, IS-IS does not support any form of automatic summarization as in RIPv2 and EIGRP. The Cisco IOS **summary-address** can be used to manually create aggregate addresses for IS-IS **[CISCTEAPAQ00]**:

summary-address *address mask* {**level-1** | **level-1-2** | **level-2**} *prefix mask*

- The parameter *address* specifies the summary address for the range of addresses being summarized.
- The parameter *mask* specifies the IP subnet mask used for the summary route.
- The **level-1** keyword indicates that only routes redistributed into Level 1 are summarized with the configured *address/mask* value.

Network Path Control and Factors That Affect Routing Table Properties   225

- The `level-1-2` keyword is used to indicate that the summary route applies both when routes are being redistributed into a Level 1-2 router, and when a Level 1-2 router advertises Level 1 routes that are reachable in its IS-IS area.
- The `level-2` keyword indicates that routes learned by Level 1 routing are summarized into the Level 2 IS-IS backbone with the configured *address/mask* value, and also summarized routes redistributed into a Level 2 router.
- The parameter *prefix* specifies IP route prefix for the destination.
- The parameter *mask* specifies IP subnet mask used for the summary route.

Chapter 2, section "IS-IS Route Summarization" of Volume 2 of this two-part book discusses in detail route summarization in IS-IS.

7.5.6 Static Route Summarization

IP routers forward packets using either routing information manually configured in their Routing Tables, or routing information learned by dynamic routing protocols. Static routes, cannot be automatically updated because they define explicit paths between two IP routers. A static route must be manually reconfigured when network changes occur. It was discussed in Chapter 4 that the use of static routes in a network, allow routers to use less CPU cycles to maintain Routing Tables unlike when dynamic routing protocols used.

Most practical networks use a combination of dynamic routing and static routes, and allow static routes to be redistributed into dynamic routing protocols. Static routes are most suitable in environments where the network design is simple, and where the network traffic is predictable and stable. Because static routes cannot react to network changes, they are not suitable for large networks with constantly changing link-states. A network may use dynamic routing to communicate between its routers, and may use a smaller number of static routes for special cases. For example, a static route can be set up as a default route, specifying a gateway of last resort (when all other routes fail or are not available).

Chapter 4, section "Summary Static Route" discusses in detail static route summarization. The Cisco IOS `ip route` global configuration command can be used to establish static routes **[CISCN3000CONFG]**:

   `ip route` *ip-prefix ip-mask* {*nh-address* | *interface*} [*distance*] [`tag` *tag-value*] [`permanent`]

- The parameter *ip-prefix* specifies the IP route prefix for the destination.
- The parameter *ip-mask* specifies the prefix mask for the destination.
- The parameter *nh-address* specifies the IP address of the next hop that can be used to reach that network.
- The parameter *interface* specifies the network interface to use.
- The parameter *distance* (optional) specifies an Administrative Distance for the static route.
- The parameter *tag-value* (optional) specifies a value that can be used as a "match" value for controlling redistribution via route maps.

- The **permanent** keyword (optional) specifies that the route will not be removed, even if the interface goes down or the next-hop IP address is not reachable.

To help understand the use of the parameters in the `ip route` configuration command, let us consider the following use cases of static routes **[CISCN3000CONFG]**:

- **Directly Connected Static Route**: In a directly connected static route, only the output interface on the router need to be specified (i.e., the interface on which the router sends all packets to the destination network). In this case, it is assumed that the destination is directly attached to the output interface, and the packet's destination is also the next hop IP address.
  o For point-to-point interfaces only, the next-hop can be specified as an interface.
  o For interfaces that attach to broadcast multiaccess media, the next-hop must be an IP address.
- **Fully Specified Static Route**: In a fully specified static route, the outbound interface on the router (i.e., the interface on which the router sends all packets to the destination network), and the next hop address must be specified. A fully specified static route can be used when the outbound interface is attached to a multiaccess network (i.e., is a multiaccess interface), and where there is the need to fully identify the next-hop IP address. The next-hop address must belong to an IP device that is directly attached to the specified (multiaccess) outbound interface.
- **Remote Next Hops for Static Routes**: The IP address of a neighbor router that is not directly connected to a particular IP router setting up a static route through some remote (non-directly attached) routers can be specified as the next hop. If an IP router has a static route through a number of remote next hops used for packet forwarding, these next hops are recursively looked up in the router's Routing Table to identify the corresponding (next) directly attached next hop that has reachability to the remote destination (see the "Traditional non-BGP Recursive Route Lookup" section in Chapter 3 of Volume 2 of this two-part book for details on recursive lookups).

7.6 ROUTE REDISTRIBUTION

We discussed earlier on that running a single routing protocol throughout a network is preferable. However, when two companies, Company A running OSPF in its network, and Company B running EIGRP, merge, the combined or merged company may end up having two separate routing domains running these two protocols. Since in many cases it is difficult to make the merged company run a single routing protocol, route redistribution becomes a very useful tool for internetworking the two separate routing domains. In some cases, a single company may have a single network with different parts under different administrative control, making route redistribution also necessary. Route distribution allows different routing protocols to exchange routing information.

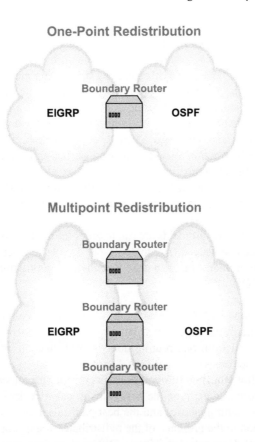

FIGURE 7.7 Route Redistribution Methods

Route redistribution techniques can be categorized as one-point route redistribution or multipoint route redistribution (Figure 7.7) **[CISCTEARDIA10]**. To perform route redistribution, at least one redistribution point (i.e., boundary router) needs to be created between the two routing domains. In OSPF, the router that performs external route redistribution is called an ASBR. Each redistribution point must have an interface to each routing domain. Route redistribution allows the boundary routers connecting the two routing domains (e.g., EGRP and OSPF) to exchange and advertise routes between them (Figure 7.7).

The boundary router runs both routing protocols and always performs route redistribution in the outbound direction. Also, this router does not change its IP Routing Table when performing route redistribution; the redistributed routes already exist in its IP Routing Table. Routes must be present in the IP Routing Table of the boundary router for them to be redistributed. In the case of EIGRP route redistribution, routes that are simply in the EIGRP Topology Table (e.g., EIGRP Feasible Successors [see Chapter 6]) are not redistributed. It is not sufficient for a route be in only the EIGRP Topology Table to be redistributed; it has to also be in the IP Routing Table. The neighbor routers of the boundary router see redistributed routes as external routes.

Assigning a routing metric to a route is one of the key considerations when performing route redistribution. As discussed in Chapter 1, each routing protocol utilizes a unique routing metric that is incompatible with other routing metric. So, a boundary router must be capable of translating the routing metric of a redistributed route into a metric that the receiving routing protocol can use. A redistributed route must be assigned a routing metric (using a default metric or a manually configured metric) that is compatible with the receiving routing protocol (see "Route Metric: Route Redistribution and The Seed Metric" section below). The methods used to configure route redistribution vary, and depend on the particular routing protocols involved in the redistribution.

We discuss in the "Route Redistribution Configuration Tools" section below, the different methods for configuring route redistribution between various IP routing protocols.

When multiple routing protocols are run in different domains of a network, routes have to redistributed between these protocols, but this has to be given careful consideration. The main problems that can arise when redistributing routes are the following [CISCTEARDIA10]:

- **Routing Loops and Route Feedback**: When more than one boundary router is used to perform route redistribution, any one of them might redistribute routes received from one routing domain back into the routing domain that originated the routes.
- **Incompatible Routing Information**: Each routing protocol uses a different routing metric to determine the best paths to network destination. The differences in the routing metrics and the best path selection algorithms, can make path selection in the presence of the redistributed routes not optimal.
- **Inconsistent Convergence Times**: Different routing protocols, even when applied to the same network, can have different convergence times. Thus, redistributing routes between two routing domains running different routing protocols that converge at different rates, can have an effect on the overall convergence time and stability of the internetwork.

Furthermore, certain differences in the characteristics of routing protocols, such as the routing metrics used, the Administrative Distance settings, the capability of using classful and classless IP addresses, can all effect route redistribution [CISCID8606REDIS12]. Network operators must give careful consideration to these differences for route redistribution to be successful.

In some cases, good network planning should be able solve a majority of the problems arising from route redistribution. However, in a majority of cases, additional network path control tools have to be used. Some problems might be solved by changing the Administrative Distances of the routing protocols, altering the routing metrics during route redistribution, and/or employing route filtering via configuration tools such as route maps, distribute lists, and prefix lists. These methods are discussed in appropriate sections below.

As discussed in Chapter 2, IP routers use the routing metric and Administrative Distance values of routing protocols to select the best paths to network destinations.

Network Path Control and Factors That Affect Routing Table Properties

The routing metric, according to the routing protocol being used, is a value assigned to a particular path between the local router and the network destination. Each routing protocol uses its own (internal protocol-specific) metric to determine the "best" path to a particular destination when multiple paths exists. When routes are redistributed one routing protocol into another, it must also be recognized that the routing metrics of the protocols involved play an important role in the route redistribution.

The Administrative Distance is used to rank the trustworthiness or believability of different routing information sources when they provide routing information to the same network destination. As discussed in Chapter 2, when more than one routing protocol provides routing information to the same destination, the Administrative Distance is used to determine which routing information source is more believability. Directly comparing routes provided by different routing protocols (which have different routing metric types) simply cannot be done. So, the Administrative Distances of the routing protocols play an important role in route selection. As will be discussed below, although the Administrative Distance helps with route selection when different routing information sources are available, a number of problems can still arise during route redistribution, such as, the formation of routing loops and route feedback, inefficient routing, or inconsistent convergence problems [CISCID8606REDIS12].

7.6.1 ONE-POINT ROUTE REDISTRIBUTION

In one-point route redistribution, only one boundary router performs route redistributing even if there are other boundary connecting two routing domains. In this scenario, the other boundary routers can route user traffic between the two routing domains but they are not configured to redistribute routes. Furthermore, one-point route redistribution can be configured to perform route redistribution in a one-way fashion (Figure 7.8, top diagram), or in a two-way (both ways) fashion (Figure 7.8, bottom diagram):

- **One-Point One-Way Route Redistribution**:
  o Routes are redistributed from one routing protocol into the other in only one direction (e.g., EIGRP into OSPF).
  o Typically, a default or static route is used in the EIGRP routing domain to allow its routers to reach the OSPF routing domain.
- **One-Point Two-Way Route Redistribution**:
  o Routes are redistributed between the two routing protocols in both directions (e.g., EIGRP into OSPF, and OSPF into EIGRP).

In Figure 7.8 (bottom diagram), the network operator sees the need to make EIGRP and OSPF advertise the routes they have learned to one another and not use a default route to the OSPF routing domain. This form of one-point route redistribution is called *mutual route redistribution*. Since, the single boundary router has one interface in the OSPF routing domain and another interface in the EIGRP routing domain, it is responsible for performing the route redistribution between the two domains (see also Figure 7.9).

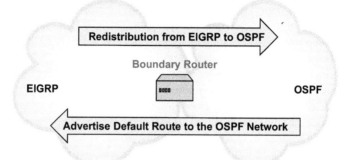
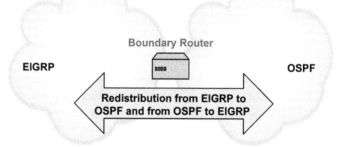

FIGURE 7.8 One-Point Route Redistribution

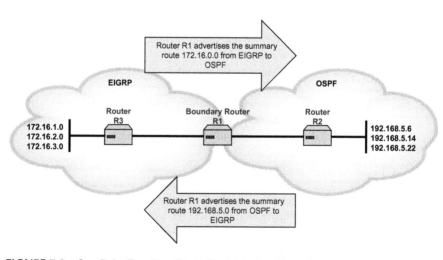

FIGURE 7.9 One-Point Two-Way Route Redistribution Example

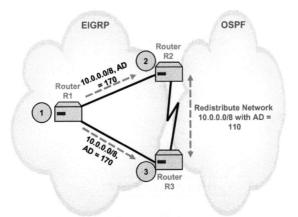

Let us assume only Router R2 is configured to redistribute an external EIGRP route into the OSPF domain.
1. Router R1 advertises the external EIGRP route 10.0.0.0/8 with an administrative distance (AD) of 170 to both R2 and R3.
2. Because only Router R2 is configured to redistribute the EIGRP routes into the OSPF domain, it redistributes the network 10.0.0.0/8 into the OSPF domain with an administrative distance of 110.
3. Although Router R3 has a direct route to R1, it will use the OSPF route via R2 to get to network 10.0.0.0/8 due to the lower administrative distance of the advertised OSPF route. This will result in suboptimal routing from R3 to R1.

FIGURE 7.10 Example Problem with One-Point One-Way Route Redistribution

When not properly addressed, one-point one-way route redistribution can create suboptimal routing when multiple boundary routers are used to connected two routing domains, and only one router is performing the one-point one-way route redistribution (see Figure 7.10). When the external EIGRP network 10.0.0.0/8 (with Administrative Distance of 170) is redistributed by Router R2 into the OSPF routing domain, its Administrative Distance is set 110 (the default Administrative Distance of OSPF). Although R3 and R1 have a direct connection, R3 will use the OSPF route through R2 to reach the external EIGRP network 10.0.0.0 because this OSPF route has the lower Administrative Distance of 110. This is one example of a scenario where one-point one-way redistribution creates a suboptimal routing.

7.6.2 MULTIPOINT ROUTE REDISTRIBUTION

In multipoint route redistribution, multiple boundary routers are used to redistribute routes between the two routing domains. The route redistribution can be done either in a multipoint one-way fashion (Figure 7.11, top diagram) or multipoint two-way (both ways) fashion (Figure 7.11, bottom diagram). Multipoint two-way redistribution tends to be more problematic than one-point redistribution. Either method, when used, has the potential of introducing routing feedback loops. Figure 7.12 describes the different techniques used for route redistribution.

One of the key concerns with multipoint route redistribution is that, it is more prone to routing loops and route feedback. In particular, two-way multipoint route redistribution can cause routing feedback loops in a network. Routing loops and possibly route feedback can occur when route redistribution is performed on more than

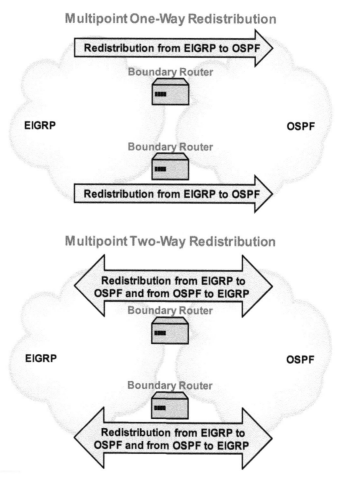

FIGURE 7.11 Multipoint Route Redistribution Examples

one router interconnecting two routing domains. As will be discussed below, route maps with route tagging is one of the methods that can be used to prevent the redistribution of routes that have already been redistributed. When using two-way multipoint route redistribution, route maps and route tags can be applied to filter routes in both direction, and on the boundary routers performing the route redistribution.

Although performing route redistribution on only one boundary router in only one direction within the network is the safest way of redistributing routes, this creates a single point of failure in the network. However, if there is the need to perform route redistribution in both directions or on multiple boundary routers, then extra steps must be taken to avoid problems such as suboptimal routing and routing loops during the route redistribution.

Multipoint one-way redistribution is more effective and works well when the routing protocol into which routes are being redistributed (i.e., the receiving routing protocol) is either EIGRP, OSPF, or BGP **[CISCTEARDIA10]**. This is because these

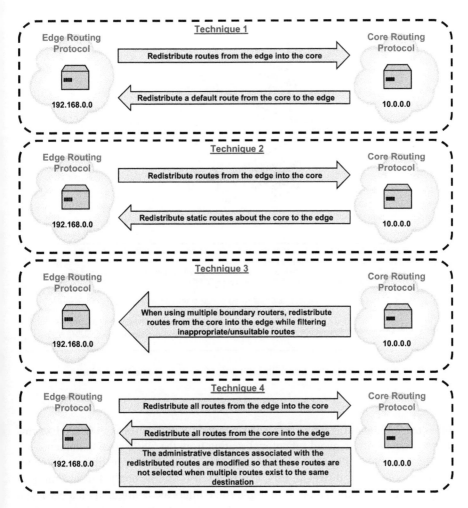

FIGURE 7.12 Route Redistribution Techniques

routing protocols support different Administrative Distances for internal and external routes (External BGP = 20, Internal EIGRP = 90, External EIGRP = 170, Internal BGP = 200).

The cost of an OSPF External Type-2 route is always the cost that came with the external route, irrespective of the interior OSPF cost to reach the ASBR that advertised that route. An OSPF External Type-1 cost is the sum of the cost that came with the external route and the internal OSPF cost to reach the ASBR that advertised that route. For the same network destination, an OSPF External Type-1 route is always preferred over an External Type-2 route.

As described in Figure 7.13, because the Administrative Distance of external EIGRP routes is higher than the Administrative Distance of OSPF routes, Routers R2 and R3 will use the suboptimal routes to destinations in the EIGRP routing domain.

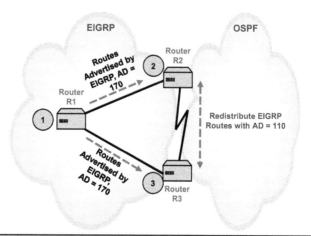

FIGURE 7.13 Example Problem with Multipoint One-Way Route Redistribution

In Figure 7.13, marking the redistributed external EIGRP routes as an OSPF External Type-1 route helps to avoid suboptimal routing from the OSPF routing domain.

The example in Figure 7.14 illustrates how a combination of filtering methods such as route maps, distribute lists, and prefix lists can be applied to incoming or outgoing routing information. If multiple route filters are configured on a particular router interface as illustrated in Figure 14, they all must permit a route that is received from a neighbor before it will be accepted into the IP Routing Table. Similarly, they all must permit an outgoing route before it is propagated to downstream neighbors. It is important to note that the use of distribute lists and prefix lists is mutually exclusive, only one of these can be configured on a particular router interface at any given time [CISCBGPCOMD19] [CISCIOSCOMD19].

7.7 PATH CONTROL TOOLS

Many networks contain redundant links and redundant devices, thereby, creating redundant paths to network destinations. In such scenarios, in addition to the use of routing protocols, the network administrator needs other tools to control how traffic flow to their destinations. This section discusses the extra tools needed to control the paths that traffic take through a network. The path control tools discussed here include distribute lists, prefix lists, Administrative Distance, route tagging, route

Network Path Control and Factors That Affect Routing Table Properties

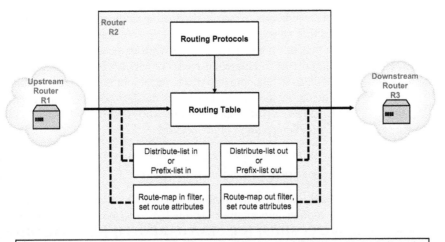

FIGURE 7.14 Using Multiple Methods to Control Routing Updates

maps, offset lists, PBR (policy-based routing), and Cisco IOS IP SLAs (Service Level Agreements).

7.7.1 The Need for Path Control Tools

All routing protocols have protocol-specific built-in mechanisms for providing network resiliency (i.e., the ability to maintain an acceptable level of service when network failures occur), improving network convergence times (i.e., the time required for the overall network to learn and settle on a new route), and providing adaptability (i.e., the ability of the network to be configured to adapt to changing conditions). Path control tools provide extra tuning knobs that can be used to improve network performance, as well as, provide some desired behaviors that otherwise cannot be provided by the built-in tools of the routing protocols.

The path control tools can be used for making more efficient use of network bandwidth, load sharing across multiple paths, influence route advertisements for specific network prefixes over other prefixes to produce certain desired network behaviors, adjust routing to provide specific services (such as quality of service [QoS], security, network optimization), and so on.

However, care and diligence must be exercised when using path control tools to manipulating routing updates, especially, when multiple routing protocols are implemented (and interworked) in a network. Otherwise inefficient routing and routing loops may result. For example, in two-way multipoint route redistribution (discussed above), careful planning and implementation is required to ensure that redistributed routes are not themselves redistributed to the originating routing domain, and traffic flows the optimal way with no routing loops.

Furthermore, other than serving as a way of improving application response times, user experience, and improving a comprehensive set of network and end-user performance parameters, the desired outcome of a path control tool must be predictable – the results should be deterministic and predictable as much as possible. Path control tools should be implemented as part of an overall network strategy, and should fit in with the routing policy implementation of the network.

7.7.2 ROUTE REDISTRIBUTION CONFIGURATION TOOLS

In this section, we describe some of the methods and commands used for redistributing routes from one routing into another (as applied to Cisco IOS) **[CISCASR-9000CR] [CISCBGPCOMD19] [CISCEMPGARROT14] [CISCHALABS00] [CISCMASCOMD14]**. Although, the router configuration commands used here (to illustrate how route redistribution works) may differ somehow from those described in current versions of router operating systems, the main goal is to describe from a high-level, the various methods available for performing route redistribution.

A special section below "Injecting Routing Information into BGP" focuses on route redistribution into BGP. Route redistribution can also be performed between two processes belonging to the same routing protocol (distinguished by unique process identifiers [IDs]), such as between two separate OSPF domains.

7.7.2.1 Redistributing Routes into RIP

The **redistribute** command can be used in the router configuration mode to redistribute routes from another routing domain into RIP. The **redistribute** command is configured under the routing process (in the local IP router) that is to receive the redistributed routes (the receiving routing protocol). The minimum command syntax is as follows **[CISCRIPCOMD18]**:

> **redistribute** *protocol* [*process-id*] [**match** *route-type*] [**metric** *metric-value*] [**route-map** *map-tag*]

- The parameter *protocol* is the source protocol (e.g., **connected, static, eigrp, ospf, isis, bgp**) from which routes are being redistributed.
- The parameter *process-id* (optional) is the EIGRP/OSPF/IS-IS/BGP instance from which routes are to be redistributed. When redistributing from OSPF (**ospf** keyword), *process-id* is an appropriate OSPF process ID from which the routes are to be redistributed; from IS-IS (**isis** keyword), *process-id* is an optional tag value that defines a meaningful name for the routing process; from BGP or EIGRP (**bgp** or **eigrp** keyword), *process-id* is an Autonomous System Number (ASN).
- The parameter *route-type* (optional) associated with the **match** keyword, applies when redistributing routes from OSPF to another routing protocol and specifies the criteria by which OSPF routes are redistributed into that protocol (**internal** – internal routes that to a specific OSPF Autonomous System;

external 1 – external routes that to the Autonomous System, but are redistributed into OSPF as Type-1 external routes; **external 2** – external routes that to the Autonomous System, but are redistributed into OSPF as Type-2 external routes).
- The parameter *metric-value* (optional) associated with the **metric** keyword, specifies the metric for the routes being redistributed. If no value is specified, and no value is specified using the **default-metric** command, then the default metric is set to 0 and interpreted as infinity (by RIP), meaning no routes will be redistributed.
- The parameter *map-tag* (optional) associated with the **route-map** keyword, is an identifier of a specific route map configured on the router. This route map is interrogated to filter the routes from the source routing protocol to RIP. If this parameter is not specified, all routes are imported.

7.7.2.2 Redistributing Routes into OSPF
The minimum command syntax for redistributing routes (e.g., connected, static, RIP, EIGRP, OSPF, IS-IS, BGP) into OSPF using the **redistribute** command is as follows **[CISCOSPFCOMD19]**:

> **redistribute** *protocol* [*process-id*] [**metric** *metric-value*] [**metric-type** *type-value*] [**route-map** *map-tag*] [**subnets**] [**tag** *tag-value*]

- When redistributing from RIP (**rip** keyword), *process-id* is not needed; from OSPF (**ospf** keyword), *process-id* is an appropriate OSPF process ID from which the routes are to be redistributed; from IS-IS (**isis** keyword), *process-id* is an optional tag value that defines a meaningful name for the routing process; from BGP or EIGRP (**bgp** or **eigrp** keyword), *process-id* is an ASN.
- The parameter *metric-value* (optional) associated with the **metric** keyword, specifies the OSPF seed metric used for the redistributed route. It is passed unaltered when redistributing routes from one OSPF process to another OSPF process on the same router, if no metric value is specified. When redistributing from another routing protocol to OSPF, the default metric is 20 when no metric value is specified.
- The parameter *type-value* (optional) associated with the **metric-type** keyword, when applied to routes from OSPF, specifies the external link type associated with the route being advertised into OSPF (Values = 1, means Type-1 External route; value = 2, means Type-2 External route). When applied to routes from IS-IS: **internal**, means IS-IS metric (that is less than 63); **external**, means IS-IS metric (that is greater than 64 and less than 128).
- The subcommand **subnets** (optional) is an OSPF parameter that specifies that subnetted IP addresses (routes) should be redistributed into OSPF. When not specified, only classful routes are redistributed.
- The parameter *tag-value* (optional) associated with the **tag** keyword, specifies a 32-bit decimal value attached to each external route to be used by OSPF ASBRs.

7.7.2.3 Default Metric for RIP, OSPF, or BGP

The **default-metric** command can be used to set a default metric for routes redistributed from another routing protocol into RIP, OSPF, or BGP:

default-metric *number*

- For redistribution into RIP, *number* is a default metric value in the range 1–15.
- For redistribution into OSPF, *number* is a default metric value appropriate for the specified routing protocol and is in the range 1–16,777,214.
- For redistribution into BGP, *number* is a default metric value applied to the redistributed route and takes a value in the range 1–4,294,967,295.

When a metric (i.e., the parameter *metric-value*) is configured in a **redistribute** command, it overrides the value in the **default-metric** command for that routing protocol. When used, the metric values specified by the **default-metric** command do not override those set by the **redistribute** command **metric** keyword.

For BGP, the **default-metric** command can be used to set the metrics of any external BGP (eBGP) route received from BGP peers, and subsequently advertised internally within an Autonomous System to iBGP peers. The value set by the **default-metric** command is the MED that a BGP router uses during the best path selection process. The MED is a non-transitive BGP path attribute that is used only within the local Autonomous System and adjacent BGP Autonomous Systems. The default metric is not set for a received BGP route if that route already has a MED value.

7.7.2.4 Redistributing Routes into EIGRP

The minimum command syntax for redistributing routes (e.g., connected, static, RIP, EIGRP, OSPF, IS-IS, and BGP) into EIGRP using the **redistribute** command is as follows **[CISCEIGRPCOMD18]**:

redistribute *protocol* [*process-id*] [**match** *route-type*] [**metric** *metric-value*] [**route-map** *map-tag*]

- Note that as stated above, for the **eigrp** or **bgp** keyword, the *process-id* is an ASN. For OSPF, *process-id* is an OSPF process ID. This value is not required for RIP or IS-IS.
- The parameter *metric-value* (optional) associated with the **metric** keyword, specifies the EIGRP seed metric for the redistributed route, expressed in terms of the EIGRP metrics bandwidth, delay, reliability, load, and Maximum Transmission Unit (MTU). If no value is specified when redistributing from another routing protocol, and no default metric has been configured, then no routes will be redistributed.

The **redistribute eigrp** command can be used in router configuration mode to redistribute routes from EIGRP into EIGRP:

redistribute eigrp *system-number* [**metric** *bandwidth-metric delay-metric reliability-metric effective-bandwidth-metric mtu-bytes*] [**route-map** *map-tag*]

Network Path Control and Factors That Affect Routing Table Properties

The parameter *system-number* is an ASN in the range 1–65,535.

7.7.2.5 Default Metric for EIGRP

The `default-metric` command can be used to set metrics for routes redistributed into EIGRP:

`default-metric` *bandwidth delay reliability loading mtu*

- The parameter *bandwidth-metric* is the minimum bandwidth of the redistributed route in kilobytes per second and can take a value from 1 to 4,294,967,295.
- The parameter *delay-metric* is the route delay in tens of microseconds and can take a value of 1, or any positive number that is a multiple of 39.1 nanoseconds.
- The parameter *reliability-metric* is the likelihood of successful packet transmission expressed as a number from 0 to 255. The value 0 means no reliability, while 255 means 100% reliability.
- The parameter *effective-bandwidth-metric* is the effective bandwidth of the route expressed as a number from 1 to 255 (where 255 is 100% loading).
- The parameter *mtu-bytes* is the MTU of the route, expressed in bytes and can be from 1 to 65,535.

The following commands shows how RIP routes are redistributed into EIGRP with following metrics values: *bandwidth-metric* = 1,000, *delay-metric* = 100, *reliability-metric* = 250, *effective-bandwidth-metric* = 100, and *mtu-metric* = 1,500:

```
router eigrp 109
network 172.16.0.0
redistribute rip
default-metric 1000 100 250 100 1500
```

Chapter 6 provides a detailed discussion of the different metrics used by EIGRP.

7.7.2.6 Redistributing Routes into BGP

The general `redistribute` command syntax for redistributing routes from one routing domain into another (including BGP) is as follows **[CISCBGPCOMD19]**:

`redistribute` *protocol* [*process-id*] {`level-1`|`level-1-2`|`level-2`} [*autonomous-system-number*] [`metric` {*metric-value* | `transparent`}] [`metric-type` *type-value*] [`match` {`internal` | `external 1`|`external 2`}] [`tag` *tag-value*] [`route-map` *map-tag*] [`subnets`] [`nssa-only`]

- The `level-1` keyword applies to IS-IS, and specifies Level 1 routes that are redistributed into the other IP routing protocol independently.
- The `level-1-2` keyword applies to IS-IS, and specifies both Level 1 and Level 2 routes that are redistributed into the other IP routing protocol.
- The `level-2` keyword that applies to IS-IS, and specifies Level 2 routes that are redistributed into the other IP routing protocol independently.
- The `metric transparent` keyword (optional) instructs the RIP metric for a redistributed route to be transported transparently to another routing

protocol. For example, when RIP routes are redistributed into BGP, the RIP metric can be passed and carried in the BGP MED attribute value. When these BGP routes are redistributed back into RIP at the remote end, the **transparent** keyword can be used to copy the propagated MED attribute value back as the RIP metric. If a *metric-value* is specified instead, that value is applied to all the routes listed.

- The **nssa-only** keyword (optional) applies to OSPF and sets the nssa-only attribute for all routes being redistributed into OSPF. See discussion on OSPF NSSA (Not-So-Stubby Area) type in Chapter 1 of Volume 2 of this two-part book.

7.7.2.7 Redistributing Directly Connected Networks and Static Routes into a Routing Protocol

Directly connected networks and static routes can also be redistributed into a given routing protocol. Although it is not common practice to redistributing directly connected networks into routing protocols, it can still be done [CISCID8606REDIS12]. Redistribution is done using the following commands, respectively [CISCIOSCOMD19]:

```
redistribute connected
redistribute static
```

Other command arguments can be used with these two commands as discussed above. Recall that routes will only be redistributed into a routing protocol if they exist in the IP Routing Table of the router performing the redistribution (the boundary router).

7.7.3 Route Metric: Route Redistribution and the Seed Metric

The primary challenge when redistributing routes from on routing protocol into another is determining the routing metric in the receiving routing protocols for the redistributed routes. As seen in Chapter 1, different routing protocols use different routing metrics for their internal route processing. For example, EIGRP uses a metric based on bandwidth, delay, reliability, loading, and MTU, while OSPF uses a metric of cost, which typically is based on bandwidth.

The issue is, EIGRP routing metrics cannot simply be converted into OSPF routing metrics (as done with monetary currencies) because the two metrics are very different, and have meaning only within their specific protocols. Thus, a solution has to be found when performing route redistribution between two different routing protocols. One option is to manually configure a metric for the redistributed route in the receiving routing protocol. However, if a metric is not manually configured for the redistributed route, the receiving protocol assigns a default *seed metric* to the redistributed route as discussed next. The seed metric serves as a "starter metric" that the receiving routing protocol can work with as soon as it receives the redistributed route.

7.7.3.1 Configuring Seed Metrics

When a route is being redistributing from one routing protocol to another, a router assigns a seed metric to the redistributed route using either the **default-metric** router configuration command, or the seed metric can be specified as part of the **redistribute** command, using either the **metric** optional keyword, or using a route map [CISCEMPGARROT14] [CISCIOSCOMD19]. The default-metric command is very useful when redistributing routes from multiple sources at the same time because, it eliminates the need to define metrics for each redistributed route separately.

For example, when routes are redistributed from RIP into EIGRP, the values used in the **default-metric** command constitute the seed metric for the RIP routes being redistributed. The seed metric is the initial value of the metric for the redistributed route used in the receiving routing protocol (EIGRP) and it must be consistent with the receiving protocol.

In Cisco IOS, when no metric is specified for redistributed routes, each routing protocol has its own default seed metric, which it assigns to the routes. The default seed metrics when a route is being redistributed into the following routing protocols are as follows:

- **RIP**: The default seed metric if no metric is specified is 0 (interpreted as infinity, unreachable) for all routes, excluding directly connected networks and static routes. Infinity means a default metric must be specified for the redistributed route to be eligible for placement in the RIP database and also to be advertised to other routers. Directly connected networks and static routes when redistributed into RIP take the default seed metric value of 1.
- **EIGRP**: The default seed metric if no metric is specified is 0 (interpreted as infinity, unreachable). This also means a default metric must be specified for the redistributed route to be eligible for placement in the EIGRP Topology Table (see Chapter 6). An internal seed metric must be specified for a directly connected network or static route redistributed into EIGRP.
- **OSPF**: The default seed metric if no metric is specified is the External Type-2 metric of 20 for all routes (including directly connected networks and static routes), except for BGP routes, which are set to the External Type-1 metric of 1.
- **IS-IS**: The default seed metric if no metric is specified is 0 for all routes (including directly connected networks and static routes). Without a specified seed metric, IS-IS does not allow a route to be redistributed. So, a seed metric must be defined for redistribution to occur.
- **BGP**: For a route learned from an IGP, the default BGP seed metric is set to the original IGP metric value of the route. Directly connected networks and static routes when redistributed into BGP take the default seed metric value of 1.

Both RIP and EIGRP have default seed metrics of "infinity", meaning these routing protocols consider any route redistributed into them to be unreachable by default, and therefore will not be advertised to any other routers (unless the default metrics are modified). Note that a RIP route with a metric of "infinity" is never entered in the IP Routing Table as discussed in Chapters 2 and 5. EIGRP is described in Chapter 6. So, a default seed metric must be defined for RIP and EIGRP during route redistribution. Note that the above default seed metrics are assigned when no metric is defined with the **default-metric** command or the **metric** keyword in the **redistribute** command. To avoid confusion or potential problems, the simplest solution is to set default metrics for all the route types being redistributed into a specific routing protocol.

Note that for OSPF, an External Type-2 metric is a metric that is assigned by the ASBR performing the route redistribution (see Chapter 1 of Volume 2 of this two-part book). An External Type-2 metric means that regardless of the number of additional routers within the OSPF Autonomous System that the external route has crossed from the ASBR that originated it, the metric remains the same (unaltered) as it was set when the ASBR redistributed it. When routes are redistributed into OSPF, those routes, by default, are set as External Type-2 routes. An External Type-1 metric is a metric that consists of the original cost assigned by the originating ASBR to the redistributed route, plus the internal Autonomous System cost required to reach that ASBR. Among the two metrics, the Type-1 is generally a more accurate metric.

After the boundary router has established the seed metric for a redistributed route, that metric is incremented normally within the receiving routing domain. The exception to this rule is OSPF External Type-2 routes, which remain unaltered in the receiving Autonomous System. A specific seed metric, different than the default metric (used when no metric is provided), can be specified for a redistributed route.

If a particular configuration uses both the **metric** keyword in the **redistribute** command and the **default-metric** command, the metric value specified by the **redistribute** command **metric** keyword takes precedence. For RIP and EIGRP, if a metric value is not specified by the **metric** keyword, and the **default-metric** command is also not used to specify a metric value, the default metric value is set to 0 (i.e., interpreted as infinity in RIP or EIGRP), except for OSPF, where the default metric or cost is set to 20 **[CISCEMPGARROT14]**.

Routes redistributed into RIP and EIGRP must have the appropriate metrics to avoid creating problems during redistribution. Redistributed routes from other routing protocols into EIGRP are tagged as EIGRP external routes. Routes that are redistributed between EIGRP processes do not need metrics to be configured. BGP uses the IGP metric of the route, while both directly connected networks and static routes redistributed into BGP are assigned an initial default seed metric value of 1.

In OSPF, route metrics for external routes can be changed at the distribution points, that is, ASBRs and ABRs, while in EIGRP, route metrics for all routes can be set at any point using the **default-metric** command, **redistribute** command, or route maps.

When routes are redistributed from BGP into OSPF (where they are tagged and treated as external OSPF routes), the default seed metric is set 1 and not 20 (as in when redistributing from other routing protocols) **[RFC1403]**. The main reason OSPF sets the seed metric of external routes provided by BGP to 1 is because, OSPF wants external traffic from the OSPF domain to take the BGP routes rather than OSPF (IGP) routes.

Note that as discussed in Chapter 2, eBGP routes have a default Administrative Distance of 20 which is much smaller than that of OSPF (which has a default of 110). By setting the seed metric of BGP routes to 1, OSPF gives preference to the BGP routes (over its own IGP routes) for external traffic. BGP is designed to provide finer/

granular routing policy implementation capabilities, and offers (using a wide range of BGP path attributes) more control over best path selection. This makes the BGP routes imported into OSPF more trustworthy and preferable. BGP is discussed in greater detail in Chapter 3 of Volume 2 of this two-part book.

7.7.4 Administrative Distance and Path Control

An Administrative Distance (also called the Route Preference) is a metric used by routers to rank routes to the same network destination that are provided by two or more routing information sources (directly connected networks, static routes, IGP routes, BGP routes). An Administrative Distance reflects the trustworthiness or reliability of the routing information source, and guides routers in the selection of routes when more than one routing information source exist, that is, when multiple routing protocols provide routes to the same destination (see "Administrative Distance and Route Selection" section in Chapter 2).

Each routing information source is prioritized in the order of most reliable (lowest Administrative Distance value) to least reliable (highest Administrative Distance value). Among multiple routes to the same destination, the route from the routing information source with the lowest Administrative Distance is installed in the IP Routing Table. Note that within any given routing protocol (e.g., RIP, EIGRP, OSPF, IS-IS, and BGP), a protocol-specific metric(s) is used to select the best path to a given network destination.

So, given that it not uncommon for (boundary) IP routers to run multiple routing protocols, the Administrative Distance comes in handy when these protocols present different routes to the same network destination. An Administrative Distance is a numerical integer that can take a value from 0 to 255. The higher the Administrative Distance value, the lower the trustworthiness of the routing information source. When the Administrative Distance is set to 255, this means the routing information source should not be trusted at all; all routes with this source must be ignored and not installed in the Routing Table.

7.7.4.1 Administrative Distance as a Path Control Tool

Before an IP router begins to run multiple routing methods and protocols, each one is configured with a default Administrative Distance value. Even though the routing protocols have default Administrative Distance values as discussed in Chapter 2, these values can be intentionally modified under certain circumstances to make some routing information sources more preferable than others. This means the Administrative Distance metric can readily be used as a path control tool because it can easily be modified to suit different networking and routing performance needs. The Administrative Distance of routes can be manipulated to influence the route selection process.

As discussed in Chapter 2, directly connected networks/interfaces have a default Administrative Distance of 0, while static routes have a default Administrative Distance of 1. This means an IP router prefers a directly connected network over static routes, and a static route over dynamic routes. The router considers a route from the routing information sources with a lower Administrative Distance to be more believable and, therefore, more preferable.

If the network administrator wants to make a dynamic route more preferable over a static route, the Administrative Distance of the static route is simply raised from the default value of 1 to a value that is higher than that of the dynamic route. For example, to make an OSPF route which has a default Administrative Distance of 110, preferable over an eBGP route with default value of 20, the Administrative Distance of the eBGP route can be set to a value greater than 110, say 120. The Administrative Distance of any dynamic route can also be raised to override a static route.

It is discussed in Chapter 4 that a floating static route is a static route that is used as a backup/standby route to a dynamic route. This is done by configuring the floating static route with a higher Administrative Distance than that of the dynamic route, so, that in the event the dynamic route fails or become unavailable, the floating static route can be used. Raising the Administrative Distance of the static route higher makes it less preferable to the dynamic route. By default, a static route router is preferred over a dynamic route because a static route has a lower Administrative Distance as discussed in Chapter 2.

In Cisco IOS, the **distance** command can be used in router configuration mode to define an Administrative Distance for routes that are inserted into the Routing Table [CISCIOSCOMD19] [CISCEMPGARROT14]:

distance *distance ip-address wildcard-mask* [*ip-standard-acl* | *access-list-name*]

- The parameter *distance* specifies the Administrative Distance required, and is an integer in the range 10–255. Values from 0 to 9 are reserved, and routes with an Administrative Distance value of 255 are ignored and not installed in the Routing Table.
- The parameter *ip-address* specifies that IP address of the router or network from which the routes are learned in the four-part, dotted decimal notation.
- The parameter *wildcard-mask* specifies a wildcard mask in the four-part, dotted decimal notation. Any bit that is set to 1 in the wildcard-mask argument instructs the routing software to ignore the corresponding bit in the IP address.
- The parameter *ip-standard-acl* (optional) specifies a standard IP ACL number that is to be applied to incoming routing updates.
- The parameter *access-list-name* (optional) specifies a named access list that is to be applied to incoming routing updates.

When the above command is applied to BGP, it sets the Administrative Distance of an eBGP route, while for EIGRP, it sets the Administrative Distance of only the internal routes of EIGRP neighbor routers.

References [CISCBGPCOMD19], [CISCEIGRPCOMD18], and [CISCOSPFCOMD19] describe specific **distance** commands for BGP, EIGRP, and OSPF, respectively. The **distance bgp** command [CISCBGPCOMD19] is used to allow a better route provided by another routing protocol to a given destination to be used over one that was actually learned via eBGP, or if some internal routes should be preferred by BGP:

distance bgp *external-distance internal-distance local-distance*

- The parameter *external-distance* specifies the Administrative Distance (in the range 1–255) for eBGP routes (routes learned from an external Autonomous System).
- The parameter *internal-distance* specifies that Administrative Distance (in the range 1–255) for internal BGP routes (routes learned from a BGP peer in the local Autonomous System).
- The parameter *local-distance* specifies the Administrative Distance (in the range 1–255) for local BGP routes. Local BGP routes are routes/networks that are listed with the **network** command. These routes are often backdoor routes that are configured for a router, or are routes/networks that are being redistributed from another routing process.

When no Administrative Distance values are specified using this command, the default values are: *external-distance* 20; *internal-distance* 200; *local-distance* 200. For example, the configuration **distance bgp 30 200 220** changes the Administrative Distance for eBGP routes to 30, internal BGP routes to 200, and local BGP routes to 220.

The **distance eigrp** command **[CISCEIGRPCOMD18]** allows the use of two Administrative Distances (internal and external Administrative Distances) that could provide a better route to a node:

distance eigrp *internal-distance external-distance*

- The parameter *internal-distance* specifies the Administrative Distance (in the range 1–255) of EIGRP internal routes (routes learned from another entity within the same Autonomous System). The default Administrative Distance for EIGRP internal routes is 90.
- The parameter *external-distance* specifies the Administrative Distance (in the range 1–255) of EIGRP external routes (routes learned from a neighbor that is external to the local Autonomous System). The default Administrative Distance for EIGRP external routes is 170.

For example, the configuration **distance eigrp 80 105** changes the Administrative Distance for internal EIGRP routes to 80, and changes the Administrative Distance for EIGRP external routes to 105.

The **distance ospf** command **[CISCOSPFCOMD19]** defines Administrative Distances for an OSPF route based on route type (external, interarea, or intra-area routes):

distance ospf {**external** *dist1* | **inter-area** *dist2* | **intra-area** *dist3*}

- The parameter *dist1* associated with the **external** keyword (optional) sets the Administrative Distance (in the range 1–255) for routes learned by redistribution from other routing domains. The default Administrative Distance value is 110.
- The parameter *dist2* associated with the **inter-area** keyword (optional) sets the Administrative Distance (in the range 1–255) for all routes sent from one OSPF area to another area. The default Administrative Distance value is 110.

- The parameter *dist3* associated with the `intra-area` (optional) sets the Administrative Distance (in the range 1–255) for all routes within an OSPF area. The default Administrative Distance value is 110.

The `distance ospf` command is used to set the Administrative Distance for an entire group of routes, rather than a specific route that passes an access list. For example, the configuration `distance ospf intra-area 105 inter-area 105 external 125` changes the Administrative Distance for both OSPF intra-area and interarea routes to 105, and changes the Administrative Distance for external routes to 125.

7.7.5 Route Tagging

Route tags are numeric values that are assigned or attached to routes to be used by IP routers for filtering routes and applying administrative policies, such as when performing route summarization and redistribution. Route Tagging can be used to tag a group of IGP or EGP routes so that they can be advertised (with their tag values) throughout a routing domain. This also allows routers to easily filter the group of routes for security purposes, or meet other routing policy and administrative requirements.

Also, in two-way multipoint routes redistribution, there is the potential of routing loop formation in the network (see "Multipoint Route Redistribution" section above). Route tagging can be used as one of the mechanisms for preventing the redistribution of routes that have already been redistributed **[CISCEMPGARROT14]**. Route tags must be applied in both directions, and redistributed routes must also be filtered in both direction and on both routers performing the route redistribution. By simply tagging routes, routers can use route maps to determine which routes have already been redistributed, and then deny them from being redistributed once again.

This is to ensure that, for example, in an EIGRP-OSPF two-way multipoint route redistribution, only routes originating from an EIGRP domain are redistributed into OSPF, and only routes originating from the OSPF domain are redistributed into the EIGRP domain. Other than route filtering and routing loop prevention, route tagging can be used for traffic engineering purposes, where specific routes can be identified and appropriate network resources reserved for them. In OSPF, route tags for external routes can be added at the distribution points, that is, ASBRs and ABRs, while in EIGRP, route tags for all routes have to be configured.

In Cisco IOS, the `set tag` command can be used in the route-map configuration mode of an IP router to set a tag value for a route in a route map **[CISCEIGRPCOMD18] [CISCIOSCOMD19]**:

> `set tag` {*tag-value* | *tag-value-dotted-decimal*}

- The parameter *tag-value* specifies a route tag value in plain decimals (in the range 0–4,294,967,295).
- The parameter *tag-value-dotted-decimal* specifies a route tag value in dotted decimals (in the range 0.0.0.0–255.255.255.255.).

In Cisco IOS, a route tag is a 32-bit value that is attached to a specific route. A tag value can be set as a plain decimal or in dotted decimals. The typical usage of the **set tag** command is to assign administrative tags to routes within a route map. The route map then uses the route tags to filter routes (see "Route Maps" section below). Route tags have no impact on routing decisions and are only used to flag or mark routes so that routing loops can be prevented when routes are redistributed between routing protocols.

The **match tag** or **match tag list** command can be used to match tagged routes within a route map so that administrative policies can be applied to the matched tagged routes. The **match tag list** command is used within a route map to match a list of route tags.

- **To Match a Route Tag in a Route Map**: In Cisco IOS, the **route-map** command can be used in global configuration mode to enable policy routing or, to define conditions for redistributing routes from one routing protocol to another:
  1. **route-map** *map-tag* [**permit** | **deny**] [*sequence-number*]
  2. **match   tag** {*tag-value* | *tag-value-dotted-decimal*} [...*tag-value* | *tag-value-dotted-decimal*]
     - The parameter *map-tag* specifies the name for the route map.
     - The **permit** keyword (optional) is used to permit only routes that match the route map to be forwarded or redistributed.
     - The **deny** keyword (optional) is used to block routes that match the route map from being forwarded or redistributed.
     - The parameter *sequence-number* (optional) specifies a number that indicates the position that a new route map statement will take in the list of route map statements that have already been configured with the same name (*map-tag*).
- **To Create a Route Tag List**: In Cisco IOS, the **route-tag list** command can be used in global configuration mode to create a route tag list:
  route-tag list *list-name* {**deny** | **permit** | **sequence** *number* {**deny** | **permit**}} *tag-dotted-decimal mask*
     - The parameter *list-name* specifies the name of the route tag list.
     - The **deny** keyword specifies packets that have to be rejected.
     - The **permit** keyword specifies packets that have to be forwarded.
     - The **sequence** keyword specifies the sequence number of an entry.
     - The parameter *number* specifies the sequence number, and has valid range from 1 to 4,294,967,294.
     - The parameter *mask* specifies a wildcard mask.
- **To Match a Route Tag List**:
  1. **route-tag   list** *list-name* {**deny** | **permit** | **sequence** *number* {**deny** | **permit**}} *tag-dotted-decimal mask*
  2. **route-map** *map-tag* [**permit** | **deny**] [*sequence-number*]
  3. **match tag list** *list-name* [...*list-name*]

7.7.6 PASSIVE INTERFACES

Configuring a router interface as passive is one way to control the advertisement of routing information in and out of a router (see also Chapter 4). To prevent other routers in a network from learning about routes dynamically, routing update messages can be suppressed from being sent through some router interfaces. The behavior of a passive interface varies and depends very much on the routing protocol running on the specified interface (RIP, EIGRP, OSPF, IS-IS).

In general, a passive interface suppresses the transmission of certain protocol-specific messages and routing updates over that interface, while allowing protocol messages and routing updates to be exchanged normally over other interfaces. By regulating the types of routing protocol messages allowed inbound and outbound, and possibly, the establishment of router adjacency relationships, passive interfaces are capable of affecting the properties of IP Routing Tables. Passive interfaces can affect the routing information a router receives and consequently, the routes a router will install in its IP Routing Table.

- **Passive Interfaces for RIP**: The `passive-interface` command can be used in router configuration mode in a RIP router to suppress the sending of RIP routing updates on an interface **[CISCEMPGARROT14] [CISCIOSCOMD19]**:

    `passive-interface [default]` [*interface-type interface-number*]

    o The `default` keyword (optional) is used to configure all interfaces on the router as passive.
    o The parameter *interface-type* (optional) specifies the interface type.
    o The parameter *interface-number* (optional) specifies the interface or subinterface number.

    When this command is configured on an interface, the RIP router will stop sending routing updates to the IP multicast address (224.0.0.9) on that (configured passive) interface. However, the RIP router will still continue to receive and process (normally) routing updates sent from its RIP neighbors on that interface. The command prevents RIP from sending routing updates out the specified interface, but allows RIP to receive updates through the interface.

    For example, the configuration `passive-interface serial0/0/0` on a RIP router sets the interface serial0/0/0 as passive, meaning the RIP router will not send routing updates out this interface. The configuration `passive-interface default` on the RIP router sets all interfaces as passive.

- **Passive Interfaces for EIGRP**: In normal operations, EIGRP routers send and receive EIGRP HELLO and UPDATE packets on all their EIGRP-enabled interfaces (see "EIGRP Message Types" section in Chapter 6). In Cisco IOS, the `passive-interface` command can be used in router configuration mode to suppress the transmission of EIGRP HELLO and UPDATE packets on an interface while still including the IP address of the interface in the EIGRP Topology Table **[CISCEIGRPCOMD18] [CISCIOSCOMD19]**:

`passive-interface [default]` [*interface-type interface-number*]

The `passive-interface` command can be used to select the interfaces over which an EIGRP router will not form EIGRP adjacencies with neighbors, but will allow the router to include the IP addresses of those interfaces in the EIGRP Topology Table.

When this command is configured, the EIGRP router adds the networks defined on the passive interfaces to its EIGRP Topology Table but suppresses the transmission of EIGRP HELLO and UPDATE packets over the interfaces. Inbound and outbound EIGRP HELLO packets are suppressed over the passive interface, preventing the link partners from establishing EIGRP adjacency relationships.

- **Passive Interfaces for OSPF**: When the `passive-interface` command is configured on an OSPF router interface, no OSPF routing information is sent or received through that router interface. The specified interface (i.e., passive interface) appears, in effect, as a stub network in the OSPF Autonomous System since the interface does not send or receive routing information **[CISCIOSCOMD19]**.

The OSPF router does not send or receive OSPF Hello packets over the passive interface. This also prevents the OSPF router from forming neighbor relationships with other OSPF routers on the specified interface. A more efficient or effective way of controlling OSPF routing updates is to create an OSPF stub area, a totally stubby area, or a NSSA as discussed in Chapter 1 of Volume 2 of this two-part book.

When this command is configured on an IS-IS router interface, IS-IS will advertise the IP addresses for the configured interface (passive interface) without actually running the IS-IS protocol on that interface **[CISCIOSCOMD19]**.

7.7.7 Static Routes

Static routes are user-defined routes that are manually entered in the Routing Table and allow packets to travel from a source to a destination along a specified path. Such routes do not change unless explicitly updated, and typically, consist of a small number of router hops. Unlike routes that are learned via dynamic routing protocols, which must be imported into the Routing Table each time a network address or prefix comes online, static routes become useable immediately as soon as they are installed in the Routing Table. Static routes are assigned the default Administrative Distance of 1, and are generally preferred over other types of routes, except directly connected networks (see Chapter 2).

Network administrators manually configured static routes for the following purposes:

- Define routes over a network to remote destinations to eliminate the need for a dynamic routing protocol.
- Define a route to a destination when there is no dynamically learned route, or when there is the need to override a dynamically learned route.

- Define specific routes to be use when two IP routing domains or Autonomous Systems are required to exchange routing information.
- Define routes for use as a gateway of last resort to which all unroutable IP packets will be sent.

However, a number of factors have to be considered when configurating static routes:

- To reduce the number of static route entries in the IP Routing Table of a router, a default static route can be defined, where appropriate as discussed in Chapter 4.
- If a router is to advertise a static route to other routers, there may be the need for the static route to be redistributed into another routing protocol.

The **ip route** command can be used to establish a static route to a destination **[CISCIOSCOMD19]**:

> **ip route** *prefix mask* {*ip-address* | *interface-type interface-number* [*ip-address*]} [**dhcp**] [*distance*] [**multicast**] [**name** *next-hop-name*] [**permanent** | **track** *number*] [**tag** *tag*]

- The parameter *prefix* specifies the IP route prefix for the destination.
- The parameter *mask* specifies the IP network mask for the destination.
- The parameter *ip-address* specifies the IP address of the next hop that is to be used to reach the destination.
- The parameters *interface-type* and *interface-number* specify, respectively, the router's interface type and interface number.
- The **dhcp** keyword (optional) causes a DHCP (Dynamic Host Configuration Protocol) server to assign a static route to a default gateway.
- The **multicast** keyword (optional) specifies that the static route being configured is a multicast route.
- The parameter *distance* (optional) specifies an Administrative Distance for the static route (in the range 1–255). The default Administrative Distance for a static route is 1 which gives it precedence over routes learned by any dynamic routing protocol. A route is considered unreachable when it has an Administrative Distance of 255. A static route with an Administrative Distance of 255 will never be installed into the IP Routing Table. The parameter *distance* allows the network administrator to give preference to the static route when multiple entries for the same network destination exist. Routes with the lowest Administrative Distance are preferred.
- The parameter *next-hop-name* associated with the **name** keyword (optional) specifies a name for the next hop route. This allows static routes to be associated with names (e.g., **name Ottawa2Toronto**). If several static routes are being configured, names can be specified that describe the purpose of each static route so that they can easily be identified.

Network Path Control and Factors That Affect Routing Table Properties 251

- The **permanent** keyword (optional) specifies that the static route will not be removed, even if the router interface on which the static route is configured shuts down.
- The parameter *number* associated with the **track** keyword (optional) specifies a number (in the range 1–500) that associates a track object (e.g., Domain Name System [DNS] server) with the static route. This parameter specifies that the static route will be installed in the IP Routing Table only if the state of the track object (e.g., a DNS server located somewhere in the network) is up.
- The parameter *tag* associated with the **tag** keyword (optional) specifies a tag value that can be used as a "match" value (in route maps) for controlling route redistribution.

Creating a static route in the IP Routing Table, requires at a minimum, the route to be defined as *static* and a next-hop IP address associated with it. In its simplest form, the **ip route** command used to configure a static route is:

   ip route *prefix mask* {*ip-address* | *interface-type interface-number* [*ip-address*]} [*distance*]

When configuring a static route, it is recommended to specify both the outbound interface and the next hop IP address for the route. In the case of a serial interface, which is a point-to-point interface, specifying only the outbound interface is sufficient. If an Ethernet interface is used as the outbound interface, then it is recommended to configure both the outbound interface and the next hop IP address **[CISCID118263STR19]**. Once a static route is created in the Routing Table, it is entered into the IP Forwarding Table when the next-hop address is reachable. The source router sends all traffic destined for the static route to the next-hop address for forwarding toward the destination.

In the following configuration example, packets for destination network 172.32.0.0 will be routed to a next-hop router at 172.32.7.8:

   ip route 172.32.0.0 255.255.0.0 172.32.7.8

When the Administrative Distance of a static route is increased to a value greater than that of a route learned by a dynamic routing protocol, the static route can be used as a standby route in the event the dynamically learned route fails. For example, routes learned by EIGRP have a default Administrative Distance of 90 for internal EIGRP routes, and 170 for external EIGRP routes. So, in order to let an EIGRP route to override a configured static route, an administrative distance that is greater than 170 can be specified for the static route. A static route with a higher Administrative Distance is referred to a *floating static route* (see Chapter 4). The floating static route is installed in the IP Routing Table only when the dynamically EIGRP learned route becomes unavailable.

In the following example configuration (an example of a floating static route), packets for destination network 10.10.0.0 will be routed to a next-hop router at 172.30.5.5 if a dynamically learned route with an Administrative Distance less than 110 is unavailable.

   ip route 10.10.0.0 255.0.0.0 172.30.5.5 110

7.7.8 Default Routes

A default route is a route in the IP Routing Table of a router that is used when no entry in the Routing Table matches the destination address of an IP packet. The default route becomes the route of last resort when the address is unknown. A default route also provides a network with a redundant path in the event of network failures and loss of connectivity (a gateway of last resort). When an IP router is not able to find a match in its Routing Table for a packet's destination address and a default route is defined, the router forwards the packet to the default route. Universally, the IPv4 default route is represented by the address and mask 0.0.0.0/0.0.0.0 (or 0/0).

Default routes, when used, can provide a small reduction in the size of the Routing Table. The need for default routes arises because, in many cases, a router may not be able to determine routes to all network destinations. To implement default routes, the common practice is to deploy some routers as "destination-aware routers" to provide full routing capabilities for a network, and configure the remaining routers with default routes to these "destination-aware routers".

The "destination-aware routers" have Routing Tables that contain routing information for the entire internetwork. The default routes can be propagated dynamically, or can be manually configured in the Routing Tables of the individual routers. Most IGPs support mechanisms for causing a "destination-aware router" to generate default routing information dynamically that is then propagated to other routers.

7.7.8.1 Setting Default Routes Dynamically

Routers can exchange the default route (0/0) via dynamic routing protocols. A router advertising a default route presents itself as a gateway of last resort to other routers in the network. Routers can learn the default route (0/0) via an IGP or BGP depending on the internetworking situation. To handle network failures and provide redundancy, routers in a network can learn default routes from multiple sources.

In BGP, the BGP Local Preference attribute can be set for different default routes to indicate which route is preferred as the primary route and which as the backup route. This allows the backup default route to be used when the primary route fails.

7.7.8.2 Setting Default Routes Statically

Network operators typically filter default routes that are dynamically learned to avoid traffic taking suboptimal (or unwanted) routes, or causing unintended routing behaviors. Thus, an Autonomous System can statically set its default routes to have more control on routing within the system. This way the network operator has more control on how the default routes are set rather than having to learn the routes dynamically which may not provide the desired routing behaviors.

The network operator may statically set a default route (0/0) to point to the IP address of the next-hop serving as the default gateway, point to a specific interface on a router, or point to a specific network address **[CISCHALABS00]**. A router can set its default route based on a network address it has learned from another router.

7.7.8.3 Configuring Default Routes

The `ip default-network` command can be used to configure a router to advertise a particular network as the default route or gateway of last resort [CISCID118263STR19]:

   `ip default-network` *network-number*

where the parameter *network-number* specifies the IP address of the network.

For example, once configured on an EIGRP router, the default route will be propagated into the EIGRP Autonomous System. Because there is no additional parameter in the command to specify the network mask, the `ip default-network` command can only be used to advertise an IP classful network. Other routers in the network will have to use their next-hop IP address to reach the advertised network (*network-number*) as their default route. The specified network (*network-number*) must be reachable before it is used to configure the default route.

The `ip route` command can also be used to configure a router to advertise a default route as the gateway of last resort [CISCID118263STR19]:

   `ip route 0.0.0.0 0.0.0.0` *interface-type interface-number* [*ip-address*]

The 0/0 static route can point to a network address, a gateway IP address, or a router interface as the default path. If there is the need to configure more than a single default route, the `ip route 0.0.0.0 0.0.0.0` command can be used several times to point to multiple networks or IP addresses. A default route can be configured using either the `ip default-network` or the `ip route` command. Details about the use of the `ip route` command are given in [CISCID118263STR19].

If a router has a directly connected interface to the network that is specified as the default route, the IGP on that router will generate a default route [CISCIPCONGUI20]. In RIP, if the router has a path to the default network, a route that links the router to the pseudonetwork 0.0.0.0 will be advertised. The network 0.0.0.0 does not really exist; RIP uses this network to implement the default routing feature [CISCIPCONGUI20]. The router will advertise the default network if a default route was learned by RIP, or if the router has a gateway of last resort and a RIP default metric is configured.

A router that generates the default route for a network might also need a default route of its own. One way to configure a router to generate its own default route, is to specify a static route to the network 0.0.0.0 through the appropriate next-hop router [CISCIPCONGUI20].

No further configuration is required when default routing information is passed through an IGP. The router simply examines its Routing Table periodically to select the optimal default network to be used as its default route. In EIGRP networks, for example, the Routing Table might contain several candidate networks that can be used as the default route. In Cisco routers, the Administrative Distance and routing metric information is used to set the default route or the gateway of last resort [CISCIPCONGUI20].

If default routing information is not being passed to the local router dynamically (via an IGP), the candidate networks for the default route can be specified using the `ip default-network` command. If the specified network appears in the router's Routing Table from any routing information source, the router will flag it as a possible choice for the default route. If the router does not have an interface on the specified default network, but does have a path leading to it, the router will consider the network as a possible candidate default route, and then will select the router that leads to the best default route to become the gateway of last resort **[CISCIPCONGUI20]**.

7.7.8.4 Generating Default Routes in RIP

The **default-information originate** command can be used to configure a RIP router to generate a default route into a RIP routing domain. The command syntax for RIP is as follows **[CISCRIPCOMD18]**:

```
default-information originate [on-passive | route-map map-name]
```

- The **on-passive** keyword (optional) causes the RIP router to send default routes only on passive interfaces.
- The parameter *map-name* associated with the **route-map** keyword (optional) specifies that the RIP routing process will generate the default route if some conditions in the route map are satisfied (see "Route Maps" section below).

The default route is a network route with which the RIP router will communicate when no other route exists in the router's IP Routing Table for a given IP packet's destination address. When the **default-information originate** is configured with the on-passive keyword specified, the RIP router will send the default route on that passive interface.

7.7.8.5 Generating Default Routes in EIGRP

The **default-information** command can be used to configure an EIGRP router to accept an exterior route or a default route into an EIGRP routing process **[CISCEIGRPCOMD18]**:

```
default-information {allowed {in | out} | in | out} [acl-number | acl-name]
```

- The **allowed** keyword configures the EIGRP router to accept default routing information.
- The **in** keyword configures the EIGRP router to accept exterior or default routing information.
- The **out** keyword configures the EIGRP router to advertise external routing information.
- The parameter *acl-number* (optional) specifies a standard access list number (in the range 1–99), or an expanded standard access list (in the range 1,300–1,999).
- The parameter *acl-name* (optional) specifies the name of a standard access list.

This command can be used to redistribute the network 0.0.0.0 into an EIGRP Autonomous System.

7.7.8.6 Generating Default Routes in OSPF

The **default-information originate** command with the following syntax can be used to configure an OSPF router to generate a default external route into an OSPF Autonomous System **[CISCOSPFCOMD19]**:

Network Path Control and Factors That Affect Routing Table Properties

```
default-information originate [always] [metric metric-
value] [metric-type type-value] [route-map map-name]
```

- The **always** keyword (optional) causes the router to always advertises the default route regardless of whether the routing process has a default route in its Routing Table or not.
- The parameter *metric-value* associated with the **metric** keyword (optional) specifies a metric to be used for generating the default route. The default metric value is 10 if a value is omitted and at the same time a value is not specified using the **default-metric** router configuration command.
- The parameter *type-value* associated with the **metric-type** keyword (optional) specifies an OSPF external link type to be associated with the default route that is being advertised into the OSPF Autonomous System: External Type-1 route or External Type-2 route. The default *type-value* is External Type-2 route.
- The parameter *map-name* associated with the **route-map** keyword (optional) specifies that the OSPF routing process will generate the default route if some conditions in the route map are satisfied.

The way OSPF generates default routes depends on the OSPF area type the default route is being injected into:

- In a standard OSPF area, the ABR does not automatically generate default routes. Instead, the **default-information originate** command must be used on the router.
- In an OSPF Stub Area and a Totally Stubby Area, the ABR automatically generates a default route via Summary Link-State Advertisements (LSAs) (Type-3 LSAs) with the Link-State ID set to 0.0.0.0 (see Chapter 1 of Volume 2 of this two-part book). The **default-information originate** command is not required for configuring a default route even if the ABR does not have a default route in its Routing Table.
- In an NSSA, the ABR can generate a default route, but not by default. Instead, the **area** *area-id* **nssa default-information-originate** command can be used to force the ABR to generate the default route.
- In a Totally NSSA, the ABR automatically generates a default route via a Summary LSAs (Type-3 LSAs).

Whenever the **redistribute** or the **default-information** command is used to redistribute routes into an OSPF Autonomous System, the redistributing OSPF router automatically becomes an ASBR. An ASBR does not, by default, generate and advertise a default route into the OSPF Autonomous System. A default static route (0.0.0.0 0.0.0.0) also needs to be configured on the router (ASBR) originating the default route.

7.7.8.7 Generating Default Routes in IS-IS

The default-information originate command can be used generate a default route into an IS-IS routing domain **[CISCISISCOMD20]**:

```
default-information originate [route-map map-name]
```

The parameter *map-name* associated with the `route-map` keyword (optional) specifies the routing process that will generate the default route if the route map conditions are satisfied. If the router on which this command is configured has a route to 0.0.0.0 (or 0/0) in its IP Routing Table, IS-IS will originate Link-State Packets (LSPs) advertising 0.0.0.0.

When a route map is not specified, the router will advertise the default route only in Level 2 LSPs. When Level 1 routing is used, Level 1 routers in an IS-IS area can use special IS-IS specific method to find the default route. A Level 1 router in an area will simply look for the closest Level 1-2 router, and then send data to this router. The Level 1 router finds the closest Level 1-2 by checking the Attached (ATT) bit in Level 1 LSPs (see Chapter 2 of Volume 2 of this two-part book).

A route map can be used for the following purposes:

- Cause the router to generate a default route in its Level 1 LSPs.
- Cause the router to advertise 0.0.0.0 conditionally.

When the `match ip address` *standard-access-list* command is used, it specifies one or more IP routes that must exist in the Routing Table before the router will advertise 0/0.

7.7.8.8 Generating Default Routes in BGP

The special section "Default Routes in BGP" below discusses in detail the methods used in BGP to advertise a default route. The `default-information originate` command (with no arguments or keywords) can be used to configure a BGP routing process to advertise network 0.0.0.0 (the default route) into an Autonomous System **[CISCBGPCOMD19]**. The `neighbor default-originate` command can be used to configure a BGP router to propagate the network 0.0.0.0 to a BGP neighbor for use as a default route.

7.7.9 ROUTE MAPS

In IP networking, route filters appear in various forms such as route maps, distribute lists, or prefix lists. Route filters can be configured in IP routers to provide specific control of routing updates in a network, and as part of security mechanisms that can hide specific network destinations from receiving routing information and traffic. Route maps and traditional ACLs are similar in function, except route maps support features that provide far more flexibility and control on filtering than ACLs**[CISCID49111RMP05] [CISCIOSCOMD19] [CISCTEARDIA10]**. The typical route map not only permits a route to be redistributed into a routing protocol, but can also modify information associated with the route when it is redistributed (e.g., routing metrics). A route map can also be used to verify if a route is an internal or an external route.

7.7.9.1 Route Map Applications

Similar to ACLs, route maps can be used for various applications as described below. However, the actual implementation of a route map varies and depends on the particular application:

- **Route Redistribution**:
  o We have already discussed above that, route filtering (via route maps) can be used during route redistribution. We saw that all IP routing protocols (RIP, EIGRP, OSPF, IS-IS, BGP) can use route maps for filtering during route redistribution (see "Route Redistribution Configuration Tools" section above).
  o The `route-map` keyword is an option that can be applied to the `redistribute` command when redistributing routes. The `redistribute` command uses the name specified by the *map-tag* parameter to reference a route map. Multiple route maps may also share the same map tag (*map-tag*). Route maps provide greater control over how routes are redistributed between routing protocols.
- **Policy-Based Routing (PBR)**:
  o Another application of route maps is to enable PBR. PBR allows a network operator to define routing policies other than that based on the standard or normal IP destination addressed-based packet forwarding. In the standard routing method, packets are forwarded based on the routing information provided by the IP Routing Table. Route maps can be used to determine which packets are subjected to policy routing. Route maps can be created on more sophisticated criteria for PBR, allowing complex routing decisions to be implemented.
  o A route map, for example, can be configured on a router to control and filter routing updates in one direction, thereby affecting traffic flowing in the opposite direction with the goal of preventing that traffic from reaching certain network destinations.
  o By using route maps with route tagging, routing priorities for specific destinations can be defined among multiple paths, allowing these prioritized paths to be used in a deterministic manner. For example, when multiple exit points exist out of a company network to the Internet connection, route maps can be used to tag and define priorities for specific destinations along those multiple exit points.
  o Routes maps can be applied to PBR using the `ip policy route-map` interface configuration command as discussed in the "Using Policy Based Routing (PBR) for Path Control" section below.
- **Network Address Translation (NAT)**:
  o Route maps can be applied to NAT to provide more control over which private IP addresses can be translated to public routable IP addresses. For example, an internal IP address (for instance a server) may be translated to one public IP address when communicating with a particular company partner, and to a different public IP address when communicating with users on the public Internet.
- **BGP Policy Implementation**:
  o Route maps play an important role when implementing BGP routing policies (importing and exporting routes). They are the primary tools for implementing inbound and outbound route filtering in BGP policies.

7.7.9.2 Defining a Route Map

In Cisco IOS, the `route-map` command can be used in global configuration mode to define conditions for redistributing routes from one routing protocol to another, or to enable policy routing **[CISCEMPGARROT14] [CISCIOSCOMD19]**:

> `route-map` *map-tag* [`permit` | `deny`] [*sequence-number*] `ordering-seq` *sequence-name*

- The parameter *map-tag* specifies the name for the route map.
- The `permit` | `deny` keywords (optional) specify the action to be taken if the route map match conditions are met; the meaning of permit or deny is dependent on how the route map is used (route advertisement or route redistribution). For example,
  - The `permit` keyword (optional) is used to permit only routes that match the route map to be passed on or redistributed.
  - The `deny` keyword (optional) is used to block routes that match the route map from being passed on or redistributed.
- The parameter *sequence-number* (optional) specifies a number that indicates the position that a new route map statement will take in the list of route map statements that have already been configured with the same name (*map-tag*).
- The parameter *sequence-name* associated with the `ordering-seq` keyword (optional) orders the route maps based on the string provided.

Figure 7.15 describes, at a high-level, the operation logic of a route map (for route distribution). A typical route map has the following structure **[CISCIOSCOMD19]**:

- A route map comprises a list of commands or statements.
- Similar to an access list, the list is processed from top to bottom.
- Sequence numbers (see parameter *sequence-number*) are used for deleting or inserting specific statements. Each route map statement is assigned a sequence number which allows the route map to easily edited.
- Each route-map command has associated with it, a list of `match` and `set` commands. The `match` and `set` commands are used to define the conditions under which routes can be redistributed from one routing protocol to another, conditions for PBR, or conditions used for BGP policy.
  - The `match` commands are used to define the conditions to be checked, for example, specify the conditions under which routes can be redistributed; these constitute the *match criteria*.
  - The `set` commands specify the particular actions to be performed if the criteria specified by the `match` commands are satisfied; these constitute the *set actions*. A specific action may add or modify the characteristics or parameters of redistributed routes that have met a match condition, such as their metrics.
  - No particular emphasis is placed on the order of the `match` commands, only that, a route must "pass" all `match` commands to be redistributed according to the *set actions* specified by the `set` commands.

Network Path Control and Factors That Affect Routing Table Properties

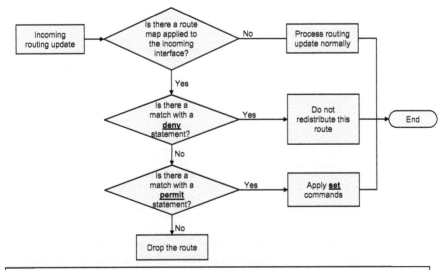

- **deny** — If the match criteria are met for the route map and the **deny** keyword is specified, the route is not redistributed.
- **permit** — If the match criteria are met for this route map, and the **permit** keyword is specified, the route is redistributed as controlled by the **set** actions. If the match criteria are not met, and the **permit** keyword is specified, the next route map with the same map tag is tested. If a route passes none of the match criteria for the set of route maps sharing the same name, it is not redistributed by that set.

FIGURE 7.15 Using Route Maps for Route Redistribution

  o A route map may have several parts (i.e., multiple references or clauses) when routes are passed through a route map, and any route that does not satisfy at least one **match** clause is ignored.
- If the *match criteria* are satisfied, and the **permit** keyword is specified in the **route-map** command, the route is redistributed according to the *set actions*. If the *match criteria* are not satisfied, and the **permit** keyword is specified, the next route map with the same map tag is processed. If none of the *match criteria* for the set of route maps sharing the same map tag (*map-tag*) are met, the route will not be redistributed.
- If the *match criteria* are satisfied and the **route-map** command includes the **deny** keyword, the route will not be redistributed.

To configure route maps for route redistribution, the following steps may be taken [CISCEMPGARROT14]:

- Define and name the route redistribution route map using the route-map command.
  o Define the match criteria (the match statements).
  o Define the set actions to be performed when there is a match (the set statements).
- Specify the route map to be used when redistributing routes.
  o Run the **redistribute** *protocol* **route-map** *map-tag* command.

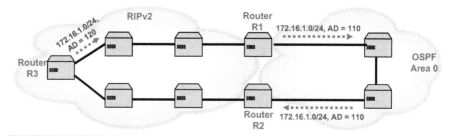

When routes are redistributed by more than one router such as in the multi point two-way redistribution configuration on R1 and R2, there is the possibility that routes can be feed back into a routing domain and cause sub optimal routing. The following explain show such a routing feed back loop can occur:
- RIPv2 on R3 advertises network 172.16.1.0 and R1 red is tributes that network into OSPF.
- OSPF then propagates this route through the OSPF domain and an OSPF router eventually advertises the network 172.16.1.0 to R2.
- R2 then redistributes 172.16.1.0 from OSPF back into the original RIPv2 network creating a routing feed back loop.

To prevent the routing feed back loop, a route map can be applied to Routers R1 and R2:
- A route map can be configured such that any route matching 172.16.1.0/24 is denied and will not be redistributed back into RIPv2.
- A route map can be configured such that all other routes are permitted to be redistributed into RIPv2 and will be assigned a RIPv2 metric of 5.

FIGURE 7.16 Route Feedback: Using Route Maps to Avoid Route Feedback

When not properly implemented, route redistribution can cause routing loops as well as route feedback in the network. Route feedback happens when a route that is redistributed from a routing protocol domain gets redistributed back into the same protocol domain from which it originated (see Figure 7.16). This route feedback can lead to suboptimal routing or routing loops in the in the overall internetwork. Route feedback and routing loops might occur when routes are redistributed by more than one router such as in a two-way multipoint route redistribution (Routers R1 and R2 in Figure 7.16).

As explained in Figure 7.16, route maps can be configured on Routers R1 and R2 to deny routes coming from a particular routing domain from being redistributed into that domain. Sometimes it is difficult to know which routing source or protocol originated a particular route, so the best way to filter redistributed routes is to use route maps with route tagging. In this case, a router sets a route tag in a route map which enables routers in the internetwork to correctly filters out redistributed routes that are being fed back into their source routing domains.

Routers R1 and R2 can tag RIP routes going into the OSPF domain, and also deny or filter these already redistributed routes from going back from OSPF into RIP. For example, let us assume a route tag number of 222 is used to identify RIP routes, and a tag of 111 to identify OSPF routes. The following statements can be configured on both Routers R1 and R2 to tag and filter redistributed routes **[CISCEMPGARROT14]**:

- Create a route map named RIPtoOSPF to **deny** the redistribution of all routes having the route tag value 111 (i.e., OSPF routes).
    ```
    route-map RIPtoOSPF deny 10
     match tag 111
    ```

Network Path Control and Factors That Affect Routing Table Properties   261

- Create a second statement for route map RIPtoOSPF to **permit** the redistribution of all other routes having the route tag value 222.
    ```
    route-map RIPtoOSPF permit 20
    set tag 222
    ```
- Create a route map named OSPFtoRIP to deny the redistribution for all routes having the route tag value 222.
    ```
    route-map OSPFtoRIP deny 10
    match tag 222
    ```
- Create a second statement for route map OSPFtoRIP to permit the redistribution of all other routes having the route tag value 111.
    ```
    route-map OSPFtoRIP permit 20
    set tag 111
    ```
- Enters OSPF configuration mode.
    ```
    router ospf 11
    ```
- Redistributes all RIP routes having route tag value 222 into the OSPF domain.
    ```
    redistribute rip route-map RIPtoOSPF
    ```
- Enters RIP configuration mode.
    ```
    router rip
    ```
- Redistributes all OSPF routes having route tag value 111 into the RIP domain.
    ```
    redistribute ospf route-map OSPFtoRIP
    ```

The result of this configuration on both Routers R1 and R2 is to ensure that only routes originating from the RIP domain are redistributed into OSPF, while only routes originating from the OSPF domain are redistributed into the RIP domain.

7.7.10 Distribute Lists

A distribute list is another way of controlling how routing updates are sent out or received into the IP Routing Table. A distribute list allows an ACL or a route map to be applied to incoming or outgoing routing updates for route filtering purposes [CISCBGPCOMD19] [CISCEMPGARROT14] [CISCIOSCOMD19]. Using distribute lists, network administrators can control which routes get distributed into and out of an IP router for security, management, and other network routing policy implementation purposes. Similar to route maps, distribute lists can also be used to filter routes in order to avoid route feedback (as discussed above).

It is important to distinguish the difference between distribution lists and ACLs. Distribute lists are used for controlling (filtering) routing updates while ACLs are used for filtering user traffic. In Cisco IOS, the `distribute-list in` router configuration command is used for filtering incoming routing updates while the `distribute-list out` command is used for filtering outgoing routing updates. An access list, a gateway, prefix list, or route map, must be defined on the IP router prior to configuring the `distribute-list` commands.

7.7.10.1 Filtering Incoming Routing Updates

The `distribute-list in` command can be used in router configuration mode in an IP router to filter routes/networks received in routing updates:

distribute-list {{*access-list-number* | *access-list-name* | **gateway** *prefix-list-name* | **prefix** *prefix-list-name*} **in** [*interface-type interface-number*] | **route-map** *route-map-name*}

- The parameter *access-list-number* or *access-list-name* specifies the number or name of the IP access list that defines which routes/networks can be received, and which can be suppressed in routing updates. The range of *access-list-number* is from 1 to 199.
- The **gateway** keyword specifies that filtering has to be done on incoming routing updates based on a gateway.
- The parameter *prefix-list-name* specifies the name of the IP prefix-list that defines which routes from the specified IP prefixes can be received, and which can be suppressed in routing updates.
- The **prefix** keyword specifies that filtering has to be done on IP prefixes in incoming routing updates.
- The parameter *interface-type* (optional) defines the type of interface from which routing updates are to be received or suppressed.
- The parameter *interface-number* (optional) specifies the interface number on which the specified access list should be applied to incoming routing updates. The access list is applied to all incoming routing updates if no interface is specified. The *interface-type* and *interface-number* arguments are applied only if an access list is specified, and not a route map.
- The parameter *route-map-name* associated with **route-map** keyword (optional) specifies the name of the route map that defines the routes/networks from which routing updates can be received or suppressed.

The **distribute-list in** command is used to filter routing updates going through the specified interface into the IP routing process under which it is configured.

When an IP prefix list is specified, it is used to filter routes based on the bit length of the IP address prefix. A single host route, subnet, entire network or supernet, can be specified. When configuring a distribution list, either a prefix list or an access list can be configured, not both. Also, this command must specify either an access list (*access-list-number* or *access-list-name*) or a route map (*map-tag*), not both.

Route filters such as distribute lists do not filter OSPF LSAs or the routing information entering the OSPF LSDB (Link-State Database). As discussed in Chapter 2, a basic tenet of link-state routing protocols like OSPF is that, all routers in an OSPF area must have identical LSDBs. This means OSPF routes or routing updates cannot, and must not be prevented from entering the LSDB. Thus, when the **distribute-list in** command is used for OSPF, it only filters and prevents routes from entering the IP Routing Table, and not LSAs from being propagated **[CISCEMPGARROT14] [CISCIOSCOMD19]**.

7.7.10.2 Filtering Outgoing Routing Updates

The **distribute-list out** command can be used to suppress routes/networks from being advertised in routing updates:

```
distribute-list {access-list-number | access-list-name} out
[interface-name | routing-process | as-number]
```

- The parameters *access-list-number* or *access-list-name* specify the number or name of a standard IP access list. The number/name refers to the list that defines which routes/networks can be advertised, and which can be suppressed in routing updates.
- The parameter *interface-name* (optional) specifies the name of a particular router interface.
- The parameter *routing-process* (optional) specifies the name of a particular routing process, or the option of using the `static` or `connected` keyword. When a *routing-process* is specified, the access list is applied to only the routes derived from the specified routing process.
- The parameter *as-number* (optional) specifies an ASN.

The `distribute-list out` command is used to filter routing updates (routes) going out of the IP routing process (under which it is configured) into the specified interface. According to reference **[CISCEMPGARROT14]**, the `distribute-list out` command filters only on the routes being redistributed by an ASBR into an OSPF Autonomous System, and applies to External Type-1 and External Type-2 routes and not to OSPF intra-area and interarea routes.

To configure a distribute list for route filtering, the following steps may be taken:

- Identify the routes/networks to be filtered using an ACL or a route map.
- Associate the distribute list with the specified ACL (*access-list-number* or *access-list-name*), or route map (*route-map-name*) using the **distribute-list** command.

7.7.11 Prefix Lists

Prefix filtering using prefix lists is a technique for marking routes for possible exclusion from the Routing Table of an IP router. This is done by matching the network address prefix in a route against a list of network prefixes that the router maintains **[CISCBGPCOMD19] [CISCEMPGARROT14]**. A prefix-list entry consists of an IP address and a network mask. The IP address can be for a single host route, a subnet, or a classful network. Prefix lists can be configured to filter routes based on an exact match of a network address, or match a range within the address.

The primary use of prefix list is filtering routing information exchanged between IP routers. Generally, prefix lists offer a more intuitive way of configuring filtering rules than traditional access list. Prefix lists (using statement sequence numbers) allow relatively more flexible and easy additions, deletions, and modification to a list without the need to completely rebuild or reconstruct the list.

The `ip prefix-list` command can be used in global configuration mode to create a prefix list or add a prefix-list entry **[CISCBGPCOMD19]**:

ip prefix-list {*list-name* [**seq** *number*] {**deny** | **permit**} *network/length* [**ge** *ge-length*] [**le** *le-length*] | **description** *description* | **sequence-number**}

- The parameter *list-name* specifies a name that identifies the prefix list.
- The parameter *number* associated with **seq** keyword (optional) specifies a sequence number for a prefix-list entry. The sequence number can take a value from 1 to 4,294,967,294. If a sequence number is not entered, the number 5 is applied to the first prefix-list entry, and subsequent unnumbered entries are incremented by 5 (default sequence numbers are in increments of 5 (i.e., 5, 10, 15, and so on).
    o A prefix list consists of statements with sequence numbers. A search for a match begins at the top of the prefix list, which is the statement with the lowest sequence number. When a match is found, the search ends and there is no need to go through the rest of the prefix list.
- The **deny** or **permit** keywords specify the action to be taken when a match is found. For example, **deny** denies access for a matching condition, and **permit** permits access for a matching condition. For example, if a route lookup yields a permit, the route is used, and if it yields a deny, the route is not used.
- The parameter *network/length* specifies the IP address (or IP prefix) to be matched and the network mask length (a number from 1 to 32).
- The parameter *ge-length* associated with the **ge** keyword (optional) specifies the minimum prefix length to be matched.
- The parameter *le-length* associated with the **le** keyword (optional) specifies the maximum prefix length to be matched.
- The parameter *description* associated with the **description** keyword (optional) specifies a descriptive name for the prefix list (1–80 characters in length).
- The **sequence-number** keyword (optional) enables or disables the use of sequence numbers for the prefix list.

A prefix list is evaluated starting with the lowest sequence number, and the processing progresses down the list (i.e., continues on higher sequence numbered statements) until a match is found. When an IP address match is matched, the permit or deny statement is applied to that route/network and the rest of the list is not evaluated. When a route is evaluated against the prefix list, the first statement that matches (either a permit or deny) ends the processing.

To increase processing efficiency, the prefix list statements that are more frequently processed (i.e., the most common [permits or denies] matches) should be assigned the lowest sequence numbers so that they can be placed near the top of the prefix list. The parameter *number* and the **seq** keyword can be used for re-sequencing statements.

Prefix lists can be used as an alternative to route maps, distribute lists, and access lists in many cases where route filtering is required. Figure 7.17 shows a networking example where route maps and prefix lists can be configured to accomplish similar

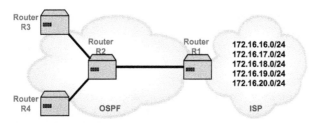

In the OSPF network, Router R1 receives route sad vertised by the ISP and in turn advertises some of these routes to Router R2. The configuration involves filtering some specific routes:
 • Router R1 advertises only routes 172.16.17.0/24, 172.16.18.0/24, and 172.16.19.0/24 to Router R2.
 • Router R3 accepts only the route 172.16.18.0/24.
 • Router R4 accepts all the routes advertised by Router R2.
 Multiple approaches exist to configure the network to meet these requirements, such as using route maps and prefix lists:
 • **Using Route Maps:**
   • Configure a route map on Router R1 and as an export policy of R1 to be used byOSPF.
   • Configure another route-map on Router R3 and as an import policy of Router R3 to be used by OSPF.
 • **Using IP Prefix Lists:**
   • Configure an IP prefix list on Router R1 and as an export policy of R1 to be used by OSPF.
   • Configure another IP prefix list on Router R3 and as an import policy of R3 to be used by OSPF.
 When compared with IP prefix lists, route maps allow route attributes to be modified and provide a more flexible way of controlling routes, but are relatively more complex to configure.

FIGURE 7.17 Example: Filtering Received and Advertised Routes using Route Maps and Prefix Lists

route filtering results. The characteristics of a prefix list include, support for incremental modifications of a list, significant performance improvement over ACLs when looking up routes in a large list, and greater flexibility in specifying network mask ranges.

7.7.12 USING POLICY BASED ROUTING (PBR) FOR PATH CONTROL

PBR (using, e.g., route maps) allows IP routers to make packet forwarding decisions using routes configured via network routing policies rather than those provided by the IP Routing Table and maintained by the routing protocols. The previous chapters show that each routing protocol has its own internal mechanisms for determining the shortest or optimal paths to network destinations. PBR allows a network operator to define a routing policy to meet internal administrative and operational needs that is different from what would be provided by a routing protocol. This section describes how route maps can be used to implement PBR. With PBR, route maps can be used to forward packets based on source and destination IP addresses, IP protocol type, and end-user applications.

The **ip policy route-map** command, can be used in addition to the **route-map** command, and the **match** and **set** commands, to define conditions for policy routing [CISCIOSCOMD19] [CISCTEARDIA10]. The **ip policy route-map** command can be used in interface configuration mode to identify a route map that can be used for policy routing on a router interface:

   ip policy route-map *map-tag*

- The parameter *map-tag* specifies the name of the route map that is to be used for policy routing. This name must match the *map-tag* specified by the **route-map** command.

To implement PBR, the following **route-map** related commands can be used:

    route-map *map-tag* [**permit** | **deny**] [*sequence-number*]
    match {*conditions*}
    set {*actions*}
    ip policy route-map *map-tag*

- The **route-map** command defines the route map conditions.
- The **match** command specifies the *conditions* under which policy routing is to be performed on the interface; these are the *match criteria*. This command defines the conditions to be checked and matched (e.g., **match ip address** command which specifies criteria to be matched using ACLs or prefix lists).
- The **set** command specifies the routing *actions* to be performed if the match criteria set by the **match** commands are satisfied; these are the *set actions*. The **set** command can be used to define various packet forwarding parameters, such as the next-hop address or interface to which an IP packet should be sent (e.g., **set ip next-hop** command which specifies the next hop IP address for matching packets). Some packets may be policy routed while other packets may be routed via the standard/normal IP packet forwarding algorithm (see Figure 7.18).

The route map statements used for PBR can be configured as **permit** or **deny**:

- If the **deny** keyword is specified, a packet meeting the *match criteria* is not policy-based routed but is forwarded through the normal destination-based IP forwarding process, using the information in the IP Routing Table (Figure 7.18).

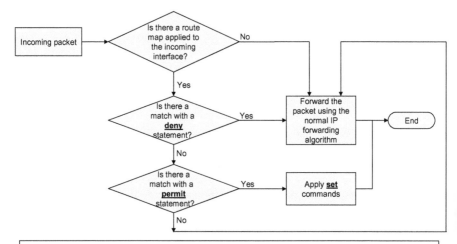

- **deny** — If the match criteria are met for the route map and the **deny** key word is specified, the packet is not policy routed, instead, the normal IP forwardng algorithm is used.
- **permit** — If the match criteria are met for this route map, and the **permit** key word is specified, the packet is policy routed as controlled by the **set** actions. If the match criteria are not met, and the **permit** key word is specified, the next route map with the same map tag is tested. If a route passes none of the match criteria for the set of route maps sharing the same name, the packet is not policy routed but forwarded through the normal IP forwarding algorithm.

FIGURE 7.18 Using Route Maps for PBR

- If the **permit** is specified, a packet meeting all the *match criteria* will have the **set** commands applied.
- If no *match criteria* are met in the route map, the packet is not dropped, instead, it is forwarded through the normal destination-based IP forwarding process.
- If there is the need to drop a packet when it does not match the specified criteria instead of using the normal destination-based IP forwarding process, a **set** statement can be configured as the last entry in the route map to route the packets to interface null 0 [CISCTEARDIA10]. The null 0 interface is a kind of "route to nowhere" interface, causing the packet to be dropped.

To implement PBR on incoming packets on a specific router interface, the following steps may be taken [CISCTEARDIA10]:

- Define and name the policy route map using the **route-map** command.
  o Define the *match criteria* (the conditions under which policy routing is allowed on the router interface).
  o Define the *set criteria* (the policy routing actions to be performed if the *match criteria* are met).
- Define the router interface to which the route map will be attached to using the **ip policy route-map** command.
- Apply the defined route map to the specified incoming interface.

It is important to note that when configuring PBR, the PBR is applied to incoming packets on a given router interface (and not on the interface on which packets are sent). PBR causes the router to use the configured route map to evaluate all packets received on the interface.

7.7.13 Offset Lists

An offset list is another mechanism for path control, but only applicable to distance-vector routing protocols such as RIP and EIGRP [CISCEIGRPCOMD18] [CISCRIPCOMD18] [CISCTEARDIA10]. An offset list is a mechanism that can be configured on a router to increase the metrics of incoming and outgoing routes learned via RIP or EIGRP. In Cisco IOS, the **offset-list** command in router configuration mode can be used to add an offset to incoming and outgoing metrics associated with routes learned via RIP or EIGRP:

offset-list {*access-list-number* | *access-list-name*} {**in** | **out**} *offset* [*interface-type interface-number*]

- The parameter *access-list-number* | *access-list-name* specifies the standard access list number or name to be applied. If *access-list-number* is 0, this indicates all networks (networks, prefixes, or routes). If *offset* is 0, no action is to be taken.
- When the **in** keyword is used, the access list is applied to incoming metrics.
- When the out keyword is used, the access list is applied to outgoing metrics.

- The parameter *offset* specifies a positive offset to be applied to metrics for IP network addresses matching the access list. If *offset* is 0, no action is to be taken.
- The parameter *interface-type* (optional) specifies the interface type on the router to which the offset list is applied.
- The parameter *interface-number* (optional) specifies the interface number on the router to which the offset list is applied.

When an offset value is configured on a RIP or EIGRP router, that value is added to the routing metrics of the routes passing through the specified interface. Based on the configured offset lists, a router can add a value to the metric of a route, either before a routing update is sent (the `out` keyword is used), or when a route is received in a routing update and is being accepted into the IP Routing Table (the `in` keyword is used). An offset list uses an ACL (i.e., parameter *access-list-number* | *access-list-name*) to match routes; any matched routes will have the specified offset applied/added to their metrics. Routes that do not match the offset list will not have their metrics changed. Figure 7.19 shows an example of a networking scenario where an offset list is used on a RIP router to make a specific route more preferable.

7.7.14   IP Service Level Agreement (SLA) Probes

In some cases, a service provider will create SLAs with some customers that define a number of service conditions they have agreed on. The SLAs form the basis for guaranteeing some specified level of service availability and performance when the

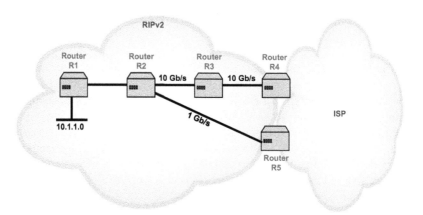

- We assume that users on network 10.1.1.0 attached to Router R1 can access the Internet through Routers R4 or R5.
- Router R2 receives routes is from each of the edge routers R4 and R5. The RIP routing metric between Routers R2 and R5 is smaller than the metric between Routers R2 and R4.
- Router R5 is only one hop away from R2 and therefore is the preferred RIP route to the ISP. However, the link between R2 and R5 is the slower speed link.
- An off set list and ACL can be configured on R2 to ensure that the preferred route to the network 10.1.1.0 will point towards Router R4.
- For example, an offset-list can be configured on R2 to add an offset of 2 to the metric of the routes learned from R5.

FIGURE 7.19   Using Offset Lists for Path Control

Network Path Control and Factors That Affect Routing Table Properties

customers use the services. An SLA may cover how the guarantees are monitored and how exception reports (i.e., failure to comply with the service guarantees) are generated. The service guarantees may include packet loss, delays, service availability, and maintenance and/outage notification times.

To monitor SLAs in real time, a service provider may use SLA probes to gather information about the performance of the network providing the service(s) to the customer. In this section, we discuss how network routing operations and path control can be based on the ability to continuously reach a target object in the network. That is, evoke network routing and path control decisions based on the reachability of the target object. Typically, the network uses probe traffic to detect the reachability of the object.

7.7.14.1 When to Use Cisco IOS IP SLA Probes

In this section, we describe how SLA probes can be used as part of some of the mechanisms used for path control [CISCIPSLAGUI18] [CISCTEARDIA10]. Figure 7.20 describes in detail an example networking scenario where path control using Cisco IOS IP SLA probes is a recommended solution. In this scenario, Router R1 is able to detect if there is a failure on its direct link to R2 (ISP 1) or the direct link to R3 (ISP 2), thus, allowing R1 to send all traffic to any reachable ISP if one these links fail. However, if a link within the infrastructure of any one of the ISPs fails, the

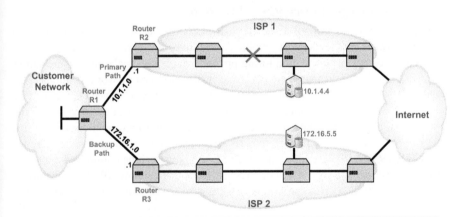

- Assume that Router R1 is multihomed to the Internet through ISP 1 and ISP 2.
- The customer network is multihomed to the two ISPs through Router R1. R1 is configured with two equal cost default static routes which enable it to load balance over the two links on a per-destination basis.
- We assume R1 can detect if there is a failure on its direct link to one ISP, and in that case use the other ISP for all traffic.
- How ever, a serious problem occurs if a link within ISP1 infra structure were to fail. This is be cause the link from R1 to R2 would still remain up and R1 would continue to use that link be cause the static default route would still be valid.
- The use of Cisco SLA probesis one effective solution to this situation.
- Static routes or PBR can be configured on R1 to be multihomed through ISP 1 and ISP 2 to the Internet, but make these routes subject to reachability tests to ward critical destinations, such as DNS servers within each ISP.
- If the DNS servers in anyone of the ISPs go down or are unreachable, R1 would remove the static route to ward that ISP.
- Configuring IP SLA probes to continuously check the reachability of a specific destination within the ISP (such as the ISP's DNS server, or any other specific destination) will conditionally announce that static route as available if the connectivity is verified to be available (using IP SLA probes).
- IP SLA probes can be configured on R1 to probe the DNS servers frequently. The IP SLA probes are attached or assigned to the static routes.

FIGURE 7.20 Using IP SLA Probes for Path Control

link (and static route) to that ISP will still remain up, and Router R1 would continue to use that link, because the static default route would still be valid.

The following are some solutions to the problem describe in Figure 7.20 but with IP SLA probes-based method offering a better solution [CISCTEARDIA10]:

- **Use a Dynamic Routing Protocol**: In this approach, a dynamic routing protocol is run between Router R1 and the ISPs, allowing R1 to learn and install the ISPs' networks in its Routing Table. With this, R1 will be aware of any link failures that occur within any of the ISPs. This solution is not deemed to be practical for smaller customer networks, and in many cases requires integration of a customer network with the two ISPs. This may be a better solution for larger customer networks or those with large traffic volumes.
- **Use Cisco IP SLA Probes with either Static Routes or PBR**: In this approach, static routes or PBR can be used from Router R1 to the ISPs, but with the routes to each ISP subject to reachability tests (by sending probe traffic to critical destinations within the ISP, such as DNS servers). Probe traffic is sent periodically to the DNS servers located within the ISPs. In this case, Router R1 would remove the static route to an ISP if the DNS servers in that ISP goes down or is unreachable. Cisco IOS IP SLA probes are applied to the static routes and are used to perform frequent reachability tests with the DNS servers within the ISPs.

7.7.14.2 Workings of Cisco IOS IP SLA Probes

As part of a path control tool, Cisco IOS IP SLAs uses active traffic monitoring to measure network performance at specific destinations in a target network. Cisco IOS IP SLAs sends performance monitoring probes (which are special messages) across a network in order to measure network performance between network locations. This feature allows network performance measurements to be taken between Cisco devices or between a Cisco device and a host (such as a network application server) with the overall goal of obtaining data that could provide a clearer picture about the service levels provided and required for IP services and applications. Reference [CISCTEARDIA10] indicates that the IP SLAs feature can be configured either via a Command-Line Interface (CLI) or an SNMP tool that supports IP SLAs operation.

Cisco IOS IP SLAs send probing messages across a network, and is able to measure network performance between any two network locations and across multiple network paths. The information collected via the Cisco IP SLAs probes can include one-way latency, network jitter (also called packet delay variation [PDV]), packet loss, response time, network resource availability, voice-quality scoring, server response time, and application performance. Other than path control, the information gathered from active monitoring has the following applications:

- **Verification of the Quality of Service (QoS) Offered to Customers**:
  o Measure the PDV, one-way and two-way network latencies, and packet loss experienced by customer traffic.
  o Provide reliable, predictable, and continuous QoS measurements.

Network Path Control and Factors That Affect Routing Table Properties   271

- **Provide Data That Can Assist Network Administrators with Network Troubleshooting**:
  o Provide consistent and reliable network performance measurements that can help immediately identify network problems and save troubleshooting time.
  o Allows the network administrator to proactively identify network issues and improve network availability. This helps the network administrator to better react to network performance metrics when failures and/or configuration changes occur in the network.
- **Provide Monitoring, Measurement, and Verification of Customer SLAs**:
  o Provide network availability monitoring and allow network administrator to validate network performance.
  o Provide better network performance profiling through active probing in a continuous, reliable, and predictable manner. These measurements and monitoring help to create a performance-aware network.
  o Allows the network administrator to verify customer service guarantees through active monitoring.
  o Monitor the performance of customer services and applications such Voice over IP (VoIP), video conferencing, real-time broadcasting, and multicasting streams.

Cisco IOS IP SLAs operations involve the use of *sources* and *responders*. The *source* is the network entity that sends probe messages to a *target* (*device*). The IP SLAs *source* (which initiates all IP SLAs measurement probe operations) can be configured either via a CLI or a SNMP tool that supports the operation of the Cisco IP SLAs [**CISCTEARDIA10**]. A *responder* in Cisco IP SLAs operations is a component that is embedded in a Cisco IOS device, that allows it to listen and respond to IP SLAs request messages sent to it.

Two types of Cisco IP SLA probe operations have been defined as described in Figure 7.21:

- In this configuration, the *target device* is a network device (such as a web server or IP host) that is not running the Cisco IP SLAs *responder* component.
- In this configuration, the *target device* is a network device (such as a Cisco router) that is running the Cisco IP SLAs *responder* component.

The accuracy of the IP SLAs probe measurement is significantly improved when the *target device* is also operating as an IP SLAs *responder*. The *source* and the *responder* use the IP SLAs Control Protocol to communicate, before the *source* starts sending probe messages. To secure the exchange of control messages, IP SLAs Control Protocol uses MD5 authentication similar to RIPv2, EIGRP, and OSPFv2.

To implement path control using Cisco IP SLAs, the following tools and processes are needed [**CISCTEARDIA10**]:

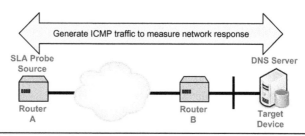

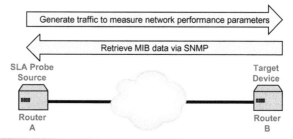

FIGURE 7.21 Types of IP SLA Probes Operations

- **Object Tracking**: This involves tracking the reachability of specific destinations (or objects) in the target network, for example, the reachability of specific objects such as DNS servers.
- **Cisco IOS IP SLAs Probes**: This involves sending different types of network monitoring probes toward the specified objects (DNS servers) in order to measure network performance. In an ICMP (Internet Control Message Protocol) Echo operation, ICMP Echo Requests are sent to a destination to check network connectivity.
- **Associate the Monitored Network Performance Results to the Routing Process**:
  o **Using Static Routes with Tracking Options**: This involves using static routes with the ability to change routing operations and control which path to use based on the ability to reach the tracked object (i.e., reachability of the tracked object). This solution is simpler and accommodates scenarios in which there is the need to choose/control which outbound or exit points (via a static route) from a network to direct all outbound traffic.
  o **Using PBR with Route Maps**: This involves using PBR with route maps to define specific traffic classes (such as voice, or specific applications) based on network conditions such as load, delay, and other network performance factors.

Network Path Control and Factors That Affect Routing Table Properties

To deploy Cisco IOS IP SLAs probes, it is very important to consider the impact of the additional probe traffic that is to be generated, including how the probe traffic affects network congestion and bandwidth utilization levels. The steps involved in configuring Cisco IP SLAs include the following **[CISCTEARDIA10]**:

- Define the Cisco IOS IP SLAs operations (or probes) to be carried.
- Define the tracking objects (e.g., DNS server) that track the state of Cisco IOS IP SLAs operations.
- Define the action associated with the tracking object.

An IP SLAs *operation* includes defining the type of protocol to be used for probe messages, frequency of probes, thresholds to trigger actions, and traps to be sent to a network management system. Each IP SLAs *operation* is specific to a *target device*. A number of tasks performed by the network administrator include configuring the IP SLAs *source* with the IP address of the *target device*, protocol, and TCP and UDP port numbers for the operation. Upon completion of the operation and the *response* has been received, the *source* stores the results in an IP SLAs MIB, which can then be retrieved using SNMP

Each IP SLAs operation can be described by the following sequence of events **[CISCTEARDIA10]**:

1. The IP SLAs operation starts with a *control phase*. At the beginning of this phase, the *source* sends a control message with the configured information for the IP SLAs operation to the *responder*.
    a. The *source* sends the control message to UDP port number 1967 which is the IP SLAs control port on the target router (the responder). Included in the control message are the protocol to be used, TCP or UDP port number, and duration of the IP SLAs operation.
    b. If MD5 authentication is enabled, the *source* computes an MD5 checksum which is included in the control message.
    c. The *responder* verifies the MD5 checksum to ensure that it is communicating with the right sender. The *responder* returns an "authentication failure" message if the authentication fails.
2. If the control message is successfully processed, the *responder* returns an "OK" message to the *source*. The *responder* then listens on the TCP or UDP port specified in the control message for a specified period of time.
    a. In the event the *responder* is not able to process the control message, it will return an error to the *source*.
    b. The *source* tries to retransmit the control message if it does not receive a response from the *responder*, and it will eventually time out if it does not receive a response.
3. If the *source* receives an "OK" message from the *responder*, it will move the IP SLAs operation to the *probing phase*.
    a. In this phase, the *source* sends one or more test packets to the *responder*.
    b. The *source* uses the response from these test packets to compute response times.

4. The *responder* receives the test packets and prepares the appropriate response messages.
    a. Based on the type of IP SLAs operation agreed upon with the *source*, the *responder* may include an "in" timestamp (for the received test message) and an "out" timestamp (for the outgoing response message) in the payload of the response packet to account for the local CPU processing time spent processing receive/test and transmit/response messages.
    b. The *source* uses these timestamps to make accurate assessments of the one-way delay and the processing time in the target device. To calculate the round-trip time (RTT), the source uses four timestamps.
    c. The *responder* then closes the user-specified TCP or UDP port after it has sent response packets to the IP SLAs test packet, or when the specified operation time has expired.

7.8 SPECIAL FOCUS: PATH CONTROL TOOLS IN BGP

BGP has a number of important features that are flexible enough, and make it possible for a network operator to control routing according defined policies. The strength of BGP for routing control lies in the path attributes and route filtering capabilities it supports. We saw in Chapter 2 that BGP path attributes are parameters that can be modified to influence how BGP routers perform best path selection. As discussed above, BGP routers can also perform filtering on a path level or network prefix level. A system of internetworks and Autonomous System can use a combination of BGP path attribute manipulation, route filtering, and defined policies to achieve the desired routing behavior. This section discusses the different methods available in BGP for path control.

7.8.1 BGP Route Filtering and Path Attribute Manipulation

In BGP, route filtering can be applied on a per-peer basis or on an IGP protocol level. Inbound filtering can be used to specify which routes a BGP router can receive from its peers. Outbound filtering specifies the routes that a BGP router can advertise to its BGP peers. At the BGP protocol level, inbound filtering specifies the routes that can be injected into BGP from a particular IGP, whereas outbound filtering specifies the routes BGP is allowed to inject into an IGP.

This section describes BGP route filtering and path attribute manipulation, which are the basic mechanisms BGP routers use to control routing information flow. Each BGP router receives UPDATE messages from its peers, and examines their contents to determine which BGP path attributes to manipulate according to the configured routing policies.

To control the routing information a BGP router learns from or advertises to its peers, route filtering can be applied to the routing updates the router receives from the peers. The process of filtering and manipulating BGP routes involve identifying specific routes, accepting or rejecting the routes, and manipulating the BGP path attributes of the routes accepted, if necessary. We discuss these actions in the following three subsections.

7.8.1.1 Identifying BGP Routes

To enable a BGP router identify which routes are to be accepted or rejected, and which routes should have their BGP path attributes manipulated, a set of criteria for differentiating or filtering routes have to be defined in the router. The criteria can be based, for example, on the IP address prefix (or NLRI) in the BGP UPDATE messages, the Autonomous System that originated the route, the number of Autonomous Systems that the route has traversed, and the BGP path attributes associated with the route.

Typically, the network administrator defines a list of criteria as part of the filtering rules used by the BGP router. The router compares the parameters of each route to the filtering rules to see if there is a match. A matching entry has associated with it the action(s) to be taken, such as accept route, reject route, or manipulate the values of some specified BGP path attributes.

Route identification can be based on the BGP AS_PATH attribute of the route and/or the NLRI of the route. The BGP AS_PATH attribute carries a list of the Autonomous Systems the route has traversed before reaching the current router. Using the AS_PATH attribute for route filtering provides a convenient method when filtering has to be done for all routes passing through the same or multiple Autonomous Systems. Using this attribute avoids having to list a large number of routes one at a time as would be done when using the IP prefix for route filtering. Using the BGP AS_PATH attribute, a filter list can be created and applied to both incoming and outgoing routing updates based on the values carried in the attribute **[CISCBGPCOMD19]**.

The network prefix and mask of the NLRI (in a received UPDATE message) can be used for route identification. The administrator may choose to define, instead, an address prefix or range of prefixes such that, if the address prefix of a route falls in that range, it can be selected for further processing. An IP prefix list, for example, can be used for route identification and filtering as discussed below.

7.8.1.2 Accepting/Rejecting BGP Routes

After the BGP router has identified a route (depending on the filtering rules defined), it can perform a number of actions on the route, such as reject it, accept it, and if necessary, accept it but modify some of the BGP path attributes associated with the route. The criteria for handling routes depend on the routing policies configured for the local Autonomous System. A route that is rejected is simply discarded without requiring further processing. A route that is accepted can be accepted but subjected to some BGP path attribute modification.

7.8.1.3 BGP Path Attribute Manipulation

When a BGP router decides to advertise a route to its peer, it determines if there is the need to modify or add more information to the BGP path attribute of the route before advertising it to the peer. A route that is accepted can have its path attributes modified (such as AS_PATH, MED), which in turn can affect the BGP best path selection process of BGP routers. Other than best path selection, BGP attribute manipulation plays an important part in the implementation of routing policies in the Autonomous System, as well as, influencing other behaviors such as load balancing in the internetwork.

7.8.2 BGP ROUTE FILTERING

Typically, BGP routers use AS-Path filter lists, prefix lists, route maps, among others, to perform route filtering, and if necessary, specify which routes would have their BGP path attributes manipulated. Route filtering can be configured to allow a network operator to permit or deny specific network prefixes sent to, or received from each BGP neighbor. This prevents BGP routers from inadvertently installing unwanted or illegitimate routes to their Routing Table. Import routing policies control the routing information the router places in its Routing Table, while export routing policies control the routing information the router advertises from its Routing Table to neighbor routers.

Route filtering ensures that the BGP routers will accept (inbound) or advertise (outbound) only the correct routes. The network operator can configure router filters to only permit known or legitimate network prefixes, and deny incorrect, unwanted or illegitimate routes that are sent to or received from each BGP neighbor (as defined by the network policy). The configured route filters can be applied to each BGP neighbor in both inbound and outbound directions.

7.8.2.1 BGP AS-Path Filter Lists

As discussed in Chapter 2, the AS-Path attribute (AS_PATH) is a well-known mandatory BGP path attribute that contains a list of ASNs that lie on the path a BGP routing update has traversed. An eBGP router prepends its ASN to the AS-Path attribute in all BGP routing updates it sends to other eBGP peers. When the update is advertised to a BGP router within the same Autonomous System (i.e., an iBGP peer), the AS-Path information is passed unaltered. When an eBGP router sends out an update to an eBGP peer, it adds its own ASN to the front (leftmost position) of the AS-Path attribute. The AS-Path lists all the Autonomous Systems that the routing update has traversed to reach its current location. The AS-Path attribute serves a very important purpose by providing BGP routers with a mechanism for avoiding routing loops in the internetwork.

BGP AS-Path access lists allow a BGP router to filter inbound and outbound network prefixes based on the BGP AS-Path attribute – permit or deny network prefixes from certain BGP Autonomous Systems. A BGP AS-Path access list can be used to filter illegitimate or unwanted network prefixes from being inserted into the BGP Routing Table. The **ip as-path access-list** command can be used in global configuration mode to configure an Autonomous System path filter to filter BGP advertisements **[CISCBGPCOMD19]**:

    ip as-path access-list *acl-number* {**permit** | **deny**} *regexp*

- The parameter *acl-number* specifies a number in the range 1–500 that specifies the AS-Path access-list number.
- The **permit** keyword permits BGP advertisement satisfying the matching conditions.
- The **deny** keyword denies BGP advertisement satisfying the matching conditions.

- The parameter *regexp* specifies a regular expression that defines the AS-Path filter. The ASN is expressed in the range from 1 to 65,535. The 4-byte ASNs are in the range 65,536 to 4,294,967,295.

For an AS-Path access list, the input string is the BGP AS-Path of the routes to which the access list is applied. The router compares each BGP route's AS-Path against each condition in the access list. If the first match is for a **permit** condition, the route is accepted or passed. If the first match is for a **deny** condition, the route is rejected or blocked. AS-path filters can be applied to both inbound and outbound BGP paths. In Cisco IOS, the first step in configuring a BGP AS-Path route filter is to create the AS-Path access list, after which the access list is linked to the desired BGP neighbor. Other ways to filter BGP advertisements are using the **neighbor prefix-list** command, the **neighbor distribute-list** command, the **neighbor filter-list** command, and the **neighbor route-map** command as discussed below.

7.8.2.2 BGP Prefix Lists

Other that the use of AS-Path filters (defined with the **ip as-path access-list** command), a BGP prefix list is another method for filtering BGP advertisements. A BGP prefix list is a technique for determining BGP routes for possible exclusion from a BGP Routing Table by matching the network prefixes of the BGP routes (carried in the NLRI field of BGP UPDATE messages) against a local list of network address prefixes. For BGP, a prefix list can be applied to inbound or outbound routing updates for a specific BGP peer using the **neighbor prefix-list** command [CISCBGPCOMD19]:

> **neighbor** {*ip-address* | *peer-group-name*} **prefix-list** {*prefix-list-name*} {**in** | **out**}

- The parameter *ip-address* specifies the IP address of BGP neighbor.
- The parameter *peer-group-name* specifies the name of a BGP peer group.
- The parameter *prefix-list-name* specifies the name of a prefix list.
- The **in** keyword indicates that the prefix list is applied to incoming advertisements from that BGP neighbor.
- The **out** keyword indicates that the prefix list is applied to outgoing advertisements to that BGP neighbor.

A BGP router can use a combination of BGP AS-Path access lists and prefix lists, where the AS-Path access list ensures that network traffic takes the desired route through an internetwork, and the prefix lists prevents BGP routers from learning illegitimate or unwanted routes. A network operator can configure prefix lists to specifically allow only those network prefixes that are permitted by the routing policy of the operator.

The **neighbor prefix-list** command and the **neighbor distribute-list** command cannot be both applied to a BGP neighbor to filter routes in any given direction (inbound or outbound) [CISCBGPCOMD19]. These two commands are mutually exclusive; meaning only one of these can be applied to filter routes in each direction at any given time. The `neighbor filter-list` command can also be used to filter BGP advertisements (see command details in **[CISCBGPCOMD19]**).

7.8.2.3 BGP Distribute Lists

The `neighbor distribute-list` command can be used in router configuration mode to filter routing information (in BGP UPDATE messages) received from, or sent to BGP neighbor as specified in an access list **[CISCBGPCOMD19]**:

> `neighbor` {*ip-address* | *peer-group-name*} `distribute-list` {*access-list-number* | *expanded-list-number* | *access-list-name*} {`in` | `out`}

- The parameter *ip-address* specifies the IP address of the BGP neighbor.
- The parameter *access-list-number* specifies the number of a standard or extended access list (in the range 1–99, and an extended access list number in the range 100–199).
- The parameter *expanded-list-number* specifies the number of an expanded access list number (in the range 1,300–2,699).
- The parameter *access-list-name* specifies the name of a standard or extended access list.
- The `in` keyword specifies that the access list is to be applied to advertisements sent from the specified BGP neighbor.
- The `out` keyword specifies that the access list is to be applied to advertisements sent to the specified BGP neighbor.

7.8.2.4 BGP Route Maps

The `neighbor route-map` command can be used in router configuration mode to apply a route map to incoming or outgoing BGP routes **[CISCBGPCOMD19]**:

> `neighbor` {*ip-address* | *peer-group-name*} `route-map` *map-name* {`in` | `out`}

- The parameter *ip-address* specifies the IP address of the BGP neighbor.
- The parameter *map-name* specifies the name of a route map.
- The `in` keyword indicates that the route map applies to incoming routes.
- The `out` keyword indicates that the route map applies to outgoing routes.

Note that the `route-map` command (discussed above in the "Route Maps" section) is used to define the conditions under which routes will be redistributed from one routing protocol into another, or used to enable PBR.

7.8.3 INJECTING ROUTING INFORMATION INTO BGP

The methods used to inject routes into BGP can affects routing stability in internetworks, especially the Internet. We discuss here the commands used in routers running Cisco IOS (and other commercial routers) for injecting routes in BGP, statically and dynamically **[CISCASR9000CR] [CISCBGPCOMD19] [CISCHALABS00] [CISCMASCOMD14]**. The discussion here includes the benefits and drawbacks of each route injection method.

7.8.3.1 Injecting Routes Statically into BGP

Routes that are statically injected into BGP stay in the BGP Routing Table regardless of the routing state of the networks introduced. This means if a statically injected route ceases to exist, BGP will still continue to advertise that route. IGP routes and aggregate routes that are statically injected into BGP to be advertised to other peers are manually defined as static routes. Predefining these routes as static ensures that they are always maintained in the IP routing and will always be advertised by BGP. However, the suitability of this method depends on the particular network scenario under consideration.

For example, advertising a failed route to a stub network to the Internet, would not pose any problems because end-users trying to access the failed route will not be successful irrespective of whether the route is advertised as static or not. However, advertising a network destination (in a given Autonomous System) from multiple points to the Internet statically may end up creating traffic blackholes. For instance, if routing problems within the Autonomous System prevents the border BGP router from accessing the statically advertised network, traffic destined to that network destination will be dropped even if there was another entry point to that destination **[CISCHALABS00]**.

1. Using the **redistribute** command: This command is used to redistribute routes from one routing domain into another. The minimum command syntax is "**redistribute** *protocol*", where, *protocol* is the protocol from which to redistribute the routes using, for example, the following keywords: **bgp, eigrp, isis, ospf, rip, connected,** or **static [CISCBGPCOMD19]**. The keyword **static** is used to redistribute static IP routes. For example, running the following commands on a BGP router in AS 65000 will redistribute **all** the static routes configured into BGP:

    ```
    router bgp 65000
    redistribute static
    ```

    This command causes the BGP router to advertise **all** static routes to its peers. To exercise more control on the advertisement, filtering (via route maps, for example) can be used to specify which routes are to be advertised (using "**redistribute** *protocol* [**route-map** *map-tag*]"). If no route map is specified, all routes are redistributed into BGP.

2. Using the **network** command: This command is used to specify the networks to be advertised by BGP. The minimum command syntax is "**network** *network-number* [**mask** *network-mask*]", where, *network-number* is the network IP address that BGP will advertise, and *network-mask* (optional) is the associated IP network or subnetwork mask **[CISCBGPCOMD19]**. For example, running the following commands on a BGP router in AS 65000 will inject only the (pre-configured) statically defined route 192.68.10.0/24 into BGP (which BGP will always advertise):

    ```
    router bgp 65000
    network 192.68.10.0 mask 255.255.255.0
    ```

    This command is used to instruct the BGP router to advertise only a subset of static routes to its peers. This provides a more controlled and finer injection of routes into BGP because the network administrator might see that total injection of static routes into the Internet may be problematic (unwanted routes may be advertised). However, injecting a large number of routes may be tedious, so, this can also be accomplished using the **redistribute** command with route maps as discussed above.

7.8.3.2 Injecting Routes Dynamically into BGP

Routes that are dynamically injected into BGP appear and disappear from the BGP Routing Table depending on the routing state of the networks announced. This means, BGP will stop advertising a dynamically injected route once it ceases to exist. The methods discussed here can be divided into those that are semidynamic (using the **network** command), and those that are purely dynamic (using the **redistribute** command) [CISCHALABS00]. In the former, only certain IGP routes are injected into BGP, while in the latter, all IGP routes are injected into BGP. This differentiation is meant to portray both the extent of user involvement and the degree of control in defining routes to be injected into BGP to be advertised.

When routes are dynamically injected into BGP, IGP routes (from RIP, EIGRP, OSPF, and IS-IS) are redistributed into BGP automatically. By allowing the IGP routes to be dynamically transferred into BGP, this method offers an easier way of configuring route injection into BGP, although it has some drawbacks.

3. Using the **redistribute** command: When this command is used, all the IGP routes are injected into BGP. For example, running the following commands on a BGP router in AS 65000 will redistribute OSPF routes into BGP:

    router bgp 65000
    redistribute ospf

    The main problem with this method is that, redistributing all the IGP routes into BGP might end up leaking unwanted routing information into BGP (such as private addresses, or unregistered/illegal internal addresses), thereby, causing other routing problems elsewhere. However, route maps can be used to filter routes to avoid unwanted routes leaking into BGP as discussed above.

4. Using the **network** command: In this method, the user manually specifies a subset of the IGP routes to be injected into BGP. Using this command involves listing all the routes to be passed into BGP. For example, running the following commands on a BGP router in AS 65000 will inject only the network 10.120.0.0 into BGP:

    router bgp 65000
    network 10.120.0.0 mask 255.255.0.0

    This method is not fully dynamic because the network address prefixes that need to be injected into BGP must be individually specified and maintained in the BGP router. These network prefixes are not automatically injected into BGP. Even though this method has the advantage of providing fine grain control of route injection into BGP, it can be impractical or tedious to use when a large number of network prefixes have to be passed to BGP.

    When this command is used, the BGP router assumes that the networks specified exist and tries to verify that by checking its IP Routing Table

[CISCHALABS00]. If the BGP router does not find an exact match for any of these networks in its IP Routing Table, it will not advertise that network. This IP Routing Table verification is to ensure that the router itself and peers in other Autonomous Systems will not mistake accept networks that do not actually exist.

7.8.3.3 Route Injection and the BGP ORIGIN Attribute

When routes are advertised into BGP using the **network** command, they are assigned the BGP ORIGIN attribute value of 0, meaning IGP, to indicate that the routes are internal to the Autonomous System (see Chapter 3 of Volume 2 of this two-part book). When BGP advertises these routes, it will include this ORIGIN attribute value in the corresponding BGP UPDATE messages.

On the other hand, when routes are injected into BGP using the **redistribute** command, they are assigned the ORIGIN attribute value of 2, meaning INCOMPLETE. This ORIGIN attribute value is used to indicate that the true origin of the route cannot be precisely known, it could have come from anywhere in the internetwork. Routes that are learned via EGP are assigned the ORIGIN attribute value of 1, meaning EGP.

The following examples explain how the ORIGIN attribute values are assigned when the **network** and **redistribute** commands are used for route injection into BGP:

- If the routes 10.10.0.0, 11.11.0.0, and 12.12.0.0 are passed into BGP using the **network** command, and 12.12.0.0 is not in the IP Routing Table, only 10.10.0.0 and 11.11.0.0 will be advertised by BGP. BGP will assign to these networks the ORIGIN attribute value of 0 (i.e., IGP).
- If the IGP route 13.13.0.0 is injected into BGP via the **redistribute** command, BGP will assign this route the ORIGIN attribute value of 2 (i.e., INCOMPLETE).
- Let us assume that the routes 10.10.0.0, 11.11.0.0, and 12.12.0.0 are defined as static routes, and are injected into BGP statically using the **redistribute** command, and 11.11.0.0 which is listed in the IP Routing Table, is also passed to BGP using the **network** command. In this case, 11.11.0.0 will have the ORIGIN attribute value of IGP, while 10.10.0.0 and 12.12.0.0 will be given the ORIGIN attribute value of INCOMPLETE.

7.8.4 BGP AS-Path Prepending: AS-Path Attribute Manipulation Using Dummy Entries

The AS-Path attribute (AS_PATH) carried in BGP UPDATE messages is sometimes manipulated to control how inter-autonomous system routing is carried out. This is possible because, during the BGP best path selection process, BGP prefers routes that have shorter AS-Paths over those with longer AS-Paths (see Chapter 3 of Volume 2 of this two-part book). Thus, to make some routes more preferable and others not, an eBGP router can modify the path information in the BGP AS-Path attribute by adding or prepending dummy ASNs to the path information. This makes the path

IP Routing Protocols

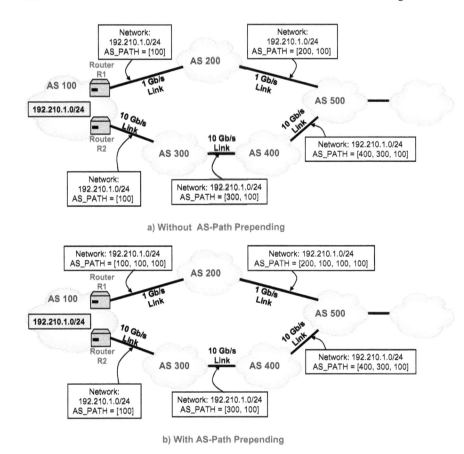

FIGURE 7.22 BGP AS-Path Prepending: AS-Path Attribute Manipulation using Dummy Entries

length indicated in the AS-Path attribute for the route longer and therefore, least preferable.

"AS-Path Prepending", as this path control method is called, involves the eBGP router adding ASNs at the beginning (i.e., leftmost position) of the BGP AS-Path attribute to make the path longer as explained in Figure 7.22. AS-Path prepending is typically used as a workaround if other path control tools such as the MED attribute are not supported, and a specific BGP route in the internetwork is required to be followed.

In practice, the eBGP router doing the AS-Path prepending uses its own ASN, or the ASN of the eBGP peer from which it learned the route, as the dummy ASNs [CISCHALABS00]. This is because using any other ASN can have unintended side effects include misleading other routers to think that these ASNs were actually traversed, or this could potentially create routing loops or traffic blackholes in the internetwork. As shown in Figure 7.22, Router R1 prepends its own ASN to avoid creating such problems.

The Autonomous System path length as indicated by the AS-Path attribute, is a BGP metric used to influence the best path selection process as explained in Chapter 2. By extending the length of the Autonomous System path as indicated in the AS-Path attribute, a BGP router can influence the best-path selection in downstream BGP peers. The **set as-path prepend** command can be used in Cisco IOS and many commercial routers to prepend an arbitrary number of ASNs to the AS-Path attribute in order to extend the Autonomous System path length of BGP routes [CISCBGPCOMD19] [CISCHALABS00].

The command syntax is, "**set as-path {tag | prepend** *as-path-string*}", where **tag** applies only when redistributing routes into BGP, and also converts the manually set tag of the redistributed route into an Autonomous System path. For example, the command used in Figure 7.22 will be, "**set as-path prepend 100 100**". Usually, the local ASN is prepended multiple times to the attribute, making the Autonomous System path length appear longer to downstream BGP routers.

7.8.5 ROUTE AGGREGATION IN BGP

BGP routers store and exchange routing information with their peers using BGP UPDATE messages (see Chapter 2). However, the amount of routing information (i.e., routes) maintained and exchanged increases as more networks and BGP routers are configured. So, BGP routers use route aggregation (or summarization) to reduce the amount of routing information exchanged. Route aggregation is the process of combining several different routes based on the BGP path attributes the carry so that only a single route can be advertised.

Using the concept of CIDR, a BGP router creates an aggregate prefix by combining several contiguous network IP addresses into one classless IP addresses that can be advertised and installed in the Routing Tables. Using route aggregation, the BGP routers only need to advertise fewer routes. Also, when aggregate routes are sent, routers have to execute the best-path selection algorithm only once, instead of running the algorithm against multiple specific routes.

With aggregate routes, the IP Routing Table size is reduced, thereby accelerating best path calculation. This also helps to conserve router memory and router resources. BGP route aggregation on eBGP routers (for Autonomous Systems that are nontransitive) reduces best path computations on routers in the core of such Autonomous Systems. Chapter 3, section "BGP Path Attributes and Route Aggregation" of Volume 2 of this two-part book discusses in detail BGP route summarization.

Unlike the **network** command which applies to routes in the IP Routing Table, BGP route aggregation applies to routes in the BGP Routing Table **[CISCHALABS00]**. The BGP Routing Table of a router contains the list of all routes learned from the BGP neighbors, plus the routes originating from the local Autonomous System. The BGP router installs the best routes from the BGP Routing Table to all known network destinations in its IP Routing Table. A BGP router can perform route aggregation if at least one more-specific route exists in the BGP Routing Table.

7.8.5.1 Performing BGP Route Aggregation

The Cisco IOS **aggregate-address** command can be used to create an aggregate route entry in the BGP Routing Table **[CISCBGPCOMD19]**:

aggregate-address *address mask* [**as-set**] [**as-confed-set**] [**summary-only**] [**suppress-map** *map-tag*] [**advertise-map** *map-tag*] [**attribute-map** *map-tag*]

- The parameter *address* specifies the aggregate IP address.
- The parameter *mask* specifies the aggregate IP network mask.
- The **as-set** keyword (optional) is used to generate Autonomous System AS_PATH attribute information (see AS_PATH segment type in Chapter 3 of Volume 2 of this two-part book). When the **as-set** keyword is used, an aggregate route entry is created and the AS_PATH segment advertised for the route will be an AS_SET consisting of all elements contained in all routes that are being aggregated.
  - o An AS_SET is an unordered list of ASNs collected from all the individual routes forming the aggregate route. By including these origin ASNs in the AS-Path of the aggregate route, it is possible ensure the integrity of BGP's routing loop prevention mechanism. To prevent routing loops, BGP does not accept a route with an AS-Path listing its own ASN.
  - o It is not recommended to use the **aggregate-address** command with the **as-set** keyword when aggregating many routes, because the aggregate route will be continually withdrawn and updated as Autonomous System path reachability information for the individual routes being aggregated changes **[CISCBGPCOMD19]**.
- The **as-confed-set** keyword (optional) performs the same function as the **as-set** keyword, except that it is used to generate autonomous confederation AS–CONFED–SET attribute information (see AS–CONFED–SET segment type in Chapter 3 of Volume 2 of this two-part book).
- The **summary-only** keyword (optional) when specified, creates an aggregate route but also suppresses all more-specific routes from being advertised to BGP neighbors.
  - o When this keyword is not specified, the **aggregate-address** command causes the BGP router to advertise the aggregated route in addition to the constituent more-specific routes.
  - o If there is the need to only suppress routes to certain BGP neighbors, the **neighbor distribute-list** command may be used, but with

caution. This is because if a more-specific route (forming part of the aggregate route) for some reason leaks out, all BGP routers (using longest prefix-matching lookup and forwarding) will prefer the more-specific route over the less-specific aggregate route generated **[CISCBGPCOMD19]**.
- The parameter *map-tag* associated with the `suppress-map` keyword (optional) specifies the name of a route map to be used to select specific routes to be suppressed. When the `suppress-map` keyword is used, the aggregate route is created but suppresses the advertisement of the selected routes.
  o The `match` clauses of the route maps can be used to selectively suppress some more-specific routes making up the aggregate route, and leave other more-specific routes unsuppressed.
  o When the `suppress-map` command and the `summary-only` configuration command are used together, the `summary-only` command does not have any effect.
  o When the `suppress-map` command is used, the more-specific routes that are suppressed will not be advertised. However, the routes that are not suppressed will be advertised in addition to the aggregated route. Reference **[CISCID5441AGGR05]** discusses in detail the impact of using the `suppress-map` command with the other configuration commands.
- The parameter *map-tag* associated with the `advertise-map` keyword (optional) specifies the name of the route map used to select specific routes that will be used to create different components of the aggregate route, such as an AS_SET segment, or BGP Communities.
  o This keyword is useful when some of the more-specific routes of an aggregate route are in separate Autonomous Systems and there is the need to create an aggregate route with an AS_SET, plus advertise the aggregate route back to some of these same Autonomous Systems. In this case, the specific ASNs must be omitted from the AS_SET to prevent the aggregate route from being dropped by the BGP routing loop detection mechanism at the receiving BGP router.
- The parameter *map-tag* associated with the `attribute-map` keyword (optional) specifies the name of the route map used to set the BGP path attribute of the aggregate route.
  o This keyword is used when there is the need to change the BGP path attributes of the aggregate route using an attribute-map route map. It is useful when one of the more-specific routes forming the AS_SET segment is configured with a BGP path attribute such as the BGP Communities NO_EXPORT attribute, which prevents the aggregate route from being advertised to eBGP neighbors.

When the `aggregate-address` command is used without any arguments or keywords, the aggregate route does not inherit the individual route attributes (such as AS_PATH or BGP Communities), which causes a loss of granularity in the resulting routing information. Also, an aggregate route is created in the BGP Routing Table if any more-specific BGP routes are present (in the Routing Table) and fall within the specified address range.

An aggregate route is treated like new BGP route with a shorter prefix length. The BGP Routing Table must contain a longer address prefix that matches the aggregate route. The aggregate route is advertised as a route originating from the local Autonomous System, and will have its BGP Atomic Aggregate path attribute set to indicate that some information might have been lost as a result of the aggregation. The BGP Atomic Aggregate attribute is set, by default, unless the **as-set** keyword is specified [CISCBGPCOMD19].

After the router has performed route aggregation, it will advertise a new network prefix with a shorter prefix length into BGP. The aggregated prefix is considered a new route, and the router performing the route aggregation becomes the originator for this new aggregate route. BGP considers aggregated routes as local routes when modifying the BGP Administrative Distance. The BGP Aggregator path attribute identifies the BGP Autonomous System and router that created an aggregate route (see Chapter 3 of Volume 2 of this two-part book).

7.8.5.2 Route Aggregation without AS_SET

When an aggregate route is created (and the **as-set** keyword is not specified), the AS-Path information from the more-specific routes before the aggregation, are not advertised. The BGP router performing the aggregation does not include BGP path attributes such as AS-Path, MED, and BGP Communities in the new BGP advertisement. The BGP Atomic Aggregate attribute is used to indicate that a loss of BGP path information has occurred.

7.8.5.3 Route Aggregation with AS_SET

The optional **as-set** keyword can be used with the **aggregate-address** command to pass the BGP path information from the more-specific to the aggregate route. As the router generates the aggregate route, it copies the BGP attributes from the more-specific routes over to it. The router stores the AS-Path settings from the original routes in the AS_SET portion of the AS-Path of the aggregate route. The writing convention is to display the AS_SET within brackets, and this counts only as one hop, even if multiple ASNs are listed in it (see Chapter 3 of Volume 2 of this two-part book).

The **as-set** keyword when used, summarizes the AS_PATH attributes of all the more-specific routes aggregated into an AS_SET. The AS_SET information in the aggregate route contains a record of the Autonomous Systems the route has passed through. An aggregate route propagated through BGP may end up going back to one of the Autonomous Systems listed in the route's AS_SET. This propagation creates a routing loop if the route is accepted by the already listed Autonomous System. A BGP router uses this information to avoid routing loops. The router does this by checking and rejecting routes if its local Autonomous System is already listed in the route's AS_SET. The routing loop detection mechanism of BGP allows a BGP router to record its own ASN in the AS_SET of the aggregate route so that the aggregate route will be rejected (preventing a routing loop) if it is propagated back to that router.

Using the **as-set** keyword with route aggregation, allows a BGP router to combine all the BGP path attributes of the original route into the aggregate route. However, in some situations, the use of the AS_SET feature can cause some problems with a routing policy:

Network Path Control and Factors That Affect Routing Table Properties   287

- For example, if one of the original routes being aggregated contains the BGP Communities NO_EXPORT attribute, the aggregate route will not be exported (see Chapter 3 of Volume 2 of this two-part book).
- Furthermore, if the AS-Path information for the individual routes being aggregated change, the aggregate route will be continually withdrawn and updated, causing route flaps. Note that the AS_SET contains information about each individual route that is summarized. This means changes in the AS-Path information of an individual route will cause the aggregate route to be updated. If the aggregate route is formed from tens or hundreds of routes and these routes have problems, there can be a constant aggregate route flap.

To resolve these types of problems, it is recommended to specify explicitly the routes from which the BGP path attributes will be copied to the aggregate route. The **advertise-map** keyword option can be used to conditionally match or deny the BGP path attributes (from the original routes) that should be permitted or denied in the aggregate route.

7.8.5.4   Changing the BGP Attributes of an Aggregate Route

The `as-set` keyword can be used to store the AS_PATH attributes of the more-specific routes being aggregated. In some cases, there may be the need to change the BGP attributes of the aggregate route, for example, to include a metric, origin, and BGP Community. A BGP Community is a logical group of network address prefixes that share some common properties. The `attribute-map` keyword can be used to manipulate the BGP attributes of the aggregate route.

BGP Communities can be used to simplify routing policies. This is done by configuring the routing information that a BGP router will accept, prefer, or distribute to other neighbors according to the community membership. When a BGP router learns, advertises, or redistributes a route, it can set, append, or modify the BGP Communities attribute of a route. When routes are aggregated, the resulting BGP aggregate route carries a BGP Communities attribute that contains all the Communities carried in all of the routes aggregated (if the aggregate route is an AS_SET aggregate).

The following situations are examples where there is the need to change the BGP attributes of an aggregate route because the aggregate contains some unwanted BGP attributes that are inherited from the more-specific routes:

- For example, if the aggregate route inherited a BGP Communities NO_EXPORT attribute from one or more of the more-specific routes, this causes the aggregate route not to be advertised to other Autonomous Systems. So, in the event one or more of the more-specific routes aggregated is configured with the BGP Communities NO_EXPORT attribute, the `attribute-map` keyword can be used to change this attribute to the BGP Communities NONE attribute, clearing the BGP Communities associated with the aggregate route. This allows the aggregate route to be advertised to eBGP neighbors.
- The BGP attributes of the aggregate route can also be changed to reflect preference for a certain aggregate route when multiple routes exist. For example, if a customer network is advertising an aggregate route through multiple links to a

service provider, the customer might want to send the aggregate route on the different links with different MED values to influence the entry point to the service provider's Autonomous System. The MED attribute (sometimes called the metric) is an optional non-transitive BGP path attribute that a BGP router uses in its BGP path selection process to determine the preferred entry point when multiple entry points exist to a neighboring Autonomous System.

7.8.5.5 Advertising the Aggregate Route Only, while Suppressing the More-Specific Routes

When the **summary-only** keyword is specified, the **aggregate-address** command advertises the aggregated route while suppressing all of its more-specific routes. However, after a particular more-specific route is suppressed, it is still possible to advertise that route to a specific BGP neighbor. This mode of route aggregation is used when the network administrator sees that advertising the more-specific routes offers no additional benefits, for instance, does not help in making better packet forwarding decisions [CISCHALABS00].

7.8.5.6 Advertising the Aggregate Route Plus All of the More-Specific Routes

Omitting the **summary-only** keyword from the **aggregate-address** command causes the BGP router to advertise the aggregated route in addition to all of the constituent more-specific routes. Usually, this mode of route aggregation is used when a customer network is multihomed to a service provider network, and the provider wants to use the more-specific routes to make better packet forwarding decisions when sending traffic to the customer network. The service provider also wants to propagate the single aggregate route toward a network access point, for example, leading to the Internet, in order to minimize the number of routes advertised [CISCHALABS00].

The more-specific routes (to the customer network) are usually stopped at the service provider network (via filtering) and not propagated beyond the provider to avoid complicating the Routing Tables of routers beyond. The service provider network can tag the more-specific routes with the BGP Communities NO_EXPORT attribute while leaving the aggregate route as is. This route tagging allows the service provider to suppress the more-specific routes so that they do not get propagated to the network access point and beyond. Only the aggregate route will be propagated to the network access point and eBGP peers. The BGP attributes are discussed in greater detail in Chapter 3 of Volume 2 of this two-part book.

7.8.5.7 Advertising the Aggregate Route Plus a Subset of the More-Specific Routes

Some networking scenarios require routes to be "leaked", where a subset of the more-specific routes in addition to the aggregate route are advertised to other routers. Route leaking can be done at the routing protocol process level by explicitly specifying the routes to suppress, or on a BGP neighbor level by specifying which routes should not be suppressed.

As discussed above, the **suppress-map** keyword can be used to explicitly list the routes that should not be advertised along with the aggregate route to BGP neighbors. If the network operator has more control over the individual routes that can form the aggregate route, it is easier to decide which BGP attributes will get passed to the aggregate route. The **advertise-map** keyword can be used to exclude a specific route (e.g., having the BGP Communities NO_EXPORT attribute) from the aggregate route; preventing the aggregate route from inheriting the NO_EXPORT attribute. With this configuration, the aggregate route can be advertised to other eBGP neighbors.

7.8.6 Default Routes in BGP

For a default route to be advertised by a BGP routing process, the default route must exist in the BGP Routing Table. The BGP **default-information originate** command can be used (with no arguments or keywords) to configure a BGP routing process to advertise the default route (i.e., network 0.0.0.0) [CISCBGPCOMD19]. For the default route to be advertised, the **redistribute** command must also be configured under the BGP routing process to complete the configuration [CISCBGPCOMD19]. For example, a default route can be redistributed from OSPF into a BGP routing process in Autonomous System 500 as follows:

```
router bgp 500
default-information originate
redistribute ospf 100
```

The **network** command can also be used to advertise a default route similar to the **default-information originate** command. However, the **default-information originate** command requires the route 0.0.0.0 to be present in the BGP Routing Table for it to be advertised, while the **network** command requires only that the route 0.0.0.0 be present in the IP Routing Table of the originating IGP [CISCBGPCOMD19].

The **neighbor default-originate** command in router configuration mode can be used to configure a BGP router to advertise the network 0.0.0.0 to a BGP neighbor to be used as a default route [CISCBGPCOMD19]:

> **neighbor** {*ip-address* | *peer-group-name*} **default-originate**
> [**route-map** *map-name*]

- The parameter *ip-address* specifies the IP address of the BGP neighbor.
- The parameter *peer-group-name* specifies the name of a BGP peer group.
- The parameter *map-name* associated with the **route-map** keyword (optional) specifies the name of the route map that allows the default route 0.0.0.0 to be injected conditionally.

The **neighbor default-originate** command does not require the default route (0.0.0.0) to be present in the Routing Table of the local router. When a route map is specified, the router injects the default route 0.0.0.0 if the route map contains a **match ip address** clause, and there exist a route that matches the IP access list exactly. The route map may also carry other match clauses. The **default-information originate** command and the **neighbor default-originate** command should not be configured on the same router at the same time. Only one of these commands should be configures at any given time.

7.8.7 IGP Routes Versus BGP Routes: Looking at Backdoors Routes

When a router receives multiple routes to the same network destination from different routing information sources (e.g., directly connected networks, static routes, IGP routes, BGP routes), the Administrative Distance (also called the Route Preference) is the parameter the router uses to determine which routing source has higher precedence, that is, which route the router should prefer. From the discussion in Chapter 2, the Administrative Distance (AD) ranks routes from those with the highest preference (low Administrative Distance value) to those with the lowest preference (high Administrative Distance value).

From Chapter 2, eBGP has an AD of 20 which is lower than that of all IGPs (EIGRP internal has AD of 90, OSPF has 110, IS-IS has 115, RIP has 120, iBGP has 200, etc.). However, there are situations where an IGP (even though having a higher AD) might provide a route that is shorter than one provided by eBGP to the same network destination. A backdoor route is one that offers an alternative path to a network destination that can be used instead of an eBGP route **[CISCHALABS00]**. In these situations, using the eBGP route to the specified network destination may not be the most efficient routing method.

Cisco IOS provides a mechanism for forcing IGP routes (with higher ADs) to take precedence over eBGP routes. The `network backdoor` command can be used to change the AD of an eBGP route to a specific destination so that IGP routes will take precedence (will be more preferred) over the eBGP route **[CISCBGPCOMD19]** **[CISCHALABS00]** **[CISCMASCOMD14]**. The command syntax is, "`network ip-address backdoor`", where, *ip-address* is the IP address of the network to which a backdoor route is being created **[CISCBGPCOMD19]**.

Figures 7.23 and 7.24 show examples of networks where IGP routes are preferred over eBGP routes the same network destination. A good feature of the `network backdoor` command is that it changes the AD of the BGP route but does not advertise the specified network in BGP updates.

7.9 UNNUMBERED INTERFACES

Each interface on a router requires a unique IP address that will be installed in the Routing Table so that routing updates and IP packets can be processed. An IP unnumbered interface allows IP processing to be carried out on a serial point-to-point interface without assigning an explicit IP address to the interface. IPv4 unnumbered interfaces are applicable only to point-to-point links and not to multiaccess networks like Ethernet. Whenever possible, it is beneficial to use IP unnumbered interfaces as it helps conserve the IPv4 address space of the network.

7.9.1 Conserving IP Addresses with Unnumbered Interfaces

Let us consider the case where IP addresses are to be assigned to the interfaces of a router using an IP Class B address range that has been subnetted using eight bits for the host identifier portion. We assume that every interface on the router requires a unique IP subnet address including point-to-point serial interfaces. Although each

Network Path Control and Factors That Affect Routing Table Properties

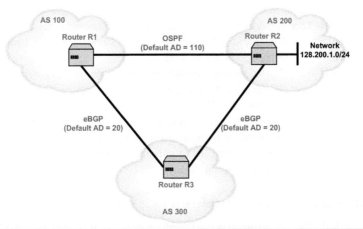

- AS 100 receives routing updates from two different routing information source about network 128.200.1.0/24, eBGP and OSPF.
- AS 100 receives routes via eBGP on the link to AS 300 and via the backdoor link running OSPF between AS 100 and AS 200.
- According to the default Administrative Distances (ADs) of the two routing protocols, the route provided by BGP should be preferred over the one provided by OSPF. The lower the AD, the higher the precedence for the route from that routing source.
- In AS 100, Router R1 that learns the route via eBGP will install it in its routing table.
- This means, traffic sent toward network 128.200.1.0/24 will take the indirect eBGP route via AS 300 and then AS 200, rather than the direct OSPF route between AS 100 and AS 200.
- To use the direct OSPF route instead, the **network backdoor** command can be used to change the AD of the eBGP route 128.200.1.0/24 from 20 to 200, which makes the OSPF route with AD of 110 to be preferred.
- A good feature of the **network backdoor** command is that it will not cause BGP to generate an advertisement for that network.

FIGURE 7.23 Illustrating an IGP Backdoor Route: Example 1

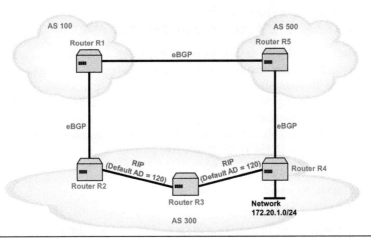

- Router R2 as a BGP router receives a route to Router R4 and network 172.20.1.0/24 through eBGP, but this route traverses at least two autonomous systems.
- Router R2 and Router R4 are also connected through RIP network (for example), and this route has a shorter path.
- RIP routes, however, have a default AD of 120, and eBGP routes have a default AD of 20, so BGP will prefer the eBGP route.
- To cause BGP to prefer the RIP route, the **network backdoor** command can be used to change the AD of the eBGP route from 20 to 200 so that the RIP route will be preferred.
- BGP treats the network specified by the **network backdoor** command as a locally assigned network, except that it does not advertise the specified network in BGP updates.
- This means that Router R2 will communicate to Router R4 using the shorter RIP route instead of the longer eBGP route.

FIGURE 7.24 Illustrating an IGP Backdoor Route: Example 2

point-to-point serial interface on the router has only two end points on the point-to-point link that need IP addresses, assigning an entire IP subnet to each serial interface uses up all the 254 available IP addresses in the subnet for that router interface although only two IP addresses are needed.

However, if an IP unnumbered interface is configured on each serial interface on the router, this would save the IP address space significantly. The IP unnumbered interface can "borrow" the IP address of a functional interface on the router. The router can then use this borrowed IP address as the source IP address for routing updates and packets sourced from the point-to-point serial interface. This allows the IP address space to be conserved. IP unnumbered interfaces are only applicable to interfaces attached to point-to-point links.

7.9.2 Configuring IP Unnumbered Interfaces

In Cisco IOS, the `ip unnumbered` configuration command can be used to enable IP processing on a serial interface without an explicit IP address being assigned to it [CISCID13786IPUN05]:

   `ip unnumbered` [*type*] [*number*]

where the parameter *type* describes the interface that is "lending" its IP address, and *number* specifies the interface number on the IP router. An IP unnumbered interface, when configured, "borrows" the IP address already configured on another interface on the router. Let us assume the interface Ethernet0 on Router R1 (Figure 7.25) is configured with an IP address as follows:

```
interface Ethernet 0
ip address 172.16.20.110 255.255.255.0
```

Traditionally, to enable IP processing on interface Serial 0 on Router R1, a unique IP address would be configured on it. However, it is also possible to configure interface Serial 0 as an IP unnumbered interface without assigning a unique IP address to it. This is done by letting Serial 0 borrow an IP address already configured on one of Router R1's other interfaces, for example, the IP address already configured on the Ethernet 0 interface. This can be done using the `ip unnumbered` interface mode command as follows [CISCID13786IPUN05]:

```
interface Serial 0
ip unnumbered Ethernet 0
```

The use of this command allows the interface on which the command is configured (Serial 0) to borrow the IP address of the specified interface (Ethernet 0). This results in these two interfaces sharing one IP address as shown in Figure 7.25. The IP address

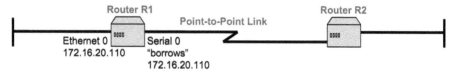

FIGURE 7.25 Configuring an IP Unnumbered Interface

configured on the interface Ethernet 0 is also assigned to the interface Serial 0, and both interfaces involved will function normally. The "unnumbered interface", Serial 0, is the interface that borrows its IP address from one of the other functional interfaces of the router.

7.9.3 Limitations of IP Unnumbered Interfaces

The only real limitation of IP unnumbered interfaces is that, it cannot be used for remote testing and management. Also, it is important to borrow the IP address of the unnumbered interface from a router interface that is always up and running. If the borrowed IP address is from an interface that is not functional, the IP unnumbered interface will not work. For this reason, it is recommended that the IP unnumbered interface takes its address from that of a loopback interface in the router since loopback interfaces are always operational as long as the router is operational. The following configuration uses the loopback interface address as the "donor" IP address for router interface Serial 0:

```
interface loopback 0
ip address 127.10.10.4 255.255.255.0
interface Serial 0
ip unnumbered loopback 0
```

It should also be noted that, an IP unnumbered interface and the `ip unnumbered` configuration command work on point-to-point links only. When the command is configured on a multiaccess interface (i.e., Ethernet) or the loopback interface, error and warning messages are displayed indicating that the command should be used on point-to-point (non-multiaccess) interfaces only **[CISCID13786IPUN05]**.

7.9.4 Receiving Routing Updates over IP Unnumbered Interfaces

When an IP router receives a routing update from a neighbor, it installs the source IP address of the routing update as the next-hop IP address in its IP Routing Table. Normally, the next-hop IP address is the address of a directly connected node on the router. This is no longer the case if an IP unnumbered interface address is configured because the serial interface on which it is configured "borrows" its IP address from a different router interface. This borrowed address could possibly be in a different IP subnet or network.

When an IP unnumbered interface is configured, routes learned through this unnumbered interface would use only the interface (identifier) as the next hop and not the "borrowed" IP address – the source IP address of the routing update, being a "borrowed" address, is not used as the next-hop, instead only the interface is used **[CISCID13786IPUN05]**. This avoids having an invalid next-hop IP address in the Routing Table due to the source of the routing update coming from a next-hop IP address that is not really directly connected but, instead belongs to a "donor" interface. This is possible because, on a point-to-point serial link, there is no confusion about which router originates a packet or which router a packet is destined to.

7.9.5 FORWARDING IP PACKETS OVER IP UNNUMBERED INTERFACES

On an IP router, every interface that is connected to a network segment is assigned an IP address that belongs to a unique IP subnet. Routers that have interfaces that are directly connected to the same network segment (or subnet) are assigned IP addresses from that IP subnet. If an IP router has a packet that is to be sent to a network that is not directly connected to the router, it makes a lookup in its IP Routing Table to determine the next-hop IP address and outgoing interface, and then forwards the packet to that directly connected next-hop node toward the packet's destination. If there is no match in the Routing Table for the packet's destination, the packet may be forwarded to a gateway of last resort. If the packet's final destination is a directly connected host or network, the router forwards the packet directly to that destination.

Each entry in the IP Routing Table contains either a network prefix (for a network using a VLSM or an aggregate route created according to CIDR rules), an IP subnet address, or the IP address of directly connected host. Each address entry may have either an outgoing interface plus a next-hop IP address, only an interface address specified (as in a static route or IP unnumbered interface), or only a next hop IP address (as in recursive route lookup).

7.10 ROUTING PROTOCOL TIMERS

Routing protocols use a variety of timers to control how a number of tasks are performed such as, how frequent routing updates are sent, the length of time before a route is declared invalid, and so on. The settings of the routing protocol timers can affect network performance and convergence. These timers can be adjusted to tune the performance of a routing protocol to better suit the needs of a particular internetwork.

For RIP, the following timer adjustments can be made (see Chapter 2):

- **Update Timer**: This timer specifies the rate at which RIP sends routing updates. The *routing-update timer* clocks the interval between periodic routing updates. The default timer setting is 30 seconds. A small random amount of time is added to the timer setting whenever the timer is reset. This is done to prevent all routers from attempting to send routing updates to their neighbors simultaneously, when their timers are reset.
- **Invalid Timer**: This timer specifies the maximum time the RIP router waits for a routing update before declaring the route as invalid. The default setting is 180 seconds and should be at least three times the update timer value. Once the timer expires, the route goes into holddown and is tagged as inaccessible, and is also advertised to neighbor routers as unreachable. When the invalid timer expires, the route is marked as invalid but is still retained in the RIP router's Routing Table until the flush timer expires.
- **Holddown Timer**: This specifies the maximum amount of time that the RIP router must wait before accepting any new routing updates for a route that is placed in holddown. The default setting is 180 seconds.

- **Flush Timer**: This timer specifies the maximum amount of time the RIP router must wait before removing a route from its Routing Table. The default setting is 240 seconds. This timer is used to purge invalid routes from the Routing Table.

RIP maintains these timers for each known route; each Routing Table entry has these timers associated with it.

For EIGRP, the following timer are used (see Chapter 6):

- **Hello Interval Timer**: This timer specifies the rate at which EIGRP sends HELLO packets. The default setting is 60 seconds for low-speed (not more than 1,544 Mb/s) and NBMA (non-broadcast multiple access network) networks, and 5 seconds for other types of networks.
- **Hold Timer**: This specifies the maximum amount of time a router will consider an EIGRP neighbor as alive when it does not receive a HELLO from the neighbor. The default Hold Time is 15 seconds, which is three times the default Hello Time (of 5 seconds). Each time a router transmits a HELLO packet, it also includes its Hold Time in it. The EIGRP router uses the Hold Timer to determine how long it should maintain the neighbor relationship without receiving EIGRP HELLO messages. Each time the router receives a HELLO message from the neighbor, it will reset the Hold Timer to the Hold Time and will decrement it until a HELLO message is received or until it expires. When the Hold Timer reaches zero, the router declares the neighbor as unreachable. The router then marks all paths through that neighbor as unusable and the neighbor relationship is torn down. The EIGRP router then runs the EIGRP Diffusing Update Algorithm (DUAL) over these destinations to determine if the route needs to be placed in the ACTIVE State (see Chapter 6).

    The Hold Timer is also used by the EIGRP Reliable Transport Protocol (RTP) as an upper bound on how long an EIGRP router should wait for a neighbor to acknowledge the receipt of an EIGRP QUERY, REPLY, or UPDATE packet. The EIGRP router tries to retransmit up to a maximum of 16, or until the Hold Time of the neighbor expires, which will cause it to terminate the neighbor relationship.
- **Active Timer**: This timer specifies the maximum time the EIGRP router waits for a reply after sending an EIGRP QUERY before declaring the route as Stuck-in-Active (SIA), and resetting the neighbor or adjacency relationship.

It also is possible to tune the different timers to enable faster convergence of the routing protocol. The desired outcome is to minimize disruptions to end users in the network in situations where quick network recovery is essential.

BGP uses two main adjustable timers to control periodic activities, such as, the rate at which BGP KEEPALIVE messages are sent (Keepalive Timer), and the maximum interval a BGP router must wait for a BGP KEEPALIVE or UPDATE messages from a neighbor before declaring it as unavailable or unreachable (i.e., Hold Timer) [RFC4271]. When a BGP connection is started, BGP will negotiate the Hold Time with the neighbor. The router uses the smaller of the two Hold Times as its Hold

Time. The router then sets its Keepalive Timer to be one third of the negotiated Hold Time **[RFC4271]**. The timers used in OSPF, IS-IS, and BGP are discussed in Volume 2 of this two-part book.

REVIEW QUESTIONS

1. Explain why a network would run multiple routing protocols.
2. What is a routing policy? Explain the differences between a routing policy and a packet filter policy.
3. What is policy-based routing (PBR)? Describe three example uses of PBR.
4. Explain how automatic route summarization in RIP and EIGRP is performed.
5. What is the purpose of a default seed metric?
6. What is difference between a RIP passive interface and an OSPF passive interface?
7. What is route redistribution? Explain some of the pitfalls of route redistribution.
8. What is the difference between one-point route redistribution and multipoint route redistribution?
9. Explain how the Administrative Distance is used as a path control tool.
10. What is route tagging?
11. What is an offset list? Which routing protocols use offset lists?
12. Explain how IP SLA Probes can be used for network path control.
13. What is a BGP AS-Path filter list?
14. What is the difference between injecting a route into BGP via the **redistribute** command or the **network** command?
15. What is BGP AS-Path prepending?
16. What is an IGP backdoor route? Why is it sometimes necessary to use a backdoor route?
17. What is an IP unnumbered interface and what benefits does it provide?
18. Name and describe the four main timers used in RIP.
19. Name and describe the three main timers used EIGRP.

REFERENCES

[AWEYA1BK18]. James Aweya, Switch/Router Architectures: Shared-Bus and Shared-Memory Based Systems, Wiley-IEEE Press, ISBN 9781119486152, 2018.

[CISCID118263STR19]. Cisco Systems, Specify a Next Hop IP Address for Static Routes, Document ID: 118263, June 4, 2019.

[CISCID49111RMP05]. Cisco Systems, Route-Maps for IP Routing Protocol Redistribution Configuration, Document ID: 49111, August 10, 2005.

[CISCID5441AGGR05]. Cisco Systems, Understanding Route Aggregation in BGP, Document ID: 5441, August 10, 2005.

[CISCID8606REDIS12]. Cisco Systems, Redistributing Routing Protocols, Document ID: 8606, March 22, 2012.

[CISCASR9000CR].	Cisco Systems, *Cisco ASR 9000 Series Aggregation Services Router Routing Command Reference, Release 4.3.x*, September 30, 2016.
[CISCBGPCOMD19].	Cisco Systems, *Cisco IOS IP Routing: BGP Command Reference*, December 1, 2019.
[CISCEIGRPCOMD18].	Cisco Systems, *Cisco IOS IP Routing: EIGRP Command Reference*, March 14, 2018.
[CISCEMPGARROT14].	Scott Empson, Patrick Gargano, and Hans Roth, *CCNP Routing and Switching Portable Command Guide*, Chapter "Configuration of Redistribution", Cisco Press, December 22, 2014.
[CISCHALABS00].	Sam Halabi, *Internet Routing Architectures*, 2nd Edition, Cisco Press, August 23, 2000.
[CISCID13786IPUN05].	Cisco Systems, Understanding and Configuring the ip unnumbered Command, Document ID: 13786, October 26, 2005.
[CISCIOSCOMD19].	Cisco Systems, *Cisco IOS IP Routing: Protocol-Independent Command Reference*, August 6, 2019.
[CISCIPCONGUI20].	Cisco Systems, *IP Routing Configuration Guide, Cisco IOS Release 15.2(7) Ex (Catalyst 1000 Switches)*, Chapter "Configuring IP Unicast Routing", January 17, 2020.
[CISCIPSLAGUI18].	Cisco Systems, *IP SLAs Configuration Guide*, Chapter "IP SLAs Overview", April 25, 2018.
[CISCISISCOMD20].	Cisco Systems, Cisco IOS IP Routing: ISIS Command Reference, June 9, 2020.
[CISCMASCOMD14].	Cisco Systems, *Cisco IOS Master Command List, All Releases*, January 27, 2014.
[CISCN3000CONFG].	Cisco Systems, *Cisco Nexus 3000 Series NX-OS Unicast Routing Configuration Guide, 9.3(x)*, Chapter "Configuring Static Routing", January 2, 2020
[CISCOSPFCOMD19].	Cisco Systems, *Cisco IOS IP Routing: OSPF Command Reference*, August 19, 2019.
[CISCRIPCOMD18].	Cisco Systems, *Cisco IOS IP Routing: RIP Command Reference*, March 14, 2018.
[CISCRIPCONGUI].	Cisco Systems, *IP Routing: RIP Configuration Guide, Cisco IOS XE Release 3SE (Catalyst 3650 Switches)*, Chapter "Configuring IP Summary Address for RIPv", January 14, 2018
[CISCTEAPAQ00].	Diane Teare and Catherine Paquet, Building Scalable Cisco Networks, Chapter "Configuring IS-IS Protocol", Cisco Press, October 27, 2000.
[CISCTEARDIA10].	Diane Teare, *Implementing Cisco IP Routing (ROUTE) Foundation Learning Guide: Foundation learning for the ROUTE 642-902 Exam*, Chapter "Implementing Path Control", Cisco Press, June 28, 2010.
[RFC1403].	K. Varadhan, "BGP OSPF Interaction", IETF RFC 1403, January 1993.
[RFC4271].	Y. Rekhter, Ed., T. Li, Ed., and S. Hares, Ed., "A Border Gateway Protocol 4 (BGP-4)", IETF RFC 4271, January 2006.

Index

A

access control lists (ACLs), 215–216
address prefix, *see* network prefix
Address Resolution Protocol (ARP), 92
adjacency table, 92
administrative distance, 3, 69–71, 97, 105,
   111–112, 117, 149, 240–246
ARP cache, 92
ARP table, *see* ARP cache
Authentication, 9–10, 117, 122–128, 134–136,
   147, 194–197
   EIGRP cryptographic authentication, 9, 147,
      195–197
   EIGRP simple password authentication, 9,
      147, 195
   RIP cryptographic authentication, 9, 117,
      124–128, 135–136
   RIP simple (or plaintext) password
      authentication, 9, 117, 122–124, 135
   routing information authentication, 9–10
autonomous systems, 11–15, 20–22, 56–57, 63
   autonomous system numbers (ASNs), 12–13,
      57, 65
   multihomed autonomous system, 15
   Private ASNs, 13
   Public ASNs, 13
   stub autonomous system, 14–15
   transit autonomous system, 15

B

black holes, 32, 34
Border Gateway Protocol (BGP), 58–62
   AS-Path attribute (AS_PATH), 65
   BGP AS-Path lists, 276–277
   BGP AS-Path prepending, 281–283
   BGP distribute lists, 278
   BGP path attribute manipulation, 274–276
   BGP prefix lists, 277–278
   BGP route, 63–64, 67
   BGP route filtering, 274–276
   BGP route maps, 278
   BGP Speakers (or routers), 63
   BGP timers, 295–296
   default routes, 256, 289
   exterior BGP (eBGP), 63
   KEEPALIVE message, 66
   interior BGP (iBGP), 63
   network layer reachability information
      (NLRI), 64
   path attributes, 66–68
   peer group, 62
   route aggregation, 283–288
   route redistribution, 239–240, 278–281
   UPDATE message, 65

C

classless inter-domain routing (CIDR), 221
control engine, *see* route processor
control plane, 78, 92–93
control plane redundancy, *see* route processor
   redundancy

D

data plane, 93
Dijkstra algorithm, *see* shortest-path first (SPF)
   algorithm
directly connected (or attached) networks, 2–3, 87
directly connected interface (or route), *see* directly
   connected (or attached) networks
distance vector routing protocols, 23, 25–31, 116
   asynchronous routing updates, 33–34
   Bellman-Ford algorithm, 23, 26, 116
   count-to-infinity, 32, 39–41
   distance-vector routing, 23
   flush timer, 36, 120
   holddown timers, 36–38, 120
   invalid timer, 34–35, 120
   periodic updates, 8, 33

299

poison reverse, 41–42, 186–187
routing loops, 31–33
split horizon, 42–44, 186–187
triggered updates, 8, 28, 30, 38–39, 47–48, 148
update timer, 35, 120

E

EIGRP authentication, 9, 147, 194–197
   cryptographic authentication, 147, 195–197
   simple password authentication, 147, 195
EIGRP flags, 163–165
   CR-Flag (0x02), 164
   EOT-Flag (0x08), 164
   INIT-Flag (0x01), 163
   RS-Flag (0x04), 164
EIGRP packet types
   HELLO packet, 149, 151–152, 155–158, 181–183
   packet size, 156
   QUERY packet, 149, 156, 159–160, 177, 183, 204–205
   REPLY packet, 149, 156, 160, 177
   REQUEST packet, 156, 161
   UPDATE packet, 149, 151, 156, 159, 177, 183
EIGRP route states
   ACTIVE State, 153, 159–160, 167, 172–175, 178, 205–206
   PASSIVE State, 153, 167, 172, 174–175, 177–178, 205–206
EIGRP timers
   active timer, 179–180, 205, 295
   hello interval timer, 29, 295
   hold timer, 152, 182, 295
EIGRP type-length-value (TLV) types, 156, 161
   Authentication Type TLV (0x0002), 161
   IPv4 Community Type (0x0104), 163
   IPv4 External Routes TLV (0x0103), 162
   IPv4 Internal Routes TLV (0x0102), 162
   Multicast Sequence Type TLV (0x0005), 162
   Parameter Type TLV (0x0001), 161
   Peer Termination Type TLV (0x0007), 162
   Sequence Type TLV (0x0003), 162
   Software Version Type TLV (0x0004), 162
Enhanced Interior Gateway Routing Protocol (EIGRP)
   advertised (or reported) distance, 152–153, 167–168, 171, 175, 184
   automatic (auto) route summarization, 147, 193–194, 223–224
   composite metric, 30, 155, 165–167
   computed distance, 152–153, 184
   default metric, 239
   default routes, 254
   diffusion update algorithm (DUAL), 30, 116, 148, 153–154, 175–179

   external routes, 185–186
   feasibility condition, 168, 174–176, 190
   feasible distance, 152–153, 168, 171, 175
   feasible successor, 153–154, 170–172, 200–201, 205–206
   internal routes, 184–185
   IP protocol number, 147
   load balancing, 147, 154, 171, 189–190
   manual route summarization, 147, 194, 223–224
   neighbor table, 30, 151–152, 198, 201–203
   poison reverse, 186–187
   reliable transport protocol (RTP), 149–151
   route redistribution, 191–193, 238
   routing table, 30, 153–154, 170–171, 199, 203–204
   split horizon, 186–187
   stuck-in-active (SIA), 174, 179–181
   successor, 153–154, 168, 170, 200–201, 205–206
   topology table, 30, 152–154, 170–171, 184, 198–199
   variance parameter, 190
   well-known multicast addresses, 8, 149, 155–156
equal-cost multipath (ECMP) routing, 69
(Exterior or External) Gateway Protocol (EGP), 21–22, 58, 60

F

forwarding architectures
   centralized forwarding, 80–81
   distributed forwarding, 80, 82–83
forwarding engine, 79
forwarding information base (FIB), see forwarding table
forwarding plane, see data plane
forwarding table, 91

I

Interior (or Internal) Gateway Protocol (IGP), 13–14, 21–22, 58
Intermediate System to Intermediate System (IS-IS)
   default routes, 255–256
   IS-IS metrics, 44–45
   Level 2 router, 57
   narrow metric, 45
   route summarization, 224–225
   wide metric, 45
IP forwarding operations, 92–95
IP routing
   aggregate routes, see summary routes
   default gateways, see default routes
   default routes, 4–5, 87, 101–102, 252–256

Index

default routing, *see* dynamic routes
default static route, 106–107
dynamic routes, 87
dynamic routing, *see* dynamic routing protocols
dynamic routing protocols, 1, 5–7, 99–101, 115
floating static route, 111–113
least-cost routing, 22–23
network prefix, 11, 71–73
network prefix length, 11, 71–73
route, 12
routing loops, 31–33, 57, 65
routing protocol, 1, 115–116, 209
standard static route, 104–105
static routes, 3, 87, 101–103, 116, 249–251
static route summarization, *see* summary static routes
static routing, *see* static routes
summary routes, 91
summary static routes, 108–111, 225–226

L

link-state routing protocols, 23, 44–55
   area, 56–57
   candidate database, 53–54
   Dijkstra algorithm, *see* SPF algorithm
   flooding process, 46–47, 50–51
   link-state database (LSBD), 23, 46, 51–53
   link-state database (LSDB) synchronization, 53, 55
   link-state advertisement, 46
   link-state packet, 46
   link-state routing, 23
   shortest-path first (SPF) algorithm, 23, 53–55
   shortest-path tree (SPT), 23, 53–56
   tree database, 53–54
longest prefix matching (LPM), 71–73, 94

M

Martian addresses, 94–95

N

network (address) prefix, 11

O

Open Shortest Path Routing (OSPF) protocol
   area border router (ABR), 57
   authentication, 9–10
   cost, 19–20, 45
   default routes, 254–255
   Hello interval, 49–50
   Hello protocol, 49
   neighbor discovery, 48–50
   neighbor table, 48
   Router Dead Interval, 49–50
   route redistribution, 237
   route summarization, 224
   routing metric, *see* cost
   routing table, 48, 54–55
   well-known multicast addresses, 8

P

path control tools
   backdoor routes, 290
   BGP AS-Path lists, 276–277
   BGP distribute lists, 278
   BGP path attribute manipulation, 274–276
   BGP prefix lists, 277–278
   BGP route filtering, 274–276
   BGP route maps, 278
   distribute lists, 136, 141–142, 261–263
   offset lists, 267–268
   passive interfaces, 100, 140–141, 248–249
   policy-based routing (PBR), 219–221, 257, 265–267
   policy routing, *see* policy-based routing (PBR)
   prefix lists, 263–265
   route filtering, 95–96, 140–142
   route maps, 256–261
   route redistribution, 14, 95, 97, 137, 191–193, 211, 214, 216–234, 236–242, 257
   route summarization, 193–194, 221–226
   route tagging, 246–247
   service level agreement (SLA) probes, 268–274
path-vector routing protocols, 57, 60–62
   speaker node or router, 60–62
periodic updates, 8, 33
prefix aggregation, *see* route summarization

R

RIP authentication, 9, 117, 122–128, 134–136
   cryptographic authentication, 9, 117, 124–128, 135–136
   simple (or plaintext) password authentication, 9, 117, 122–124, 135
RIP message types
   message size, 119
   request message, 118–122
   response message, 118–122
RIP timers
   flush timer, 36, 91, 120, 295
   holddown timer, 36–38, 120, 295
   invalid timer, 34–35, 91, 120, 294
   update timer, 35, 120, 294
route aggregation, *see* route summarization
route preference, *see* administrative distance
route processor, 77–79

route processor redundancy, 84–85
route redistribution, 14, 95, 97, 191–193, 211, 214, 216–234, 236–242, 257
    multipoint route redistribution, 231–234
    one-point route redistribution, 229–231
route summarization, 91, 108, 193–194, 221–226
routing domain, 13, 211
routing engine, *see* route processor
routing information base (RIB), *see* routing table
Routing Information Protocol (RIP)
    automatic (auto) route summarization, 222
    default routes, 254
    manual route summarization, 222
    route redistribution, 236–237
    routing table, 120
    UDP port number, 117
    well-known multicast addresses, 8, 33, 117
routing loops, 31–33, 57, 61, 65
routing metrics, 15–20, 68–69, 96–97
    bandwidth, 17
    cost, 19–20
    default metric, 238–239, 241–243
    delay, 18
    hop count, 15–16, 25, 28
    infinity hop count, 17
    limitations of the hop count, 17
    maximum hop count, 16–17, 39, 117
    network diameter, *see* maximum hop count
    reliability, 18–19
    routing metric translation, 96
    seed metric, 240–243

    traffic load, 18
routing policy, 11, 61, 136, 212–219
routing prefix, *see* network prefix
routing protocols
    asynchronous routing updates, 33–34
    convergence, 10
    periodic updates, 8, 33
    routing information authentication, 9–10
    routing loops, 31–33, 57, 61, 65
    triggered updates, 8, 28, 30, 38–39, 47–48, 148
routing table, 1, 85–91, 116, 209
routing updates, 8

S

shortest-path first (SPF) algorithm, 23, 53–55
stub networks, 101
supernetting, *see* route summarization

T

triggered updates, 8, 28, 30, 38–39, 47–48, 148

U

unnumbered interfaces, 290–294

V

Variable-Length Subnet Mask (VLSM), 221
Virtual Routing and Forwarding (VRF), 89–90, 139

Milton Keynes UK
Ingram Content Group UK Ltd.
UKHW021840161024
449632UK00006B/93